Philipp H. Hoff

Greentech Innovation and Diffusion

GABLER RESEARCH

Schriften zum europäischen Management

Herausgeber/Editor:
Roland Berger Strategy Consultants – Academic Network

Die Reihe wendet sich an Studenten sowie Praktiker und leistet wissenschaftliche Beiträge zur ökonomischen Forschung im europäischen Kontext.

This series is aimed at students and practitioners. It represents our academic contributions to economic research in a European context.

Philipp H. Hoff

Greentech Innovation and Diffusion

A Financial Economics and Firm-Level Perspective

With a foreword by Prof. Dr. Thomas Berndt

GABLER

RESEARCH

Bibliographic information published by the Deutsche Nationalbibliothek
The Deutsche Nationalbibliothek lists this publication in the Deutsche Nationalbibliografie;
detailed bibliographic data are available in the Internet at http://dnb.d-nb.de.

Doctoral thesis, University of St. Gallen, St. Gallen, Switzerland, 2011

1st Edition 2012

All rights reserved
© Gabler Verlag | Springer Fachmedien Wiesbaden GmbH 2012

Editorial Office: Marta Grabowski | Sabine Schöller

Gabler Verlag is a brand of Springer Fachmedien.
Springer Fachmedien is part of Springer Science+Business Media.
www.gabler.de

Cover design: KünkelLopka Medienentwicklung, Heidelberg
Printed on acid-free paper

ISBN 978-3-8349-3600-4

Dedicated to my parents,
Brigitte and Heinrich Hoff

FOREWORD

The terms greentech, cleantech, sustainable technology, renewable energy are often used as synonyms for a common sector that is currently a major focus of public, political and economical interest. Yet amidst the euphoria, surprisingly little attention has been paid to the issue of innovation, sector emergence and their financing from a firm level perspective.

Philipp Hoff's Ph.D. dissertation, published here, tackles this issue head on. Taking a broad, interdisciplinary approach, he investigates the related areas of innovation, growth and financing. In so doing, he attempts to determine whether the diffusion of green technology is constrained by the financial situation of greentech firms.

This is a remarkable study, for three reasons. First, the review of literature presents an outstanding analysis of the debate about innovation in the area of greentech and the diffusion of new technology. This forms the starting point for developing a conceptual framework that is solidly grounded in theory and practical in its application.

Second, by applying the familiar innovation and organizational lifecycle concept to greentech firms, the author achieves interesting – and somewhat surprising – insights into the impacts of "pull" and "push" instruments. In particular, he finds that while extending existing push instruments would reduce the financial constraints on young firms, pull instruments do not necessarily foster investment in more developed firms.

Most studies would stop at this point. Not so this book. Instead, the author undertakes an empirical analysis based on a survey of more than 500 German greentech firms. This is the first time that a study has been carried out in this field on such a scale. The author also uniquely employs the variable "capital endowment" in his analysis. By this means, he is able to find answers to many practical questions that have puzzled previous investigators.

This study is not only an excellent reference work for regulators and financers in the greentech industry. It is also a highly valuable source of inspiration and material for further research. I heartily recommend it to anyone working in the field of corporate finance – not just specialists in venture capital, but others, too – to CEOs and CFOs of greentech companies, members of regulatory bodies and other researchers. Its excellent structure, many illustrations, clear figures and precise summaries make it a treasure-trove of information. It is my sincere hope that this book has the practical influence and achieves the broad readership that it deserves.

Prof. Dr. Thomas Berndt

ACKNOWLEDGEMENTS

Every undertaking such as this one requires sources of inspiration and support, as well as challenge. I am glad to have found such sources who, together, formed a very fruitful environment that enabled me to pursue this dissertation and fulfill a personal goal I had had for many years. I would like to acknowledge these here.

First and foremost, I would like to thank my doctoral advisor Professor Thomas Berndt for the opportunity to work on this research topic and for the many valuable and encouraging discussions along the way. I am deeply grateful for the considerable trust he placed in me and for his willingness to tread new paths and supervise a work in an emerging stream of literature. I very much appreciate his constructive guidance and the way I was welcomed at his institute. I also want to express my gratitude to Professor Rolf Wüstenhagen, who agreed to supervise this study in the role of co-advisor. Professor Wüstenhagen, a recognized authority on clean energy, gave valuable guidance to my work, particularly at the outset when I designed the outline of this book. My thanks also go to the many teachers and research fellows whom I have met at the University of St. Gallen over the past two years. Every course and seminar I attended enhanced my knowledge and methodological skills, broadened my perceptions and challenged my own thoughts. During my time at the HSG, I not only learned a great deal but was also able to extend my social and professional network.

I also want to thank the many greentech firms who demonstrated their interest in the topic by agreeing to participate in the empirical study. Their contribution has been priceless. They made this study relevant, statistically representative and hence a success.

My profound gratitude also goes to Professor Torsten Henzelmann, who made this work possible by allowing me to take an 18-month leave of absence in order to complete it. He has been a tremendous mentor over the past five years. He has assisted me by challenging my research ideas and providing insights into their practical relevance based on the in-depth industry knowledge he has gained from working as a Senior Partner at Roland Berger Strategy Consultants. In this context, I would also like to thank the firm for supporting my work financially – especially Christian Krys, who helped to organize my leave. My thanks also go to Nigel Robinson for carefully checking and improving the language and Aleksandar Dzhamurov, who assisted me in programming the questionnaire.

Finally, I want to thank my family and all my friends who helped me to keep a mental balance and thus overcome the many hurdles associated with such a project. Working on this thesis was not always easy and took a great deal of self-discipline, stamina and motivation. I am fortunate to have parents such as Brigitte and Heinrich Hoff, who have always stood by me

X

and enabled me to become the person I am. Finally, I want to express my gratitude to Sandra Kobialka, my dear fiancée, who not only shared the successes but also endured the setbacks with me. She has become a sustainable part of my life and I am looking forward to our marriage.

St. Gallen, October 2011   *Philipp H. Hoff*

TABLE OF CONTENTS

FOREWORD.. VII

ACKNOWLEDGEMENTS IX

TABLE OF CONTENTS XI

LIST OF FIGURES XV

LIST OF TABLES... XVII

LIST OF NOTATIONS AND ABBREVIATIONS XIX

ABSTRACT XXI

ZUSAMMENFASSUNG XXIII

1   INTRODUCTION.. 1
1.1   Motivation for this research 1
1.2   Research questions and contributions... 3
1.3   Organization of research 4
1.4   Scope and limitations 5

2   REVIEW OF EXISTING KNOWLEDGE ABOUT GREENTECH 7
2.1   Definitions and terminology.. 7
       2.1.1   Greentech 7
       2.1.2   Clean energy versus greentech 8
2.2   The five perspectives covered by greentech business research 9
       2.2.1   The sustainability perspective 10
       2.2.2   The entrepreneurial perspective 13
       2.2.3   The innovation and diffusion perspective 14
       2.2.4   The policy and regulation perspective 15
       2.2.5   The finance perspective... 18
       2.2.6   Conclusion – A plea for a financial economics and firm-level analysis 21

3   REVIEW OF OTHER RELEVANT LITERATURE 24
3.1   Innovation and diffusion – A focus of technology 24
       3.1.1   Definitions and terminology.. 24
               3.1.1.1   Invention and innovation 24
               3.1.1.2   Innovation and diffusion 26
       3.1.2   The evolutionary process of technology innovation and its diffusion 26
       3.1.3   Drivers of technology innovation and diffusion 33
       3.1.4   Empirical evidence about technology innovation and diffusion........ 35

3.1.5 Conclusion – The parallels between technology and firm development 38

3.2 **Innovation – A basis for corporate development** **39**

3.2.1 Theory of innovative firms 41

3.2.2 Empirical evidence about innovative firms 45

    3.2.2.1 The measurement problem 47

    3.2.2.2 The determinants of innovation 47

3.2.3 Conclusion – Characterizing innovative firms 53

3.3 **Growth – A measure of corporate development** **57**

3.3.1 Patterns of growth 57

3.3.2 Theory of firm growth 60

    3.3.2.1 The corporate lifecycle concept 61

    3.3.2.2 Critical appraisal of the corporate lifecycle concept 64

3.3.3 Empirical evidence about growing firms and the corporate lifecycle 68

    3.3.3.1 The measurement of growth 69

    3.3.3.2 Empirical evidence about the corporate lifecycle concept 71

    3.3.3.3 The determinants of growth 73

3.3.4 Conclusion – Characterizing growth firms 79

3.4 **Financial economics – A necessity for corporate development** **83**

3.4.1 Definitions and terminology 84

    3.4.1.1 Capital structure and financing instruments 84

    3.4.1.2 Financial constraints 85

3.4.2 Corporate finance and capital structure theory 85

    3.4.2.1 The capital structure irrelevance theorem 85

    3.4.2.2 Asymmetric information and the principal-agent problem 86

    3.4.2.3 The "capital structure puzzle" and the quest for a universal theory 87

3.4.3 SME finance and the importance of firm demographics 88

    3.4.3.1 The financial growth cycle and firms' capital structure 88

    3.4.3.2 The availability of funds and the concept of financial constraints 92

3.4.4 Empirical evidence about SME finance, capital structure and
firm demographics 93

    3.4.4.1 Measurement of capital structure, financial constraints
and the diversity of capital instruments 93

    3.4.4.2 Empirical evidence about the financial growth cycle of the firm 95

    3.4.4.3 The determinants of capital structure 96

    3.4.4.4 The determinants of financial constraints 104

3.4.5 Conclusion – Characterizing financial structures and constraints 108

3.5 **Public policy – Bypassing financial constraints** **112**

3.5.1 The gap theory and the one-firm, one-technology case 113

3.5.2 The status quo for greentech policy measures in Germany 114

3.5.3 Empirical evidence on the effectiveness of firm-level support programs 116

3.5.4 Conclusion – Policy effectiveness at the level of the firm 118

4 **THEORETICAL FRAMEWORK** **119**

4.1 **Interdisciplinary summary of the main findings from previous chapters** **119**

4.2 **The chicken and egg problem –
A causality dilemma in interdisciplinary research** **120**

4.3	**Formulating an integrated framework**	122
	4.3.1 Innovation, growth and capital demand	123
	4.3.2 Innovation, growth and capital supply	125
	4.3.3 Merging capital demand and supply	127
	4.3.4 Implications and derivation of research hypotheses	132
4.4	**Conclusion – Proposition of an integrated framework**	134

5 EMPIRICAL ANALYSIS **137**

5.1	**Research methodology**	137
	5.1.1 Overall empirical strategy and approach	137
	5.1.2 Sample and response	137
	5.1.3 Questionnaire development	139
	5.1.4 Measures and variables	140
	5.1.4.1 Main variables of interest	140
	5.1.4.2 Control variables	144
	5.1.5 Conclusion – A mixed method approach for exhaustive coverage	145
5.2	**Comparative analysis**	145
	5.2.1 General sample characteristics and the non-response bias	145
	5.2.2 General innovation activity of greentech firms	148
	5.2.3 General growth of greentech firms	149
	5.2.4 General financial constraints faced by greentech firms	150
	5.2.5 Comparative analysis of greentech firms	151
	5.2.5.1 Methodology	151
	5.2.5.2 Firm characteristics	152
	5.2.5.3 Innovation activity and growth	154
	5.2.5.4 Financial structure	156
	5.2.5.5 Financial constraints and applications for capital	161
	5.2.5.6 Alleviation of financial constraints	164
	5.2.6 Conclusion – Theoretical framework supported	170
5.3	**Regression analysis and testing of the research hypotheses**	171
	5.3.1 Overall empirical model – Financing greentech innovation and growth	171
	5.3.2 Descriptive statistics for relevant measures and variables	173
	5.3.3 Financing innovation of greentech firms (model 1)	175
	5.3.3.1 Specification of the regression model	175
	5.3.3.2 Model performance	177
	5.3.3.3 Underlying model assumptions and validity	180
	5.3.3.4 Presentation of results	181
	5.3.4 Financing growth of greentech firms (model 2)	185
	5.3.4.1 Specification of the regression model	185
	5.3.4.2 Model performance	187
	5.3.4.3 Underlying model assumptions and validity	189
	5.3.4.4 Presentation of results	191
	5.3.5 Determinants of financial constraints for greentech firms (model 3)	194
	5.3.5.1 Specification of regression model	194
	5.3.5.2 Model performance	196
	5.3.5.3 Underlying model assumptions and validity	199
	5.3.5.4 Presentation of results	201

    5.3.6   Conclusion – Main research hypotheses confirmed 208

6   FINAL DISCUSSION AND IMPLICATIONS 210

6.1   Overall summary of results 210

6.2   Theoretical contributions 213

    6.2.1   Contributions to general business research 213
    6.2.2   Contributions to greentech research 214
        6.2.2.1   Contributions to the innovation and diffusion perspective 214
        6.2.2.2   Contributions to the policy and regulation perspective 215
        6.2.2.3   Contributions to the finance perspective 216

6.3   Managerial implications 217

6.4   Policy implications 218

6.5   Limitations and outlook 221

7   EXECUTIVE SUMMARY 223

APPENDIX 227

REFERENCES 241

LIST OF FIGURES

Figure 1-1: Organization of research 5

Figure 2-1: Current perspectives in greentech business related research............... 10

Figure 2-2: Greentech market development... 11

Figure 2-3: Policy instruments and actions 17

Figure 2-4: VC investment in greentech in Germany (environmental and energy category).. 19

Figure 3-1: Diffusion of innovations... 28

Figure 3-2: Diffusion of innovations and adopter categories.............................. 30

Figure 3-3: Modified adopters and innovation diffusion curve 33

Figure 3-4: Percentage of all corn acreage planted with hybrid seed in
different regions in the US 36

Figure 3-5: Innovation diffusion in various industries... 37

Figure 3-6: The one-firm and one-technology innovation and diffusion case........ 39

Figure 3-7: Industry sales and firm population in the US automobile industry...... 40

Figure 3-8: Determinants of innovation – A synthesis 54

Figure 3-9: The multiple-firm and multiple-technology innovation diffusion case......... 59

Figure 3-10: Exploratory study results on growth patterns................................... 60

Figure 3-11: Determinants of growth – A synthesis 81

Figure 3-12: The corporate lifecycle – A modified growth model 82

Figure 3-13: The one-firm, one-technology case – A financial perspective.......... 83

Figure 3-14: Firm continuum and sources of finance 89

Figure 3-15: Determinants of capital structure and financial constraints – A synthesis 111

Figure 3-16: The need for and impact of policy intervention on firm development –
The one-firm, one-technology case... 114

Figure 3-17: Number of greentech-related policy instruments available
on the German market, by type 116

Figure 4-1: The circularity and inconsistency problem in integrating
financial constraints, innovation and growth 121

Figure 4-2: The corporate innovation and growth cycle 125

Figure 4-3: The financial growth cycle and capital supply................................ 126

Figure 4-4: Capital demand, supply and the existence of financial constraints..... 128

Figure 4-5: Effect of innovation activity, growth, policy and
firm measures on financial constraints... 131

Figure 4-6: The corporate financial innovation and growth cycle –
An integrated framework 134

Figure 5-1: Technology lines and strategic focus 147

Figure 5-2: Sustainability orientation and internationality 148

Figure 5-3: Innovation output and R&D effort 149

Figure 5-4: Average sales growth 150

Figure 5-5: Existence of financial constraints and main reasons 150

Figure 5-6: Key demographics: Employment and profit 153

Figure 5-7: Key demographics: Sales and firm age 153

Figure 5-8: Innovation, R&D and growth 155

Figure 5-9: Aggregated financing baskets 157

Figure 5-10: Equity financing 158

Figure 5-11: Debt financing 160

Figure 5-12: Financial constraints 162

Figure 5-13: Applications for equity capital – Rejection and success rates 162

Figure 5-14: Applications for debt capital – Rejection and success rates 164

Figure 5-15: Share of government-supported firms by instrument type 166

Figure 5-16: Share of government instruments within capital structure 167

Figure 5-17: Share of firms with pull policy-dependent business models 167

Figure 5-18: Bank relationship satisfaction and average interest rate paid 169

Figure 5-19: Collateral and the number of bank relationships 169

Figure 5-20: Empirical approach 171

Figure 5-21: Overview of three models 173

Figure 6-1: The greentech financial innovation and growth cycle (modified) 210

Figure 6-2: Empirical model (confirmed) 212

LIST OF TABLES

Table 2-1: Greentech industry categories 8

Table 3-1: Selected empirical research on the determinants of innovation 46

Table 3-2: Selected empirical research into the determinants of growth 69

Table 3-3: Selected empirical research on the determinants capital structure 99

Table 3-4: Selected empirical research into the determinants of financial constraints 105

Table 4-1: Comparison of empirical study results in the various research areas 120

Table 5-1: Overview of measures and variables 142

Table 5-2: Descriptive statistics for measures and variables 174

Table 5-3: Predictive performance of the specified model 178

Table 5-4: Model summary and evaluation 179

Table 5-5: Correlation of predictors and residuals 180

Table 5-6: Empirical results of the first model on innovation 181

Table 5-7: Empirical results of the first model on process innovation 184

Table 5-8: Model summary and performance 188

Table 5-9: Correlation of predictors and residuals 190

Table 5-10: Empirical results on growth 192

Table 5-11: Predictive performance of the specified model 197

Table 5-12: Model summary and performance 199

Table 5-13: Correlation of predictors and residuals 200

Table 5-14: Empirical results of submodel (a) on financial constraints 203

Table 5-15: Empirical results of submodel (b) on financial constraints 204

Table 5-16: Empirical results of submodel (c) on financial constraints 205

Table 5-17: Empirical results of submodel (d) on financial constraints 206

LIST OF NOTATIONS AND ABBREVIATIONS[1]

BA	Business angel capital
BMU	Bundesministerium für Umwelt
	(German Federal Ministry for the Environment)
bn	Billion
CEO	Chief Executive Officer
CF	Cash flow
CFO	Chief Financial Officer
CHP	Combined heat and power (cogeneration)
D_S	Debt supply
EEG	Erneuerbare Energien Gesetz (feed-in law for renewable energy)
e.g.	Exempli gratia; for example
EQT_S	Equity supply
et al.	et alii
etc.	et cetera
EU	European Union
EUR	Euro (currency)
$EXTC_D$	External capital demand
GDP	Gross Domestic Product
GMM	General Method of Moments
GOV_S	Government support programs
H-L	Hosmer-Lemeshow
HSG	University of St. Gallen
i.e.	Id est
IHK	Industrie und Handelskammer (chamber of commerce and industry)
IPO	Initial public offering
KPI	Key performance indicator
LED	Light-emitting diode
LTD	Long-term debt
m	Million
Max.	Maximum value
Min.	Minimum value
MTD	Medium-term debt
n	Number of observations
n.a.	Not applicable
No.	Number

---

[1] Does not contain a list of the variables used in the empirical section (see section 5.1.4 for an overview)

NPV	Net present value
OLS	Ordinary least squares
p.	Page(s)
p.a.	Per annum
PE	Private equity
R&D	Research and development
RAM	Random access memory
S.E.	Standard error
SME	Small and medium-sized enterprise
STD	Short-term debt
TC_S	Total capital supply
UK	United Kingdom
UMFIS	Umweltfirmeninformationssystem (environmental firm database)
US	United States of America
VC	Venture capital
Y	Sales
ZRESID	Standardized residual

ABSTRACT

The world is going green. Sustainable technologies, such as renewable energy and electric vehicles, are increasingly becoming part of our daily life. However, this development cannot be taken for granted, as persisting institutional systems pose substantial barriers to it. Accordingly, recent research has increasingly made an effort to improve our understanding of how environmental innovation can be generated and diffused throughout the overall market space. The focus has, however, mostly been on technology from a policymaker's perspective, not on the actual instrument of change: "the greentech firm". This dissertation fills the ensuing gap by providing an insight into the emerging German greentech industry, one of the largest in the world. It develops an integrated and interdisciplinary theoretical framework in which to assess the relationships between innovation, growth and financing from a firm-level perspective; it then tests this framework empirically. It thereby overcomes contradictions in contemporary business research in the fields referred to, as well as demonstrating that firms bound by financial constraints do not fully exploit their potential. In essence, the study finds that: (1) Innovative activity and corporate growth depend heavily on the availability of capital. At the same time, it appears that particularly innovative firms are more likely to face financial constraints. (2) A lack of funds is very apparent for around a quarter of the firms investigated and seems most severe in the early part of the growth state, where firms focus on commercializing existing products. (3) Government support programs only partially offset these effects. Demand-pull programs, such as Germany's feed-in law for renewable energy (EEG), actually evidence a contrary impact. (4) The conclusion is that government instruments should focus on firms and not technologies in order to make support programs more effective and avoid technology "lock-in". The best thing a firm can do is to engage in relationship banking and apply for various capital sources more frequently.

ZUSAMMENFASSUNG

Der blaue Planet wird grün. Nachhaltige Technologien wie erneuerbare Energien und Elektrofahrzeuge gehören immer mehr zu unserem täglichen Leben. Allerdings ist diese Entwicklung keineswegs sicher, denn träge institutionelle Rahmenbedingungen und technologische Pfadabhängigkeiten stellen wesentliche Barrieren dieser Entwicklung dar. Die aktuelle Forschung versucht daher verstärkt, ein besseres Verständnis davon zu erlangen, wie sich Innovationen im Umweltbereich generieren und im Markt durchsetzen lassen. Der Fokus ist dabei in der Regel auf die Technologieebene aus Sicht der politischen Entscheidungsträger und nicht auf das eigentliche Instrument des Wandels gerichtet: dem Greentech-Unternehmen. Die vorliegende Dissertation schließt diese Lücke, indem sie Einblicke in die aufblühende – und weltweit führende – deutsche Greentech-Landschaft gibt. Sie entwickelt einen integrierten und interdisziplinären theoretischen Rahmen für die Beurteilung der komplexen Beziehungen zwischen Innovation, Wachstum und Finanzierung aus Firmenperspektive. Im Anschluss wird das Modell empirisch validiert. Damit gelingt es dem Autor, die Inkonsistenzen der aktuellen Innovations-, Wachstums- und Finanzierungsforschung zu überwinden und außerdem zu zeigen, dass Unternehmen, die in der Finanzierung eingeschränkt sind, ihr Potenzial nicht voll entfalten können. Die wichtigsten Ergebnisse der Untersuchung lauten: (1) Innovationsvermögen und Unternehmenswachstum sind in hohem Maße von der Kapitalverfügbarkeit abhängig. Zugleich scheinen gerade innovationsfreudige Unternehmen häufig vor finanziellen Engpässen zu stehen. (2) Rund ein Viertel der betrachteten Unternehmen leidet unter einer deutlichen Unterfinanzierung, die gerade in der Frühphase des Wachstums, in der der unternehmerische Schwerpunkt auf der Kommerzialisierung vorhandener Produkte liegt, am stärksten ausgeprägt zu sein scheint. (3) Öffentlichen Förderprogrammen gelingt es nur teilweise, hier Abhilfe zu schaffen. Nachfrageorientierte Programme wie z.B. das deutsche Erneuerbare-Energien-Gesetz (EEG) entfalten sogar eine gegenteilige Wirkung auf der Unternehmensebene. (4) Fazit: Staatliche Instrumente sollten auf Unternehmen und nicht auf Technologien ausgerichtet sein. So sind Förderprogramme effektiver und es kann ein technologischer Lock-In-Effekt vermieden werden. Die Empfehlung an Unternehmen lautet dabei: Stärkere Konzentration auf ein gutes Verhältnis mit den Hausbanken, Nutzung verschiedener Finanzierungsinstrumente und Intensivierung der Mittelakquise.

1   INTRODUCTION

The world we have created is a product of our thinking;
it cannot be changed without changing our thinking
Albert Einstein

The world is going green. Sustainability is increasingly becoming a part of our daily life. This revolutionary process is being accompanied by a *"tectonic shift taking place in the business world as a result of the megatrends of population growth, globalization, urbanization and climate change."* (Oltmanns (2011)) Recent disasters, such as the explosion of the nuclear power plant in Fukushima and the Deepwater Horizon oil spill in the Gulf of Mexico, have fostered this process by changing our perception of the customs and standard practices that allowed them to occur. Although the need for change is typically associated with higher economic cost, it actually harbors tremendous economic potential and infinite opportunities for innovative firms to engage in developing and marketing green technologies. These technologies will increase efficiency, substitute fossil resources and enhance the world's ability to recover from negative externalities such as CO_2 emissions. Clean energy is a hallmark of the emerging discipline that is increasingly being recognized as an industry in its own right and is often referred to as green technology or "greentech". However, one major concern of policymakers is that barriers imposed by today's economic system hinder its progress and ultimately block a more sustainable technology trajectory. This is because externalities are often not priced in – and hence not appropriately integrated in – the institutional system we currently know. With Germany leading the way, many countries have therefore devoted considerable financial resources to supporting innovation in and the diffusion of green technologies, hoping to set sustainable development in motion while at the same time sparking industry and fueling job creation. It is therefore crucial to understand the issues of technological innovation, firm growth and its financing, and how the two are interrelated. Interdisciplinary research combining these elements is scarce, however; and greentech evidence at firm level is literally non-existent. This book therefore addresses the existing research gaps by not only developing an integrated framework, but also providing one of the first empirical and firm-level studies in this field, centering around the German greentech industry, which has become one of the largest in the world.

1.1   Motivation for this research

Historically, it was thought that environmental degradation is an unavoidable by-product of wealth generation. Only more recent studies and practices have integrated environmental and business-related aspects (e.g., Christmann (2000); Hart (1995); Sharma and Vredenburg (1998); Shrivastava (1995)), showing that economic growth and the reduction of negative externalities can be achieved at the same time (e.g., King and Lenox (2002); Klassen and McLaughlin (1996); Rivera (2001)). One specific field in this line of thought is the concept of "greentech", which can be considered a sub-category of literature on ecological innovation. (Rennings (2000)) Greentech comprises an emerging technology-oriented industry that generates economic rents by developing and selling technologies or services that reduce or eradicate environmental pollution. Its practical importance is therefore extraordinarily great, since greentech offers the means to achieve sustainability while at the same time generating wealth and economic growth. Despite the promise that the greentech industry holds for sustainable development, however, academic research is still quite rare, lagging behind the attention afforded in practice (O'Rourke (2009)). This dissertation is motivated by four main considerations: First, most existing research has adopted the perspective of the policymaker, assessing the effectiveness of support mechanisms in terms of the degree to which technology is diffused (e.g., Enzensberger, Wietschel, and Rentz (2002); Jacobsson and Lauber (2006); Loiter and Norberg-Bohm (1999)). However, since financing is key to both innovation and diffusion (e.g., Beck and Demirguc-Kunt (2006); Börner and Grichnik (2005); Brush, Ceru, and Blackburn (2009); Carpenter and Petersen (2002); Gompers and Lerner (2001)), research should perhaps take a step back and move in the direction of financial economics and corporate finance (Dinica (2006)). Second, research into greentech that uses evidence acquired at firm level is very limited. Indeed, hardly any study to date has empirically investigated the financial characteristics and potential constraints of greentech firms. Third, there is scant research that elaborates on the role of finance in greentech innovation and diffusion and that integrates relevant political aspects. Finally, existing research indicates an implicit funding gap when greentech moves from the research stage to commercialization. (e.g., Burtis (2006); Chertow (2000); Foxon et al. (2005)) Although there is only anecdotal evidence to support the existence of this gap and very limited theoretical validation of the conditions under which it occurs, a potential financing gap is of great practical significance considering the extent of governmental funding programs and the question of their effective allocation. This dissertation aims to increase knowledge about greentech innovation and diffusion, its financing and relevant policies from a firm-level perspective. Drawing on existing business research and contemporary empirical evidence, it formulates propositions that lead to an integrated theoretical framework that overcomes the causal antagonisms of interdisciplinary research efforts. Finally, hypotheses are derived and tested empirically.

1.2 Research questions and contributions

Although the volume of research has increased in recent years, knowledge about how greentech is emerging as a firm-driven market space is still limited. This dissertation aims to address precisely this issue. It seeks to diminish the research gap by developing and introducing a financial economics and firm-level perspective. Moreover, it elaborates on the way in which existing public support programs affect a firm's financial situation and presents empirical evidence to test theoretical assertions and implications. The leading research question is therefore:

Is innovation in and the diffusion of green technologies constrained by the financial situation of greentech firms?

As the above research question is rather broad, the following sub-questions operationalize the overall objective of this thesis and contribute to addressing the identified research gaps. The first question aims at assessing technology innovation and diffusion and its implications for greentech firms. In particular, it seeks to identify factors that affect firms' ability to innovate and grow:

- What affects firms' ability to innovate and grow and, hence, the emergence of the greentech market space?

The second question relates to the role of finance. While existing research has focused strongly on venture capital (VC) and VC investors (e.g., Diefendorf (2000); Kenney (2009); Moore and Wüstenhagen (2004); Randjelovic, O'Rourke, and Orsato (2003); Wüstenhagen and Teppo (2006)), the present study aims to broaden the scope of analysis by integrating finance theory and empirical evidence that is mainly derived from small and medium-sized enterprises (SMEs). This analysis is accompanied by a theoretical and empirical validation of the aforementioned funding gap in the greentech innovation and diffusion process based on an investigation of the financial constraints faced by firms. The dissertation is therefore ultimately guided by the following question:

- Do greentech firms face financial constraints and, if so, what are the main determinants thereof?

The third sub-question assesses the role of government support systems. The dissertation argues that, contrary to the majority of existing research that adopts a technology perspective (e.g., Chertow (2000); Grubb (2004); Loiter and Norberg-Bohm (1999)), analysis must focus on the firm. Doing so will allow for a more coherent assessment of policy actions covering every aspect of progress in the industry. Moreover, it takes a look inside the "technology innovation and diffusion black box" by assessing the very instruments – the greentech firms themselves – that are driving the emergence of this market. It will also enable researchers to broaden their focus from renewable energy technologies – where diffusion can be measured in terms of installed capacity – to other greentech sectors in which adequate measures of diffusion are less obvious. The third question addressed is:

- How do greentech-oriented government support programs affect financing and, hence, innovation and diffusion?

The fourth research question builds on the first three sub-questions. It aims to draw implications for two groups of key stakeholders: greentech firms and policymakers. The question can be formulated thus:

- What implications for greentech firms and policymakers can be drawn from analysis based on the aforementioned research questions?

In answering the above questions, this dissertation simultaneously contributes to several bodies of literature. First, it contributes to general business research by proposing an interdisciplinary framework that integrates the findings of literature on innovation, growth and finance. Second, it sharpens our understanding of how greentech firms innovate, grow and, in doing so, drive a shift in our institutional and technocratic system. Third, it provides a fresh, firm-level perspective on how government support programs impact the greentech market space. Finally, it nudges financial research on greentech firms in the direction of financial economics. As such, it broadens the scope of analysis from venture capital to other financing sources and constitutes one of the first rigorously empirical analyses in its field.

1.3   Organization of research

The research structure is shown in figure 1-1. The dissertation is composed of five main sections. First, it introduces the topic and defines both key terms and research questions (chapter 1). This introduction is complemented by an overview of existing knowledge in greentech business and economic research, plus a clear definition of the overall research context (chapter 2). Second, the dissertation reviews relevant literature at technology and firm level that covers the main pillars of the theoretical framework: technology innovation and diffusion, firms' innovative activities, firms' growth and development, financial economics, and policy support programs in the greentech industry (chapter 3). Third, it proposes a theoretical framework that integrates and derives hypotheses from the literature streams analyzed in chapter 3 (chapter 4). Fourth, it empirically tests the theoretical framework and provides quantitative evidence regarding the greentech industry. Finally, it draws conclusions and issues recommendations to greentech firms and policymakers, as well as discussing the major contributions to this study (chapter 6). An overall summary of the findings concludes the dissertation (chapter 7).

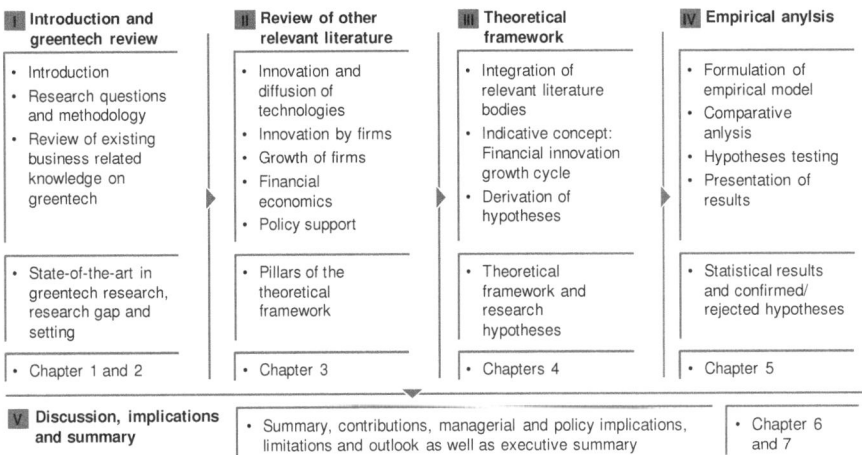

Figure 1-1: Organization of research

Although their lengths vary, all five parts are equally important and pursue different objectives. The first part summarizes existing knowledge by identifying the main research perspectives adopted to date. It demonstrates that firm-level evidence is scant, but that it deserves in-depth attention from researchers, particularly in the fields of innovation, corporate growth and financing. The second part aims at theory building, preparing fertile ground for the theoretical framework and, hence, for empirical testing. The third part seeks to integrate the findings of the second part into a single comprehensive framework. The theoretical framework thus developed is new and is one of the major contributions made by this dissertation, given that existing literature fails to incorporate the partially contradictory results of the distinct literature streams. The fourth chapter aims at validating the theoretical framework and providing insights into greentech firms' innovative, growth and financing activities. The fifth part formulates recommendations for policymakers, greentech firms and researchers. It also elaborates on how the findings of the dissertation add to existing knowledge in the discussed literature bodies.

1.4 Scope and limitations

This dissertation focuses on the greentech industry, its development, its financing patterns and the effect of government initiatives on individual firms. The mixed-method approach combines existing knowledge from various bodies of literature, develops a theoretical model and validates the formulated propositions empirically. A broad definition of the greentech industry along technology lines opens the door to a profound analysis that goes beyond clean energy on which many researchers in the field have focused hitherto.

However, the analysis does have certain natural limitations. The empirical work presented in this dissertation is restricted to Germany. The firms it covers are of German origin, or at least have branches in the German market. Its analysis of government support examines programs that are available to the German market and are focused on either green technologies or technology-oriented environmental projects. The sources of these programs may be European, federal or regional. Analysis in this dissertation is also limited to the level of the individual firm. Accordingly, the decision patterns of equity investors, banks and other financial institutions and how these groups perceive the industry are not a primary subject of this study. The dissertation also has technical shortcomings in that its line of argumentation is based on several assumptions. These assumptions are discussed in chapter 6.5 in order to identify potential areas of future research.

2 REVIEW OF EXISTING KNOWLEDGE ABOUT GREENTECH

"Cleantech has the potential to help propel our economic system towards a more environmentally sustainable future. However, the nascent cleantech industry is now at a critical juncture – it could become a large growth industry, fall back into a niche investment category, or simply dissipate into the many different industries to which it connects"
O'Rourke (2009)

In the statement quoted above, O'Rourke reminds us that greentech (see below for a comparative definition of "greentech" and "cleantech") has reached a point in its evolution at which future development will be shaped, and at which the industry's fate hinges on decisions taken by stakeholders and investors in particular. Researchers are laying the foundation for these decisions by providing knowledge about greentech from various perspectives. At the same time, however, greentech remains a young, multidisciplinary and interdisciplinary field of research in which many practical and highly relevant issues have yet to be touched by academic researchers (Russo (2003)).

2.1 Definitions and terminology

2.1.1 Greentech

The term "greentech" stands for green technologies and competes with other symbolically isomorphic terms such as cleantech, sustainable technology, etc. Definitions of greentech still vary across a broad range of technology lines. (Kenney (2009)) One often cited definition is that used by the US National Science and Technology Council, for which greentech *"is a technology that advances sustainable development by: 1) reducing risk, 2) enhancing cost effectiveness, 3) improving process efficiency, and 4) creating products and processes that are environmentally beneficial or benign"* (US National Science and Technology Council, 1994, cited in Chertow (2000)).[2] In this dissertation, the term greentech refers to technologies that reduce environmental degradation by optimizing the use of natural resources while at the same time generating economic value. However, the problem with all greentech definitions is that they do not clearly state which technologies they include or exclude. For this reason, many scholars and practitioners have drawn up their own industry categories (e.g., Büchele et al. (2009); Burtis (2004); O'Rourke (2009)). This dissertation draws on the six industry categories defined by Büchele et al. (2009) and shown in table 2-1.

---

[2] For a detailed analysis of the term and its emergence see O'Rourke (2009)

Technology line	Sub-technologies
Environmentally sound energy technologies	Efficient power plant technologies (e.g. combined heat and power, steam power plants) Emission reduction at power generation (e.g. flue gas cleaning, carbon capture and storage) Renewable energy technologies (e.g. solar pannels, wind turbines) Energy storage technologies (e.g. compressed air, geothermal and hydrogen storage) Hydrogen technologies and fuel cell applications (e.g. fuel cell)
Energy efficiency	Insulation materials (e.g. polyurethane building pannels, insulated glazing) Energy-efficient cooling systems (e.g. solar cooling) Energy-efficient production technologies (e.g. pumps, compressed air systems) Energy-efficient products (e.g. home appliances, lamps) Energy services (e.g. contracting, energy consulting)
Sustainable water management	Efficient-use technologies (e.g. flood control, appliances) Water recovery technologies (e.g. ultra filtration, desalination)
Sustainable mobility	Alternative motor drives (e.g. electronic engines) Traffic management systems (e.g. smart traffic control) Energy efficient automotive components (e.g. LED lamps, light weight materials) Efficient transport systems (e.g. freight networks)
Material efficiency and sustainable resources	Material efficient products and machinery (e.g. recycable paper) Renewable substitutes (e.g. organic plastics) New materials (e.g. white biotech)
Waste management and recycling	Ex post technologies (e.g. Recycling, autoclaving) Ex ante technologies (e.g. waste minimizing machinery)

Table 2-1: Greentech industry categories
Source: Büchele et al. (2009)

2.1.2   Clean energy versus greentech

Clean energy can be considered a sub-category of greentech. It comprises energy technologies and services that reduce environmental degradation, are socially acceptable and are economically beneficial. According to Moore and Wüstenhagen (2004), clean energy breaks down into four main categories: renewable energy, distributed energy systems, natural gas and demand-side energy efficiency. In this dissertation, clean energy is defined as environmentally friendly power and energy efficiency technologies. As the review of existing knowledge on greentech clearly shows, clean energy is the focal point of the majority of existing research. There are two reasons for this: First, greentech spans many different industries whose greentech activities are difficult to isolate from other business segments. For example, a firm that produces parts for combustion engines may also market components for electric vehicles. Clean energy, on the other hand, can be narrowed down to a certain set of firms that consider clean energy to be their core competence. Second, the performance and visibility of greentech tends to be more readily apparent in clean energy applications than in many other valid cases. For example, the diffusion of wind power technologies is not only highly visible but also easy to measure in terms of installed power generation capacity. By contrast, sustainable water technologies literally remain below the visibility surface, while their impact too is more complex to measure.

This dissertation argues that research must widen its scope and cover the entire greentech industry if we are to understand the role technology could play in achieving sustainable development. Weight is lent to this argument by the facts that clean energy makes up less than half of the overall greentech market volume (Büchele et al. (2009)), and that many existing government programs do not aim solely to support clean energy. A broader research focus would strengthen the perception of greentech as a coherent industry, provide decision-makers with guidance on effective policy instruments and thus increase the likelihood of a technology-embracing trajectory toward sustainable development.

2.2   The five perspectives covered by greentech business research

The following review of existing knowledge seeks to prepare the ground for the theoretical framework and empirical analysis. The first exhaustive review of existing business-related knowledge on greentech outline's the industry's practical relevance and provides a comprehensive overview of current research issues. It further aims to synthesize existing knowledge about greentech in order to validate the identified research gaps, gain a new perspective, identify links between ideas and concepts, understand the methodologies that have been used and place the proposed research in its overall context. Accordingly, the review is based on recommendations made by Randolph (2009) regarding the objectives and structural components of best-practice literature reviews.

Research into greentech is a truly multidisciplinary and interdisciplinary field. To accommodate this fact, the following literature review is drawn from five main perspectives to reflect the status quo in economic and business-related greentech literature as shown in figure 2-1: (I) A sustainability perspective; (II) an entrepreneurial perspective; (III) an innovation and diffusion perspective; (IV) a policy and regulation perspective; and (V) a finance perspective. These five perspectives were identified by an exhaustive literature review process with selected citation focusing on peer-reviewed journal articles. The review process excluded technological and engineering-oriented studies, focusing instead exclusively on business and economics publications.

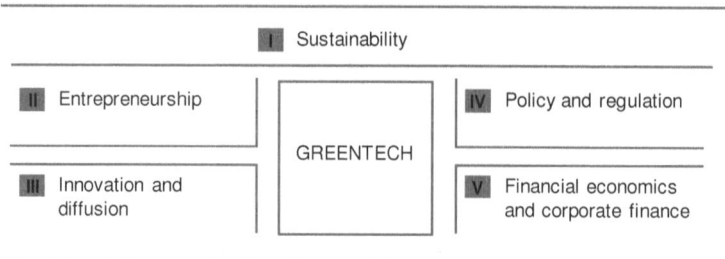

Figure 2-1: Current perspectives in greentech business related research

The literature search was conducted using the EBSCO database in early 2010. Searches were made for greentech and its assorted synonyms (such as greentech, cleantech, clean tech, envirotech). A total of 2,171 articles were found, of which 826 had been published in academic journals. After excluding technology-focused and non-business-related papers, a mere 69 relevant articles were identified. The main journals were Research Policy, the Journal of Business Venturing and Energy Policy, for instance. A further 43 relevant publications were identified by means of citation tracking. The five perspectives were developed non-exclusively based on the main streams of thought, e.g. the innovation value chain as an aspect of the innovation and diffusion perspective. Additionally, common notions were aggregated and literature patterns summarized.

2.2.1   The sustainability perspective

Greentech is increasingly regarded as a "silver bullet" to achieve sustainable development without harming economic and social welfare. The term "sustainable development" first became established at the United Nations Conference on the Human Environment in 1972. The term rose to prominence through the "Our Common Future" report written by Norwegian Prime Minister Gro Harlem Brundtland (Brundtland (1991)) and presented to the United Nations by the World Commission on Environment and Development. This report also established the commonly accepted definition of sustainable development: *"Sustainable development is development that meets the needs of the present generation without compromising the ability of future generations to meet their own needs" (Brundtland (1991)).*[3] In line with this definition, sustainable development implies that the use of resources must not

---

[3] Ongoing debate surrounds the practicability and operational definition of sustainable development. In 1998, the German council of environmental advisors therefore formulated a more action-oriented definition that covers nine common areas of sustainability problems. With respect to the environmental dimension of sustainability, these areas are: (1) greenhouse gas emissions, (2) ozone layer depletion, (3) acidification, (4) eutrophication, (5) toxic impacts on ecosystems, (6) toxic impacts on humans, (7) loss of biodiversity, (8) use of land and (9) use of resources. The council further identified energy, mobility and waste as the main economic sectors in which to tackle these problems. (SRU 1998, cited in Rennings (2000))

exceed their regeneration or substitution. In other words, the ratio of resources used to resources regenerated must be equal to one. In practice, this can only be achieved by leveraging renewable resources and increasing their regeneration capabilities, as well as excluding non-renewable resources from current consumption patterns.

In this context, greentech may serve as a major tool to support sustainable development in various ways. There can be little doubt that fighting environmental problems necessarily involves the rigorous application and diffusion of technologies. (e.g., Grubb (2004); Huber (2004)) Consequently, many authors advocate the benefits of greentech by outlining its potential for "green jobs" and wealth creation, coupled with the ability to simultaneously reduce negative environmental externalities. (e.g., Büchele et al. (2009); Burtis (2006); Horbach (2008b); Stern (2008); WWF (2009)) Greentech is expected to satisfy the three dimensions of sustainability – social, environmental and economic – often referred to as the "triple bottom line". (Elkington (1998)) According to Germany's national bureau of statistics and as shown in figure 2-2, the German greentech industry grew from EUR 20 bn in 2007 to EUR 24.7 bn in 2008, involving 2,914 and 3,024 firms respectively. The employment effects are substantial, with the firms concerned giving work to more than 80,000 people. (Destatis (2009))

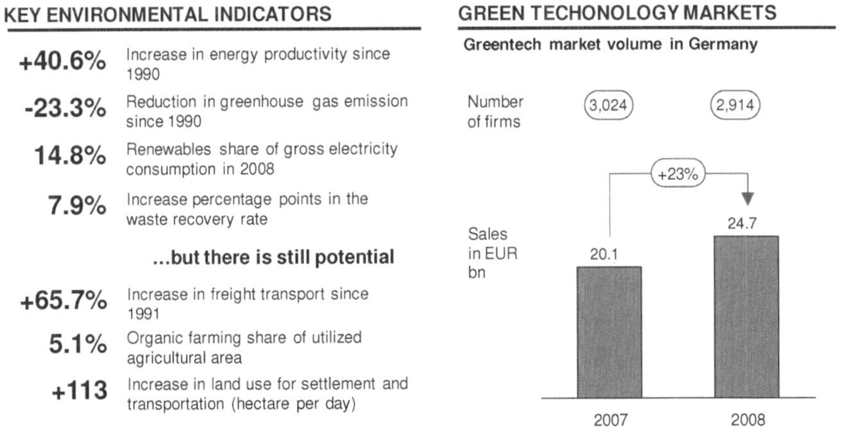

KEY ENVIRONMENTAL INDICATORS

+40.6%	Increase in energy productivity since 1990
-23.3%	Reduction in greenhouse gas emission since 1990
14.8%	Renewables share of gross electricity consumption in 2008
7.9%	Increase percentage points in the waste recovery rate

...but there is still potential

+65.7%	Increase in freight transport since 1991
5.1%	Organic farming share of utilized agricultural area
+113	Increase in land use for settlement and transportation (hectare per day)

GREEN TECHONOLOGY MARKETS

Greentech market volume in Germany

Number of firms: (3,024)   (2,914)

+23%

Sales in EUR bn: 2007 = 20.1   2008 = 24.7

Figure 2-2: Greentech market development
Source: Destatis (2009), Federal Environment Agency (2009)

The figures shown above can be regarded as rather conservative estimates. The industry segments covered include water conservation, climate protection, noise reduction, air purification and nature protection. Other technology lines subsumed under the definition of this dissertation, such as sustainable mobility technologies and technologies that improve material and resource efficiency, are not considered. Other authors believe that the market is significantly larger. For example, Büchele et al. (2009) estimate that the greentech market accounted for approximately 8% of Germany's GDP and employed 1.8 million people in 2007, while WWF (2009) calculate that, in Germany, another 800,000 jobs may be created by 2030. Compared to the economic and social impact of the greentech industry, its environmental performance is more difficult to assess. According to the German Federal Environment Agency (2009), national energy productivity has increased by 40.6% since 1990, greenhouse gas emissions have been reduced by 23.3% since 1990, the share of gross electricity consumption from renewable energies reached 14.8% in 2008, and the waste recovery rate has increased by 7.9% since 1990. It is thought that the improvement in these KPIs is attributable primarily to the growing utilization of green technologies in Germany. The energy sector in particular has made significant structural changes to improve its overall environmental impact.

The German greentech industry is today one of the largest in the world. According to Büchele et al. (2009), German firms supplied up to 30% of global market demand in 2007. Specifically, German firms' had the following global market shares: 30% for environmentally sound energy; 12% for energy efficiency; 6% for material efficiency; 10% for sustainable water management; 18% for sustainable mobility; and 24% for resource and waste management technologies. Nevertheless, it should be noted that greentech alone can never lead to sustainable development. It must be accompanied by significant political and social changes. Moreover, the technologies that are available today only enable resource consumption to be reduced and environmental degradation to be decelerated. Neither can be halted entirely, however. Some researchers even claim that the gains achieved by adopting green technology are offset by the increase in resource utilization due to increased production. (e.g., Müller-Christ (2001))

Whichever view is closer to the truth, there are ultimately three implications that cannot be avoided. First, greentech has the long-term potential to fulfill the three dimensions of sustainability. However, further radical innovation seems necessary because, as we have just seen, the technologies available today can only reduce but not eliminate environmental degradation. The interrelation between sustainable development and the use of green technologies will therefore continue to attract the keen interest of researchers, practitioners and society at large. In the area of sustainability research, room for further studies is unlimited. One of the most urgent research gaps seems to be the one mentioned above – a fact that leads to the second implication: Further research is needed to clarify the effect of using green technology on sustainable development. Third and finally, whatever impact greentech may have in the long run, its present economic development is already of extraordinary practical importance today.

2.2.2 The entrepreneurial perspective

Entrepreneurship is widely recognized as a means to transform societies and economies by introducing innovations and new technologies to existing markets. Following a robust stream of literature on entrepreneurship, some researchers investigate the ability to recognize business opportunities derived from social and environmental market failures. (e.g., Cohen and Winn (2007); Dean and McMullen (2007); Hall and Vredenburg (2003); Hart and Milstein (1999)) The principal idea is largely based on market failure theory and environmental economics. Following this line of thought, negative externalities caused by market participants cannot be traced to their origins and thus cannot fully be internalized. The economic costs that arise from these externalities are instead spread among all participants, such that market mechanisms fail to yield an equilibrium state. While environmental economists argue that these costs must be allocated to the originator and that government interaction is necessary in order to achieve sustainable development (Dean and McMullen (2007)), entrepreneurship researchers see substantial business opportunities in precisely this situation. Where a market fails to yield an equilibrium state, new firms can engage in arbitration activities to exploit market imperfections. This leads to entrepreneurial rents on the one hand, but may also change the rules of the game on the other hand. Applying the concept of "creative destruction" (Schumpeter (1943)), these authors therefore claim not only that market failures create entrepreneurial opportunities, but also that exploiting these opportunities may solve the problem of negative externalities. In other words, entrepreneurial activity could reduce environmental degradation.

Cohen and Winn (2007), for instance, identify four market failures that violate perfect market assumptions: (1) firms are not perfectly efficient; (2) externalities exist, (3) pricing mechanisms work imperfectly; and (4) information is not distributed perfectly distributed. These market failures provide entrepreneurs with windows of opportunity that may be either disruptive or incremental in nature (Hart and Milstein (1999)). They may also be driven by market mechanisms (such as profit and revenue) or policy actions (such as energy security and wealth creation) (Hall and Vredenburg (2003)). However, opportunities are exploited not simply because market failures exist, but as the outcome of entrepreneurial action and its interrelation with institutional systems. Entrepreneurial action in the greentech industry hinges on the willingness of individuals to pursue opportunities and on individuals' orientation towards sustainability. (e.g., Choi and Gray (2008); Kuckertz and Wagner (2010)) Greentech entrepreneurs in particular possess considerable potential to bring about institutional change that could lead to greater sustainability. This is because greentech follows well established norms and economic principles, presenting significant financial upside potential while at the same time accommodating environmental considerations. (Basu, Osland, and Solt (2009); Randjelovic, O'Rourke, and Orsato (2003)) When entrepreneurs with a focus on sustainability exploit a given opportunity, they begin to shape their institutional system and vice versa (Anderson and Leal (2001); Meek, Pacheco, and York (2010)) under favorable socioeconomic conditions over time (York and Venkataraman (2010)). During this process, new technologies can develop from niche markets into industry standards, thereby challenging incumbent firms and other market players. (Hockerts and Wüstenhagen (2009)) The entrepreneurial firm thus triggers a process of technology innovation and diffusion.

Although greentech and sustainability have so far received little attention in entrepreneurship literature (Hall, Daneke, and Lenox (2010)), three major conclusions can be derived. First, environmental problems open up significant opportunities for entrepreneurial action in general and for greentech in particular. Second, the exploitation of opportunities depends on the characteristics and persuasions of the entrepreneur. To put that another way, sustainability-oriented individuals are more likely to become greentech entrepreneurs. Finally, the exploitation of opportunities impacts the market space concerned. Socioeconomic and institutional systems will thus change, creating new opportunities for new entrants.

2.2.3   The innovation and diffusion perspective

The innovation and diffusion process for products and technologies has for many years attracted considerable attention in academic research. (e.g., Schumpeter (1943); Levitt (1965); Rogers (2003)) Based on this research stream, greentech – and especially renewable energy technology – are increasingly being recognized as a real-time case study for innovation and diffusion theory. Many scholars have adopted an evolutionary perspective to analyze innovation and diffusion patterns. (e.g., Balachandra, Kristle Nathan, and Reddy (2010); Chertow (2000); Grubb (2004); Neuhoff (2005); O'Rourke (2009); Wüstenhagen, Markard, and Truffer (2003)) Indeed, the perception that green technologies evolve by means of a time-dependent process seems to be commonly accepted. Within this process, technologies achieve or pass certain stages of development such as R&D, commercialization and diffusion. Although literature provides us with no mutually exclusive, collectively exhaustive and widely agreed definition of these stages, there seems to be a consensus that they exist and that they characterize a certain development situation in a given technology trajectory. However, when it comes to stage transition in greentech, the development stages of technology lines differ significantly. While some technologies (such as fuel cells and solar heating panels) are still at an early stage of development, others (such as sewage systems and wind power applications) have already reached a point of large-scale diffusion. (e.g., Dinica (2006); Jacobsson and Lauber (2006); Neuhoff (2005); Srinivasan (2007)) The commonly held belief is that, from the moment of initial innovation, technologies undergo a process of learning and optimization, gradually gaining market acceptance and, hence, market share. This process is not necessarily linear and may well alternate between steady incremental improvements and disruptive changes. (Garud and Karnøe (2003); Lund (2007))

Due to greentech's powerful influence on public welfare, some authors argue that policy must support the innovation and diffusion process. The underlying idea is that free market mechanisms alone (demand, supply and pricing, for example) are not strong enough to move green technologies from one stage of development to another because of structural barriers such as the availability of financing, of the inability to internalize externalities and persisting conventional technology regimes. (e.g., Chertow (2000); Grubb (2004); Rennings (2000)) Moreover, the commercialization of new technologies requires collaboration between the players involved and detachment from their traditional counterparts Hellström (2007). It can thus take decades to overcome the barriers raised by existing technology trajectories Dismukes, Miller, and Bers (2009). Even technologies that are already cost effective, such as wind power applications Cowan and Daim (2009), may be locked out because of deeply rooted social and institutional practices. Greentech innovation and diffusion therefore involves the support of various stakeholders that oppose the existing lobbies whose preference lies with conventional technologies. (Jacobsson and Johnson (2000); Unruh (2000)) Greentech innovation and diffusion thus implies the co-evolution of sociotechnological systems (e.g., O'Rourke (2009); Russo (2003); Tsoutsos and Stamboulis (2005)) as well as certain social and cultural factors (Brechin and Steven R. (2003); Sovacool (2009); Wüstenhagen, Wolsink, and Bürer (2007)). The sociotechnological system involves all direct stakeholders, such as governments, educational institutions, distribution channels, etc. These stakeholders have a strong impact on technology development by supporting or ignoring technologies that enter the market space. The sociotechnological system is to some extent influenced by social and cultural factors that are reflected in the public opinion and perceptions of the environment, technology, economy, etc.

Two main implications can be derived from existing research into greentech innovation and diffusion. First, different green technologies have currently reached different stages of innovation and diffusion; and these stages determine their respective characteristics and development paths. Second, in the case of greentech, free market mechanisms are weak and may not be sufficient to overcome persisting structural barriers. Policy action is required to deal with market failures, overcome barriers to innovation and diffusion and thus shape the sociotechnological system.

2.2.4   The policy and regulation perspective

"The problem that environmental regulation seeks to correct, pollution, is a negative externality. What we seek to protect – air, water, land – are commons. If polluters picked up

*the costs of environmental degradation on their own, such regulations would not be
necessary. But, historically, the commons have been too available and the cost, therefore, of
proper treatment of non-product output, viewed as too high. Thus market fails on its own to
get the socially desired outcome. Without regulation, followed by enforcement, individuals
and firms will pollute too much."*

Chertow (2000)

Chertow's statement reveals that greentech market development has to a large extent been driven by policy action and regulation. The ratio of benefits to the environment and society to economic returns is high. Policy intervention is seen necessary for three main reasons. First, failure to internalize environmental externalities (e.g. where the environmental cost of air pollution caused by a factory owner is shared by the entire local community) leads to failure in the market mechanisms that would otherwise cause greentech to be used more widely. (Tsoutsos and Stamboulis (2005)) Second, consumers nowadays are not prepared to pay a premium for a public good from which everyone benefits (such as non-subsidized energy from renewable technologies). Free-riding thus remains one of the key challenges to sustainability. (Menanteau, Finon, and Lamy (2003)) Third, spill-over effects combined with insufficient R&D market signals and incentives for greentech firms create substantial barriers to technology innovation. For example, R&D investments are risky and are therefore to some extent avoided by risk-averse managers. Even where firms do innovate, other firms will learn from their experience and skim off some of the returns. (e.g., Nemet (2009); Rausser and Papineau (2010)) Policy makers therefore try to strengthen market mechanisms to the benefit of green technologies in order to overcome these market failures. (e.g., BMU (2009)) The majority of research to date has thus focused on policy instruments, actions and regulations promoting greentech in general and renewable energy technologies in particular. Figure 2-3 provides an overview of the main instruments used in the greentech context worldwide.[4]

---

[4] It should be noted that the terms push and pull policies are used widely but not consistently. In most cases, push policies refer to government activities and support programs that improve conditions for technology development and testing. Conversely, pull policies refer to activities that foster existing technologies by creating technology demand. For a detailed analysis of categorization approaches in literature, see Taylor (2008).

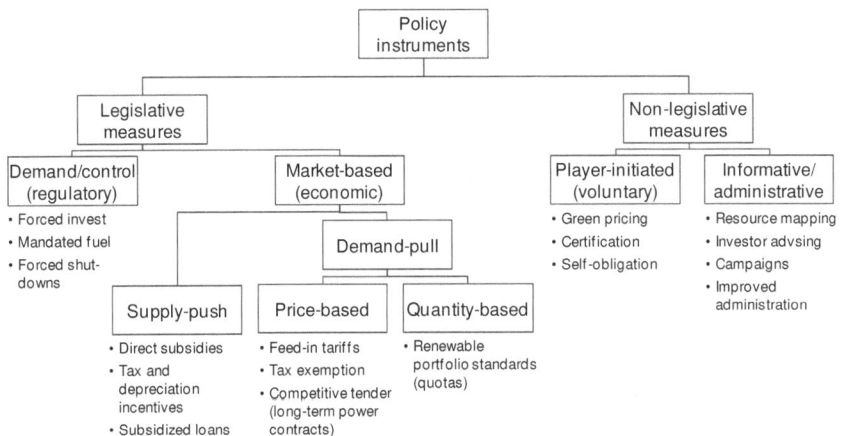

Figure 2-3: Policy instruments and actions
Source: Adapted from Enzensberger, Wietschel, and Rentz (2002)

Early studies suggested that implementing a polluter-pays principle would be enough to foster greentech innovation and diffusion (Huber (1986); Kemp and Soete (1992)), and that technologies would quickly become competitive due to learning effects (Clulow (1999)). Since then, however, debate has intensified regarding which policy mix is the most effective and efficient in promoting green technologies. Many observers assess policy effectiveness in terms of the degree of diffusion, i.e. installed capacity for renewable energy technologies. The cost of implementing the chosen policy is thus seen as a measure of efficiency. In this context, demand-pull policies – especially feed-in tariffs – seem to yield the best diffusion results (e.g., Couture and Gagnon (2010); Lipp (2007); Loiter and Norberg-Bohm (1999); Menanteau, Finon, and Lamy (2003); Mitchell, Bauknecht, and Connor (2006); Wüstenhagen, Markard, and Truffer (2003)), although not necessarily at the lowest cost (Lund (2007)). However, others claim that demand-pull instruments are not sufficient to stimulate (disruptive) innovation. These authors argue that demand-oriented systems lock in existing technologies, and that innovation must be considered as an important variable when assessing policy effectiveness and efficiency. Policy-driven programs must therefore take effect at the very beginning of the diffusion path, i.e. during the R&D and technology demonstration stage. Appropriate actions include supply-push instruments such as government R&D programs and grants. (Nemet (2009); Rausser and Papineau (2010); Taylor (2008)) Combining these views would imply that policymakers should implement a policy mix that uses multiple instruments and programs. Some authors therefore suggest that policy-driven actions must be aligned with the dynamics of innovation and the stages of development that make up a specific technology. Policymakers should focus on the transitional evolution of technologies along the innovation value chain. Based on innovation and diffusion theory, it is proposed that new technologies are best addressed by supply-push instruments, while more mature technologies should be supported by demand-pull programs. Fuel-cell technologies can best be supported by means of R&D subsidies, for example, whereas wind power technologies should be flanked with demand-push incentives. (Chertow (2000); Foxon et al. (2005); Grubb (2004); Krozer and Nentjes (2008); Neuhoff (2005); Nill and Kemp (2009)) The disadvantage of this approach, however, is that innovation is still regarded as a "black box" and that interaction between the actors is largely disregarded. Accordingly, some authors propose to broaden the analysis framework by introducing institutional theory (Hoffman and Ventresca (1999)), considering all relevant stakeholders (Enzensberger, Wietschel, and Rentz (2002)) and integrating competitive forces in the policy logic (Hockerts and Wüstenhagen (2009)). Another measure of policy effectiveness and efficiency that should be considered is the policy's impact on private investment and a firm's financing situation. Because policy programs are implemented through the direct (e.g., grants) or indirect allocation of public funds (e.g., tax breaks), there is a strong connection to a firm's financing decisions and greentech investment markets. In this context, policymakers should consider policy-related financing effects to avoid crowding out and/or substituting for investments that would have occurred anyway. (Dinica (2006); Kasemir, Toth, and Masing (2000)) Broadly speaking, whatever policy mix is implemented, there seems to be a general consensus that legislative and financial support should be flanked by "soft" programs such as education, information, mediation, etc., in order to reduce information asymmetries, increase awareness of sustainability issues and foster innovation. (Horbach (2008a); Lewis and Wiser (2007); Wüstenhagen and Bilharz (2006))

Although many issues are still the subject of discussion, three main findings can be condensed from existing research. First, there are no common evaluation criteria that appropriately assess the effectiveness and efficiency of policy initiatives. Second, there is growing agreement that there is no one best policy. On the contrary, a complex mix of policies is needed to adequately address the needs of greentech innovation and diffusion. Finally, policymakers should always consider the impact of their actions on financial markets and firms' financing decisions.

2.2.5   The finance perspective

There are two main financial research streams in literature that are of relevance to greentech: (1) corporate finance, in which almost all studies are limited to venture capital (VC); and (2) project finance. Although both constructs seek to finance greentech development, their research objects are quite different. Corporate finance sees the firm as the main source of insight and thus centers on green technology and service firms. Project finance takes greentech projects (such as wind parks or solar power plants) as its focal point and therefore restricts itself to the application or end-customer segment in the value chain. Both constructs are important to the development of the industry. On the one hand, venture capital can provide funds to enable technology firms to grow and invest in R&D that will ultimately bring innovations to market. On the other hand, project finance provides funds for the implementation and use of existing (project-compatible) technologies, thereby enabling their diffusion. This section focuses on corporate finance issues.[5]

As demonstrated earlier, corporate finance issues in the greentech industry have been researched largely from an industry development point of view, with the majority of existing surveys limiting their scope to VC. In this context, ongoing debate has questioned whether VC can indeed bring about greentech growth and development, or whether other financing mechanisms need to step in – or even be developed in the first place. VC has proved to be an important driver of innovation, the professionalization of firms and economic growth in industries such as biotechnology and the Internet. Some authors thus believe that VC harbors the same potential for greentech. (Burtis (2006); Moore and Wüstenhagen (2004); O'Rourke (2009)) However, until 2006, worldwide VC investments in this industry were virtually non-existent. VC investment in greentech only began to pick up in 2007; and ·even now, it still does not represent a significant amount compared to other industries. Figure 2-4 traces this development in Germany. Although the investment volume in 2007 increased by more than 50 times compared to 2006, VC's share of overall greentech investment does not exceed 4%. Nor has this changed in the meantime – a fact that above all affects early-stage and innovative green technologies. (e.g., Burtis (2004); Burtis (2006); Chertow (2000))

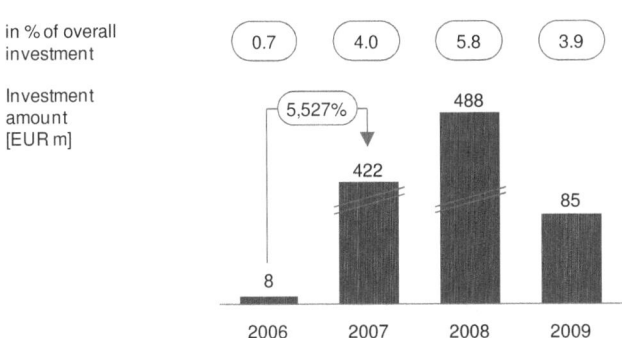

Figure 2-4: VC investment in greentech in Germany (environmental and energy category)
Sources: BVK (2007), BVK (2008), BVK (2009) and BVK (2010)

---

[5] A discussion of project finance and its implications is beyond the scope of this dissertation. The interested reader may wish to refer to Enzensberger, Fichtner, and Rentz (2003).

The greentech market is highly fragmented and associated transaction costs are high. (Kammlott and Schiereck (2009)) In addition, there seems to be a mismatch between greentech applicants' characteristics and current VC firms' acceptance criteria. (Hamilton (2009); Kenney (2009)) One problem that exacerbates this potential mismatch is the lack of deal experience among greentech firms and VCs alike. In other words, greentech firms do not know how to source VC and VC firms do not know how to find the right investment targets. (Christensen (2009); Randjelovic, O'Rourke, and Orsato (2003)) Furthermore, the track record of successful exit stories is very limited, while empirical evidence is rather indicative. Even more problematic is the fact that what little evidence there is does not seem to promise the same returns in greentech as in, say, the biotech or Internet industries. (Moore and Wüstenhagen (2004); Teppo (2006)) Some authors also see greentech's dependence on regulatory and policy instruments as a major cause for investors' reservations. In this context, Diefendorf (2000) was one of the first scholars to recognize that venture capital firms tend to avoid industries that depend on government support. At the same time, other authors suggest that government needs to support VCs to bridge the investment gap in the innovation value chain, as seen above. (e.g., Burtis (2006); Chertow (2000); Murphy and Edwards (2003))

However, more recent evidence suggests that VC investors do in fact appreciate certain government support systems as they improve investment economic conditions. (Bürer and Wüstenhagen (2009); Bürer (2008)) Moreover, there is a growing consensus that policy action does not necessarily crowd out private investment, but could actually increase it. (e.g., Lewis and Wiser (2007)) As indicated above, one major requirement for this to happen is that policy action must take account of financial considerations (Dinica (2006); Hamilton (2009)) and take long-term steps to improve market stability (Hamilton (2009)). Finally, some authors argue that the valuation methods used to assess greentech ventures are inadequate to account for the unique specifics of this industry. With respect to renewable energy technologies, conventional engineering-based valuation approaches ignore both the value of energy security and managerial options. (Awerbuch (2000)) Indeed, the engineering-based valuation techniques used today were developed at a time when green technologies did not exist. It is therefore perhaps only natural that they favor conventional technology solutions. (Rave (1998)) Overall, the availability of capital is essential if an industry is to develop. (Tsoutsos and Stamboulis (2005)) This issue will therefore be a major pillar of greentech innovation and diffusion. Practitioners have apparently got ahead of research. For instance, the German Federal Ministry for the Environment recently launched a program to foster investment in greentech, arguing that only 2% of Germany's private capital is devoted to sustainable investments. (Pohl (2010)) Although research is generally scarce, the knowledge already available lays a firm foundation for further research efforts.

Four main implications can be derived. First, distinctions must be drawn between two types of greentech finance: corporate finance and project finance. Second, there seems to be a lack in VC; and indicative evidence suggests that firms face financial constraints when trying to take technologies from early-stage development to commercialization. Third, policy action directly impacts investment markets and firms' financing decisions, although we currently only have little knowledge about how this impact materializes in practice. Finally, greentech seems to possess properties and attributes – regarding risks and investors' perception of returns, for example – that are very different from other industries. Such differences may justify different financing mechanisms.

2.2.6 Conclusion – A plea for a financial economics and firm-level analysis

Even though greentech is already of great practical significance, only future development will show if it could truly become the "silver bullet" for which politicians hope. The dissertation has analyzed existing research efforts in the greentech business from five perspectives. These perspectives provide a coherent overview of existing knowledge and knowledge gaps, describe the main characteristics of the greentech industry, flesh out the most important findings from present-day research, and enable relevant bodies of literature to be identified in order to develop the theoretical framework for this dissertation. From an entrepreneurial and sustainability perspective, greentech provides the means to exploit business opportunities while solving environmental problems and offering economic rents to the entrepreneur. In this context, opportunities arise from market failure, causing negative externalities that environmental entrepreneurs may recognize and address by developing greentech solutions. Adopting an innovation and diffusion perspective shows that green technologies grow as opportunities are recognized in a time-dependent process composed of different stages of development. Moreover, the innovation and diffusion process is embedded in a sociotechnological system that is influenced by its stakeholders and institutional interaction. These factors can either accelerate the process of innovation and diffusion or raise substantial barriers to it. Policy action plays a major role, for two reasons. First, market mechanisms that promote greentech are weak, while inherited sociotechnological systems favor conventional technologies. Second, government policy has an increasing interest in spreading greentech as it reduces environmental degradation without harming economic welfare and development. On the other hand, the policy and regulatory perspective shows that taking proper actions to promote greentech innovation and diffusion is difficult. Whether or not an action is both effective and efficient depends on the selected evaluation criteria and on the overall policy objective. It appears that there is no one best policy instrument, but rather a most suitable policy mix that must be aligned with the given industry setting. Moreover, policy action is directly connected to private finance and investment markets, because it affects the availability of funds to greentech firms and the profitability of certain greentech projects. The finance perspective thus implies that the effect of government policy on private finance gives a good indication of the effectiveness and efficiency of a given policy mix. A good policy mix will therefore overcome a greentech firm's potential financing constraints and thereby advance innovation in and the diffusion of green technologies. It will also positively affect private investment markets, such as VC, without crowding out investments that would have been made in any case. Going forward, the efficiency and effectiveness of government support programs remains the subject of substantial discussion.

This literature review lays the foundation for subsequent analysis in several ways. First, this dissertation accepts that green technology innovation and diffusion are promising steps towards a sustainable development. Second, technological innovation and diffusion is believed to be brought about by firms that invent new technologies, bring them to market (innovation) and capture a large-scale market space (diffusion). Third, the availability of financing is seen a necessary requirement to enable R&D and innovation and, ultimately, to allow green technologies to be commercialized. Finally, greentech seems to be a market that is still strongly driven by government action. Government action is therefore appreciated as an important force that directly impacts firms' financial situation, the availability of capital to certain market segments and thus the innovation and diffusion process. Given what we do know about the importance of greentech innovation and diffusion for sustainable development, the key role played by greentech firms in this process, the dependence of greentech firms on financing and the powerful influence of government action, it appears puzzling that we actually know so little. First, there is scant research on what determinants genuinely impact a greentech firm's innovation activity and growth. Nor can we say whether the innovation and diffusion process is indeed subject to financial constraints, because we do not know whether the greentech firms that actually invent and market technologies are experiencing these constraints. Second, though research into government pull instruments with respect to technology diffusion results is extensive, the assessment of push instruments has barely received any attention from researchers. However, given that the use of push instruments such as grants or loans is connected to firms' overall financing arrangements, it would be necessary to examine whether the firms receiving government support are indeed outperforming their unsupported competitors. Third, we have no knowledge about the forms of financing that greentech firms actually use (apart from VC, which is only available to a mere fraction of firms). As a result, we cannot know whether government push programs ease the burden on greentech firms and help to reduce potential financial constraints.

To answer the research questions discussed above and deal with the identified knowledge gaps, the dissertation must seek to integrate three major fields of study: innovation, growth and finance. It also has to introduce a new perspective on greentech research – the firm-level perspective – across all three fields. It thus shifts the focus of analysis away from a broad technology-oriented view toward a more specific firm-level view. In doing so, it also responds to the call to move greentech research in the direction of financial economics (Dinica (2006)), broaden the scope of analysis from VC to various capital sources, and pick up the theme that policy action should target improvements in the financing of greentech firms (e.g., Burtis (2006); Murphy and Edwards (2003)). Moreover, to connect firm-level findings with existing research, the dissertation must also take due account of the evolutionary nature of greentech development, as evolutionary frameworks have previously been used by several scholars exploring this subject from a technology perspective. (e.g., Chertow (2000); Foxon et al. (2005); Grubb (2004); Krozer and Nentjes (2008); Neuhoff (2005); Nill and Kemp (2009))

3   REVIEW OF OTHER RELEVANT LITERATURE

This chapter lays the foundation for the theoretical framework by reviewing literature in the relevant areas, i.e. innovation, growth and financial economics. It is organized as follows: First, it evaluates theoretical models of technology innovation and diffusion and presents related empirical evidence on various industries. Second, it reviews implications for firms and analyzes existing empirical evidence. Third, it summarizes the findings throughout corporate lifecycles as a basis for integrating innovation, corporate finance and policy research. This section also reviews the main determinants of firm growth. Fourth, it discusses finance theory and the corresponding empirical findings, focusing on capital structures and financial constraints. Finally, it analyzes greentech programs launched by the German government and draws implications at the level of the firm.

3.1   Innovation and diffusion – A focus of technology

"It is now accepted that the development and diffusion of new technologies are central to the growth of output and productivity. But our understanding of the innovation process, and its economic impact, is still deficient."
OECD (2005)

As indicated by the OECD statement above, innovation and diffusion are central drivers of economic development. In the case of greentech, they are also enablers of sustainable development. Innovation – the process of bringing new products and services to market – has thus become one of the most important issues in business research. Innovation affects our way of doing things. It improves existing products and services and reduces prices. It changes market structures, challenging large incumbents while opening up new market spaces for new entrants. And it reduces both negative externalities and environmental degradation. Innovation is a competitive force in the global environment. Firms that do not innovate might get overtaken by their competitors and thus lose market share, sales and profitability. Innovation is therefore an imperative not only for firms, but also for policymakers.

3.1.1   Definitions and terminology

3.1.1.1   Invention and innovation

This dissertation regards inventions as sources of economic growth and a means to overcome environmental degradation. In the context of greentech, an invention is a new device, service or process derived from study and experimentation that reduces negative external effects. Invention is the basis for innovation. Invention is the initial idea for a new product or process, while innovation is the attempt to put that idea into practice. Property rights to inventions can be protected by patents granted by national patent offices. However, many inventions do not have sufficient technological and/or economic value to become greentech innovations. Innovation is a term derived from the Latin meaning 'to launch something'. This dissertation uses the OECD's definition of innovation, where innovation centers around either products or processes: *"A technological product innovation is the implementation/commercialisation of a product with improved performance characteristics such as to deliver objectively new or improved services to the consumer. A technological process innovation is the implementation/adoption of new or significantly improved production or delivery methods. It may involve changes in equipment, human resources, working methods or a combination of these."* (OECD (2005)) As noted above, an innovation is an invention that delivers technological and economic value to the customer. It can be a technology, a product or a service. In practice, as pointed out above, it is also helpful to differentiate between product and process innovations. There have been several theoretical attempts to assess the nature of innovations. One often cited concept goes back to the early works of Joseph Schumpeter, who distinguished between five major sources of innovation: (1) new products, (2) new methods of production, (3) new sources of supply, (4) the exploitation of new markets and (5) new ways to organize business. This broad differentiation reflects the complexity of how innovation comes about in practice. For example, a solar panel manufacturer might develop a new thin film technology, manufacture standard panels using a new, cost-efficient production method, source other input substitutes, introduce existing product lines into developing markets or develop a business model that hires out solar panels for a fee instead of selling them. All of these actions can properly be described as innovations.

A more product-oriented approach to differentiation is suggested by Sternberg, Pretz, and Kaufman (2003), for whom innovation can be realized in eight distinct ways: (1) replication, (2) redefinition, (3) forward incrementation, (4) advance forward incrementation, (5) redirection, (6) reconstruction/redirection, (7) reinitiation and (8) integration. Depending on the actual product development stage for a certain technology, the simplest way to innovate is to copy that particular technology and add it to a given product portfolio (1). Incremental improvements, such as increases in the efficiency of a wind turbine, fall into category (2). If they go beyond what people are ready to use, as is the case with decentralized wind power applications, then they fall into category (3). Developing a substitute or a technology that changes the way of doing things, such as switching from solar power panels to solar heating applications, is subsumed under categories (5) and (6), while reconstruction involves going back to a development stage in the past. A similar construct is category (7), for example where wind turbines are developed by taking windmills as a starting point – i.e. going back to the basic principle of using wind as a source of energy – and applying the fundamental characteristics of windmills to a new application, i.e. generating electricity. The final category (8) comprises innovations that result from the combination and integration of different technologies, such as the development of combined-heat and power (cogeneration) applications.

Picking up on the practical notion of the distinction between process and product innovations, Carayannis, Gonzalez, and Wetter (2003) elaborates a framework that breaks innovation down into four components: process, content, context and impact. Summarizing the major innovation typologies in literature, the author suggests that process innovations be divided into evolutionary and revolutionary ones, where the first is incremental and the second non-incremental. The second component, "content", denotes the nature of the innovation and distinguishes between incremental, generational, architectural and radical innovations. The third component, "context", can be seen as a longitudinal measure that categorizes innovations as continuous or discontinuous. The final component concerns itself with the impact of the innovation, which may either be disruptive or incremental. As was shown, there have been several attempts to assess the nature of innovation. In the interests of simplification, this dissertation primarily distinguishes between incremental and non-incremental innovations and between product and process innovations. Generational and architectural innovations are subsumed under the incremental heading and are accordingly based on existing products. It is also assumed that non-incremental innovations may have a disruptive impact on their conventional substitutes.

3.1.1.2 Innovation and diffusion

A single innovation will only have a significant impact if it achieves commercialization on a large scale. In other words, an innovation must create market demand and spread (be diffused) into various market segments. Innovation diffusion can thus be defined as "*the process of market penetration of new products and services, which is driven by social influences. Such influences include all of the interdependencies among consumers that affect various market players with or without their explicit knowledge.*" (Peres, Muller, and Mahajan (2010)) The diffusion of innovations follows a path of technology adoption. This dissertation analyzes theoretical models and frameworks that describe this path in respect of the greentech industry.

3.1.2 The evolutionary process of technology innovation and its diffusion

Despite the increasing importance of innovation in business research, it has not always received the attention it deserves from scholars. In fact, innovation was long seen as a random phenomenon that could hardly be assessed by theoretical modeling[6] as it was believed to occur more by chance. Joseph A. Schumpeter was one of the first scholars to oppose this common belief, formulating the notion of "creative destruction" as a fundamental concept of capitalism: *"The essential point to grasp is that in dealing with capitalism we are dealing with an evolutionary process [...] that incessantly revolutionizes the economic structure from within, incessantly destroying the old one, incessantly creating a new one. This process of Creative Destruction is the essential fact about capitalism."* (Schumpeter (1943)) Schumpeter's ideas and concepts are still very present in innovation literature as a whole. Creativity *per se – "the ability to flexibly produce work that is novel (i.e. original, unexpected), high in quality, and useful, in that it meets task constraints"* (Sternberg, Pretz, and Kaufman (2003)) – has also remained an underlying concept in economic theory. Creativity is the basis for innovation and, hence, for non-incremental economic development and competitive advantages among firms. Moreover it is still common to view innovation as an evolutionary process that drives not only economic change but also the co-evolution of social systems.[7]

As pointed out earlier, the notion that innovations shape their economic and social environment was introduced by Schumpeter and is now well established in literature. The evolutionary process from invention to innovation and to large-scale diffusion has since attracted a great deal of attention from researchers, especially in the technology management and marketing domain. Accordingly, diffusion models have been developed and analyzed with different purposes for some decades. In technology management, for example, there is a strong focus on the process of technology adoption by firms. In marketing, however, the focus lies more with the adoption of products by consumers. The innovation and diffusion process is usually described as an S-curve that characterizes the cumulative adoption of a certain product or technology over time. The underlying idea is that adoption follows a bell-shaped or normal-shaped curve indicating the frequency distribution of buyers or adopters. Most existing research goes back to the influential work "Diffusion of Innovations" by Rogers (2003), originally published in 1962 and now in its third edition. Rogers focused on technologies and their diffusion, assessing how new products and services spread through social systems over time.

---

[6] Although in early literature economists accepted the innovation process to be a "black box", there was still an agreement that R&D investment is a major driver. How this understanding has altered and developed over the last decades is reflected in the generations of innovation models seen in literature. The interested reader may refer to a Marinova and Phillimore (2003) for an exhaustive literature review.

[7] It is worth noting that, with respect to innovation, five areas of research have evolved. The innovation and diffusion process and its prescriptive value is one of them. The others are: consumer response to innovation; organization and innovation; market entry and innovation; and outcomes from innovation. Please refer to Hauser, Tellis, and Griffin (2006) for an exhaustive literature review.

The diffusion of an innovation starts with the commercialization of an invention and may take different development paths, depending on the individual perception of each adopter. As shown in Figure 3-1, some innovations develop faster than others. For example, Innovation II is initially adopted faster than Innovation I, meaning that more people or firms are willing to adopt Innovation II early in the technology lifecycle than is the case for Innovation I. Rogers defines diffusion as *"the process by which an innovation is communicated through certain channels over time among members of a social system"* (Rogers (2003)).

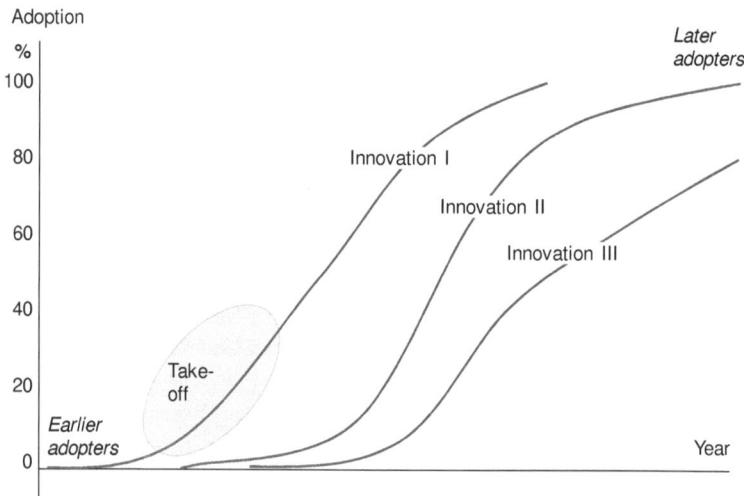

Figure 3-1: Diffusion of innovations
Source: Rogers (2003)

Diffusion is thus comprised of four main constituents: (1) the innovation itself, (2) communication channels, (3) time and the (4) social system. The characteristics of an innovation determine its rate of adoption. For instance, if the relative advantage compared to existing solutions is high and if the new solution is comparatively easy to use, the adoption of the innovative technology is, *ceteris paribus*, likely to be relatively fast. Communication channels are another important feature of diffusion as they influence potential users' willingness to adopt innovations and, hence, their adoption patterns. Communication channels involve both direct interpersonal communication and marketing and mass media instruments. Time is a mediating variable. Time enhances the process of learning through experience and allows multiple potential adopters to observe an innovation's performance. Over time, uncertainty surrounding an innovation is reduced and production costs are lowered, making the innovation attractive to a broader set of potential adopters. However, the final element in the equation – the social system – is probably the most complex one. Rogers defines a social system as "*a set of interrelated units that are engaged in joint problem-solving to accomplish a common goal. A system has structure defined as the patterned arrangements of the units in a system, which gives stability and regularity to individual behavior in a system*" (Rogers (2003)). The social system could thus support the diffusion of innovations if they meet the demands of opinion leaders. However, the social system may also hinder diffusion if a given innovation has a disruptive effect on the wealth situation of important system units. The individual preferences of the above-mentioned constituents will increase the adoption rate over time, steepening the slope on the typical S-curve. The S-curve can be segmented into adopter categories. The first individuals or firms to adopt are referred to as innovators (the first ~2.5% of all adopters), followed by the early adopters (the next ~13.5% of all adopters), the early majority (the next ~34% of all adopters), the late majority (the next ~34% of all adopters) and the laggards (the last ~16% of all adopters).

These categories were first established and roughly quantified by Rogers (2003) and are still widely accepted and used by many scholars from multiple disciplines. (e.g., Bass (1969); Mahajan, Muller, and Srivastava (1990); Agarwal and Gort (2002)) Mahajan, Muller, and Srivastava (1990) validated the categories both analytically and empirically. He found that category sizes are indeed approximately comparable to those suggested by Rogers. Each category reflects the innovativeness of a homogeneous group of adopters with respect to a certain product or technology, i.e. their propensity to buy a certain product or adopt a certain technology. Figure 3-2 shows the innovation diffusion curve and the various adopter categories and segment sizes. The *y*-axis denotes the degree of adoption. The *x*-axis stands for the point in time. Out of a pool of potential adopters, maximum penetration is achieved when the degree of adoption reaches 100%. The first graph shows cumulative adoption by adopter categories. The second graph shows the distribution of individual adopters in each group. This simply means that, when early adopters start to purchase the product or adopt a certain technology, they follow the group of innovators that have already adopted it. According to Rogers, the innovators segment makes up 2.5% of the overall population of potential adopters. The first graph therefore indicates the overall adoption rate, which is equal to the area of the second graph that shows the individual number of adopters over time. The diffusion model defined by Rogers offers several advantages. It is easy to use, has mutually exclusive and collectively exhaustive categories, conceptualizes earlier empirical findings (e.g., Mansfield (1961); Fourt and Woodlock (1960); Griliches (1960)) and exhibits good predictive value. However, the model has also been subject to criticism pointing out four major drawbacks. First, some authors (e.g., Bass (1969)) argue that the given characteristics and structures are rather indicative, i.e. that they lack rigorous empirical and analytical evidence. Moreover, it is argued that the proposed model does not hold for all innovations. Mahajan and Peterson (1996), for instance, provides examples in which the model does not hold. Second, successive technology generations cannot be mapped, although they are a very

common phenomenon in practice (Islam and Meade (1997)). Third, the model offers no
explanations for turning points that indicate the transition from one phase to the other. This
assertion not only reduces the predictive value of the model, but also limits the implications
that can be drawn for the purposes of proactive technology or product management. (e.g.,
Golder and Tellis (1997); Golder and Tellis (2004)) Finally, some authors (e.g., Kline and
Rosenberg (1986)) claim that the innovation and diffusion process is not linear and is a lot
more complex than Rogers suggests.

Adoption

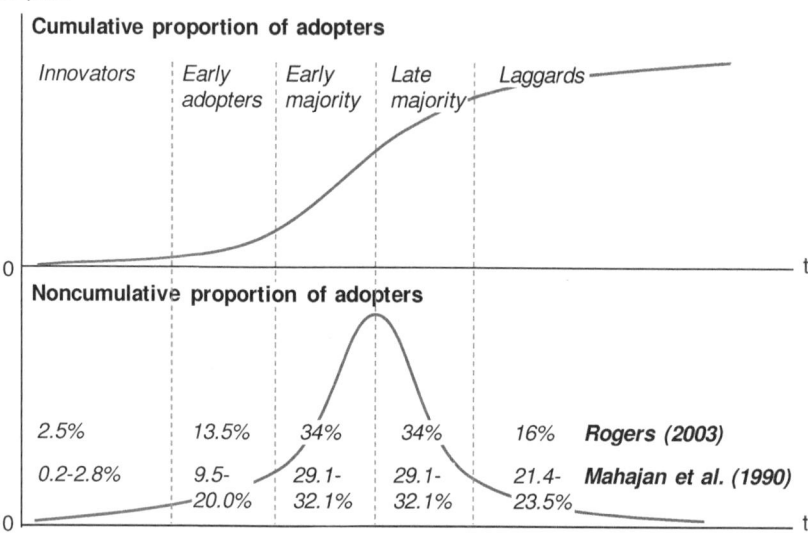

Figure 3-2: Diffusion of innovations and adopter categories
Source: Mahajan, Muller, and Srivastava (1990)

In response to the initial criticism, several models were constructed to back Rogers' findings
with analytical and empirical evidence. One of the first and most influential models was
developed by Bass (1969). Bass constructed a mathematical model that comprised the adopter
categories defined above, assuming the typical S-curve for cumulative adoption. Although
competing models existed at the time, an exhaustive literature review covering all of them is
beyond the scope of this dissertation.[8] Instead, the section below critically reflects on the

---

[8] The interested reader may refer to Islam and Meade (2006), who has reviewed analytical diffusion models over
the last 25 years and identified eight basic models (including the Bass model), most of which were developed
in the 1970s.

literature discussed herein, assesses the main paradigms that explain the S-curve properties and drivers, and presents some brief empirical evidence.

The second disadvantage has nourished a research stream dedicated to technology generations and their inclusion in existing frameworks. Generational innovations are of considerable practical importance. In the mobility sector, for example, the steam engine first substituted for horses, before itself becoming obsolete and being replaced by gas engines. In the future, we may now see gas engines being edged out by fuel cells or electric motors. The introduction of a new technology generation has a significant impact on the diffusion process for the earlier technology. It will ultimately diminish the latter's potential, leading customers or firms that would have adopted the earlier technology to now adopt the new generation instead. Existing users may switch too, and the potential adopter base may increase or decrease depending on the features of the new-generation technology. (Norton and Bass (1987)) The introduction of a new generation thus leads to a new innovation and diffusion process that is not necessarily similar to the previous one. In their recent review of literature on innovation and diffusion models, Peres, Muller, and Mahajan (2010) point out that, although there is agreement that generational diffusion approximately follows the typical S-curve, literature on technology generations provides contradictory answers to several key questions. It is therefore not clear whether a new generation accelerates the diffusion process, shortens the time to phase transition or motivates users to cannibalize new generations or even leapfrog them to future generations. For the purposes of this dissertation, however, the important point is that a new technology generation implies the initiation of an additional diffusion process.

The third point, as discussed above, addresses the issue of turning points and thus, indirectly, the predictive value of the diffusion model. The first turning point that needs to be addressed is when a technology or product is introduced to the market and initial growth is fostered. This point is essential because it necessitates substantial investment in production and distribution. In the interests of clearer analysis, Golder and Tellis (2004) suggest four main stages: (1) introduction, (2) growth, (3) maturity and (4) decline. These stages occur due to certain events within the product lifecycle. Two main turning points that have increasingly been the subject of research are known as the "takeoff" and the "saddle" points. Takeoff occurs when the innovation's price is reduced and meets the price sensitivity of a larger adopter group (early adopters). Additionally, Moore (2006) argues that "*cracks in the bell curve*" hinder the transition from one adopter group to the next. Essentially, the largest potential gap is between initial takeoff and commercialization. He presents several examples of technologies for which the market suddenly declined following an initial rise. He referred to this "saddle" in the adoption curve as a "chasm", a concept that was later formalized and tested by scholars such as Goldenberg, Libai, and Muller (2002) and Mahajan and Muller (1998).

The fourth critique is not as easy to resolve and requires further assessment. Markets are social constructs. An invention may be economically and technologically feasible and even preferable to existing substitutes, but may still fall at the commercialization hurdle due to information asymmetries and market constraints (Carayannis, Gonzalez, and Wetter (2003)). In such cases, inventions may never achieve diffusion and could fail to plot the expected S-curve. In other words, the S-curve may be limited to a certain degree of adoption – or may even stop abruptly. The S-curve model also implies a certain set of consecutive steps in which an invention is developed by R&D activity, becomes an innovation through market launch and is diffused thanks to production and marketing activities. Although it is true that a technology may go through all these steps, they are not necessarily consecutive. Innovations may arise during the production of a related product or as a result of learning effects from marketing activities. Essentially, feedback loops in the process may alter the course of an innovation, shifting back to a phase that was assumed to be finished. Arguing that the implicit assumption of linearity in the innovation diffusion process is false, Kline and Rosenberg (1986) thus develop a chain-linked model that includes feedback loops and interdependencies between development steps, marketing and research activity. His model indirectly links the innovation and diffusion of technologies to the firm. It also contends that innovation does not necessarily depend on research activity, but can also depend on factors such as skilled labor. While acknowledging the existence of exceptions from the S-curve model, this dissertation nevertheless accepts it as a useful tool with which to understand and assess a broad range of innovations in the greentech industry. This position aligns with that of Hall (2007), who, based on a review of literature on innovation diffusion models, argues that the linear model, though often criticized, holds true in many cases. Without invention, for example, there would hardly be any kind of diffusion. Figure 3-3 summarizes the modifications proposed to resolve the above criticisms to the S-curved technology innovation and diffusion model.

Innovation begins with the market launch of an invention. The technology is first adopted by innovators, who start the diffusion process. As demonstrated above, the process may experience interruptions or irregularities that materialize in the form of short take-off periods and development "saddles". Moreover, the process of innovation may abruptly stop at any time as indicated by the dotted lines. However, when a technology is finally adopted by larger groups of followers, the diffusion process gathers momentum and the adoption curve is represented by a convex shape (take-off). As penetration increases, the adoption curve turns concave and finally decreases. The cumulative diffusion curve (S-curve) thus grows continually and could be reinforced by the introduction of new generations (Mahajan and Muller (1996)).

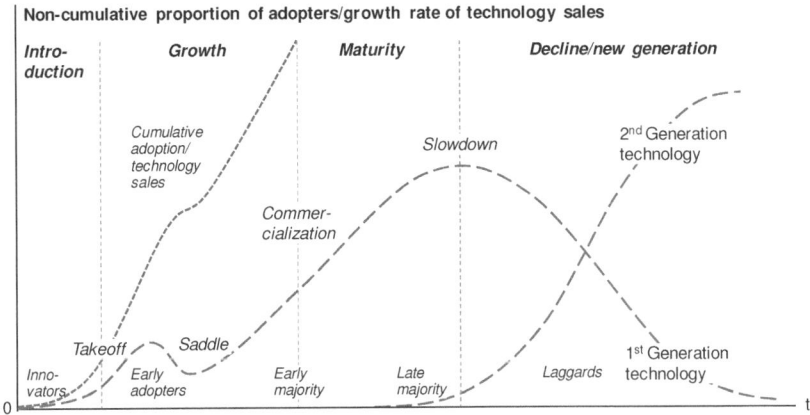

Figure 3-3: Modified adopters and innovation diffusion curve
Source: Adopted from Mahajan, Muller, and Srivastava (1990)

3.1.3 Drivers of technology innovation and diffusion

The process of innovation and diffusion is rather complex and often involves significant time lags. Besides the fact that only a small number of inventions actually enter the market (thereby becoming innovations), the time between the original invention and its commercialization often exceeds 15 years (Grübler (1996)) – or may never happen at all. This section deals with the main drivers of technology innovation and takes a closer look at the explanations research has offered to date.

In his literature review, Hall (2007) identifies two fundamental assumptions as the two main explanations for adoption provided by relevant literature: (1) user homogeneity and (2) user heterogeneity. The basic implication is that adoption depends either on how information is distributed or on the characteristics of the individual adopter. The models that assume user homogeneity are often called epidemic models (e.g., Geroski (2000)). The thinking behind them is that every potential adopter has the same utility function. Given perfect information, all potential users would adopt at the same time, assuming that performance of the innovation is superior to that of existing substitutes. The factor that drives diffusion is therefore the availability of information. In the beginning, when the innovation is brought to market, only few potential users (the innovators) know about it. As they adopt the technology, others learn from them and follow their example. Basic adoption models are thus formally defined as a function of the number of previous adopters. (e.g., Bass (1969); Mansfield (1961)) Models that assume user heterogeneity take a different approach, requiring examination of the decision to adopt that is taken at the level of the individual or the firm. The fundamental belief here is that potential users have differing utility functions, and they adopt in line with changes in the characteristics of innovations or in the related environment. The logic of adoption is summarized by Hall (2007) as follows: Adoption takes place when the expected benefit of the innovation exceeds the initial investment because the costs associated with the old technology are sunk costs and therefore no longer influence the decision. While the new technology can thus be considered to be in a state of absorption, the expected benefit is subject to significant uncertainty that determines the propensity to adopt.

Nelson, Peterhansl, and Sampat (2004) combine both approaches and point out that information availability also determines the adoption decisions of heterogeneous user groups. If there is no clear feedback on the performance of certain innovations, social constructs may evolve that favor inferior technologies. This is, for example, the case when a superior innovation is outpaced by an inferior one due to perceived associated risks caused either by limited information and uncertainty about the superior innovation, or by the strong preferences of opinion leaders for conventional substitutes that ultimately prevent an innovation from taking off. This contrasts with the classical view that, in the long run, clear and objective information will be available to all market participants. (e.g., Mansfield (1961); Mansfield (1968)) Moreover, when a specific innovation is adopted – even though it may not necessarily be superior to available substitutes – information cascades may affect follower groups and thus create something of an adoption "bandwagon" (Geroski (2000)). Another related effect that has been investigated in literature is what is called the network effect. (e.g., Hauser, Tellis, and Griffin (2006); Peres, Muller, and Mahajan (2010)) This effect implies that the more potential users adopt an innovation, the greater its benefit will be. This situation could lead to technology lock-out in which the inferior technology becomes the industry standard even though a better substitute is available. Additionally, as adoption progresses, the cost of switching increases for the existing user base. One very simple example is that of the telephone, where the original invention was first disregarded as irrelevant and the telegraph network was improved incrementally. Not until inventor Bell gained a reputation in the telegraphing community was the telephone too successfully adopted. (Sternberg, Pretz, and Kaufman (2003)) This example also highlights the strong bias within the system. The beliefs of inventors and agents regarding a given technology trajectory determine its actual development process. Such beliefs and expectations can therefore act as self-fulfilling prophecies. (Geroski (2000))

A more general set of drivers is defined by Rogers (2003): (1) decision level (e.g., individual, group), (2) information channels (e.g., direct communication, mass media), (3) social system (e.g., political system, regulation) and (4) change agents (e.g., opinion leaders, early adopters). Some factors are thus internal, coming from within the innovation diffusion process (such as adoption decisions taken by groups that first need to formulate a common opinion), while others are external, coming from the outside (such as change agents that explicitly disseminate information about a certain technology to drive technology take-off). Internal and external factors have been an important subject in marketing literature, with researchers trying to assess how marketing activities such as advertising impact the innovation diffusion process. (Mahajan and Peterson (1996)) With respect to high technology, Moore (2006) points out that, without active support, many technologies may not even achieve commercialization and will therefore never pass the commercialization stage. Activities that influence the innovation diffusion process can be summarized and divided into supply-push and demand-pull instruments, as shown in chapter 2. (e.g., Geroski (2003)) The underlying view is that technologies stem from two main sources of stimulus. Push stimulus comes from the development side of the process. A typical example is that of Post-Its®, for which the original goal was to develop a very strong (not weak) glue. Post-Its® were not developed on purpose, but were still the result of a technology-focused R&D program. Supply-push stimulus, or technology-push stimulus, holds out significant potential for non-incremental innovations that may create a completely new and as yet unknown market. The pull stimulus, on the other hand, comes from the market side and involves the desire to meet existing demand. Demand-pull is a rather feeble driver for radical innovation, because users must be able to articulate it and innovation can only occur when demand is formulated.

3.1.4 Empirical evidence about technology innovation and diffusion

Empirical evidence about the S-curve phenomenon collected in exploratory surveys dates as far back as to the 1960s. (e.g., Fourt and Woodlock (1960); Griliches (1960); Mansfield (1961)) Probably one of the first technology-oriented studies was conducted by Griliches (1960) in 1960. In his study, the author concerned himself with the patterns of diffusion for hybrid corn among farmers in the US. He found that both the speed of diffusion and adoption of the innovation depended heavily on the region concerned. In regions with higher average income, innovations were adopted both earlier and faster adopted than in poorer regions. The overall adoption process was mainly driven by farmers' profit expectations. Larger fields and better technologies gave richer areas greater benefits from adopting hybrid corn compared to poorer areas. The data presented established the S-curve phenomenon, since all adoptions in all regions revealed a comparable pattern (Figure 3-4). Although this survey is rather old, there is an important parallel to today's greentech industry. As in the US in the 1960s, green technologies are today being adopted by different regions with different income levels and different structural characteristics. It should therefore follow that greentech will be adopted earlier and faster in developed regions such as Germany, where firms' profit expectations from the adoption of greentech is relatively high due to the relative scarcity of this resource and the high perceived value of environmental protection. On the other hand, innovation adoption and diffusion speed could be expected to be lower in regions such as Brazil, where the adoption of greentech yields less competitive value, natural resources are rich and the problem of environmental degradation is not perceived as urgent. The implication is that policy support costs would be comparatively higher in Brazil than in Germany, because greentech market mechanisms are comparatively weaker. In his literature review, Hall (2007) presents innovation and diffusion data on several industries, as shown in Figure 3-5. Compared to the patterns in Figure 10, the data presented by Hall for different industries shows that variations in adoption time are accompanied by pronounced variations in diffusion rates. These differences arise from the different benefits yielded as products improve and network effects grow. Discrepancies also exist in respect of complementary investments. For example, a computer often involves implementation costs (for training and software). Similarly, a phone needs a monthly network provider schedule in order to function. In light of the research question posed by this dissertation, financing may be another significant cost block. Mansfield (1968), for instance, points out that adoption of the diesel locomotive depended on the liquidity of railway firms because the external cost of capital was higher than the internal cost. By analogy, green technologies will only be diffused if the firms that develop and commercialize them receive sufficient financing and achieve sufficient sales.

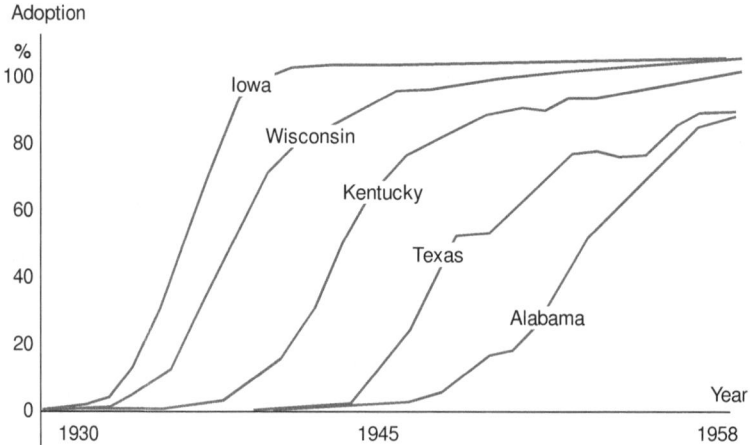

Figure 3-4: Percentage of all corn acreage planted with hybrid seed in different regions in the US
Source: Griliches (1960)

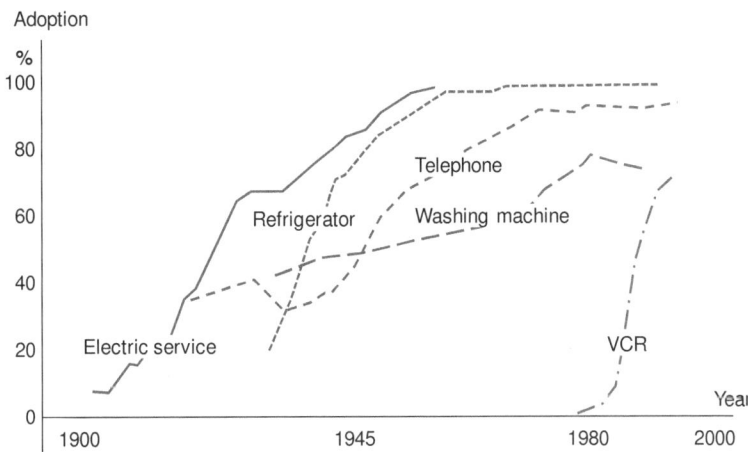

Figure 3-5: Innovation diffusion in various industries
Source: Hall (2007)

Besides descriptive statistics, a wide range of literature has empirically tested the above innovation and diffusion models. For example, one of the first deterministic models, developed by Mansfield (1961), was mainly based on a review of the adoption of 12 innovations in various, mostly heavy goods industries. Also, Bass (1969) uses empirical data from 11 consumer durable industry sectors to test his formal model. In his survey, he runs regression models to evaluate the fit of the model to the given data. Results show that the model explains up to 95% of the overall variance. He concludes that the S-curve model, as formalized in his survey, can properly be used to engage in long-term forecasting. In a later article, Norton and Bass (1987) modify the previous model by introducing technology generations. The modified model is tested using data from the semiconductor industry and different generations of RAM chips. The authors find that the general S-curve pattern remains, but that market volumes increase with new generations. (Islam and Meade (1997)) also use the Bass model as their framework. However, this work shows that the assumption of constant coefficients of innovation and imitation does not hold, and that changes in coefficients significantly improve model performance. The authors use data on two and three mobile phone generations respectively. They too conclude that the general S-curve pattern holds true for follow-up generations. They claim that model coefficients change because attitudes toward a given technology change as well. Goldenberg, Libai, and Muller (2002) also test the parameters of the Bass model and find evidence for asymmetric regularity – i.e. a saddle – in the S-curve. An initial rise in adoption rates is followed by a downturn when the overall diffusion curve decreases somewhat. It is only after this dip that adoption rates pick up again and industry sales grow exponentially thanks to the increasing number of previous adopters. The authors explain this phenomenon by positing the existence of two markets: the early market and the late market. The early market is composed of more professional users, i.e. users who need the innovation for their work. Examples include stock brokers, who needed mobile phones in the early 1990s. By contrast, the late (or main) market is composed of private users who are influenced by the popularity of the technology. Golder and Tellis (1997) explicitly analyze the take-off points at which innovative technologies begin to achieve large-scale diffusion. According to the authors' findings, take-off occurs after the initial price has been lowered to around 60%, diffusion has reached around 2% of potential adopters and the innovation has been on the market for around six years. The list could go on indefinitely. Overall, it seems that the general S-curve pattern is broadly accepted and remains a reasonable approximation of the innovation and diffusion process of technologies. (e.g., Grübler (1996); Freeman and Soete (2004)) Although it may vary in terms of the time to adoption, diffusion speed, take-off time and even shape, the fundamental notion that inventions become innovations and move through a time-dependent process remains true.

3.1.5   Conclusion – The parallels between technology and firm development

Although a technology perspective is useful to assess innovation and growth in a given market space, and although there is strong consensus in literature that the market structure and the technological environment within which a firm operates *"play decisive roles in shaping the relationship between firm size and technological change"* (Audretsch and Acs (1991)), only a firm-level perspective can truly integrate financing in the overall framework. Additionally, Love and Ashcroft (1999) point out that the market structure and industry determinants are less relevant than firm-internal factors to innovations. The innovation and diffusion process thus only stakes out the contours for this dissertation. However, an understanding of how technologies emerge and grow is fundamental to the intention of assessing the innovativeness and growth potential of firms. We have seen that only a limited number of inventions actually enter the market and become innovations. Once innovations do enter the market, they might gain momentum and become large-scale substitutes for conventional solutions, or they might create completely new market spaces. On the other hand, they may never gain momentum and attempts to commercialize them may fail. To be successful, new technologies must be adopted. A green technology will thus only spread if it satisfies a specific market demand. Diffusion accelerates with an increasing number of adopters, through incremental technology improvement, information cascades, reductions in the cost of adoption, network effects, and so on. The diffusion process may also face irregularities where early and late markets, other substitutes and/or new technology generations exist. For their part, policymakers can launch government support programs to foster innovation by providing the required resources or accelerate diffusion by fueling market demand.

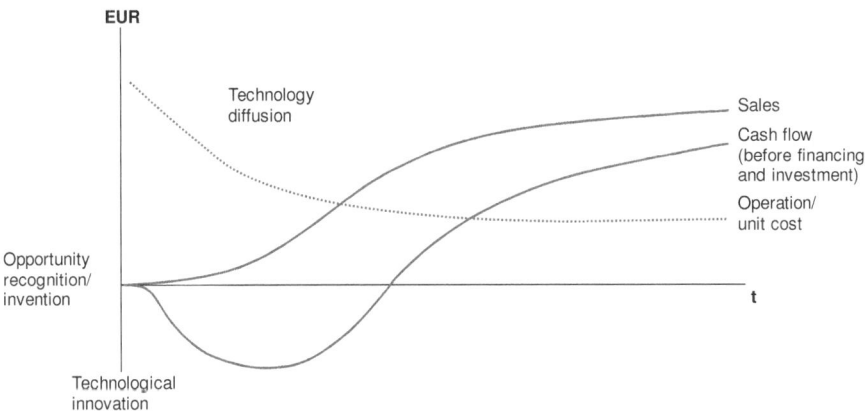

Figure 3-6: The one-firm and one-technology innovation and diffusion case

However, to understand how technologies are actually developed and how they are brought to market, one must shift from a technology perspective to a firm perspective. To start with, let us assume a market with only one technology and one firm. The implications for the firm can be derived fairly easily, as shown in figure 3-6. At first, the firm would need to engage in R&D and create a product prototype. Since the firm is not yet generating revenues, it incurs a negative cash flow. Once the product has been successfully developed and brought to market, it becomes an innovation. If the innovation satisfies demand, the firm will start to make revenues and receive compensation for its investment. If more users adopt it and the selling price stays constant, revenues will grow in the S-curve shape and the firm will ultimately reach profitability. Nevertheless, before the firm can generate sufficient internal funds, it must expand capacity considerably if it is to cope with the exponential increase in demand. After production has been expanded successfully, the firm grows along with the diffusion of its technology. At the same time, unit costs decrease because fixed costs can be spread over the increased number of products sold. The firm learns and improves its processes, which also lowers its variable costs. In a one-firm and one-technology world, the attributes of a firm's development are thus very similar to those elaborated in this chapter.

3.2 Innovation – A basis for corporate development

The one-firm and one-technology case demonstrates that firms develop in line with the technology innovation and diffusion process. However, in the real world, the assumptions made do not hold. The real world is much more complex, as there are many firms with varying sets of resources and many products at different stages of development.

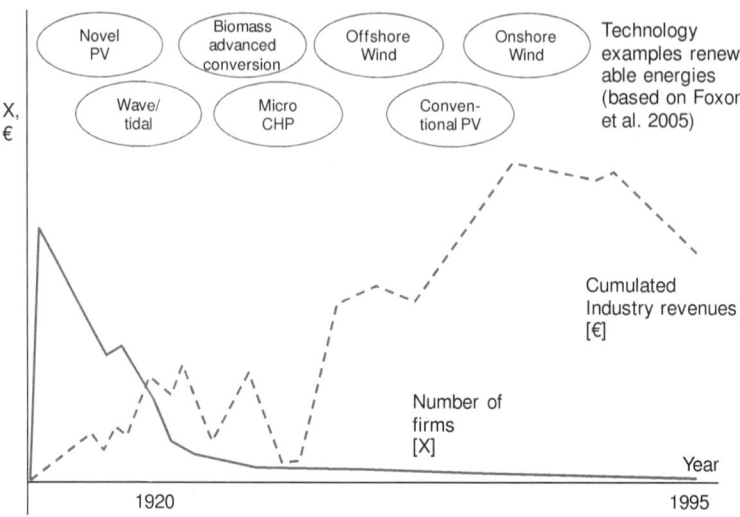

Figure 3-7: Industry sales and firm population in the US automobile industry
Source: Stylized and adapted from Geroski (2000) and Foxon et al. (2005)

As demonstrated above, it is not unlikely that the overall market for a certain product will develop in the form of an S-curve. However, firms may enter a market, grow, fail or exit. As claimed by Geroski (2000) and as shown in Figure 3-7, technological evolution is accompanied by a certain pattern of firms entering and exiting the market. The automobile industry provides a good example. In the early 19th century, several different automobile types were invented and a large – exponentially increasing – number of firms competed to find the best technology. With a large number of firms and technologies on the market, production costs for each car were very high. Car prices were therefore high as well, such that only a small number of wealthy individuals could afford to own one. It was only in the 1920s that the internal combustion engine asserted its domination and became the industry standard. Actual diffusion was then driven by the improved mass production technology introduced by Henry Ford. The price of cars decreased tremendously and industry sales started to increase exponentially. The drop in prices was driven by the larger number of products sold. Firms that could not lower their cost structures left the market space. Accordingly, competition increased over time, leaving only the most competitive firms still in operation.[9] Another problem is that

---

[9] One might be tempted to think that greentech is at a rather early stage of development. However, the actual stage of development varies considerably depending on the technology line. Figure 13 shows the differing developmental stages for selected technologies. For example, new PV and tidal power generation technologies are fairly young and have a limited track record. On the other hand, wind power technologies have already reached a point of large-scale diffusion. Wind power can in fact look back on a more than 100-

the product/technology lifecycle is not necessarily linked to a firm's growth cycle as was the case in the one–firm, one-product world. A young firm may, for example, enter a very mature market, while a mature firm may enter a very new market. Overall, the implications drawn from the one-firm case are very useful to gain an initial understanding of how firms develop with the emergence of a technology. Nevertheless, further investigation is needed if we are to fully understand the internal factors that actually drive not only a firm's innovation activity, but also its growth trajectory. The section that follows therefore looks at a body of research that has concerned itself with innovation at the level of the individual firm.

3.2.1 Theory of innovative firms

Schumpeter originally developed modern innovation theory. He complemented Adam Smith's principles of land, labor and capital by introducing entrepreneurship and technology into the overall framework. Challenging equilibrium theory, he underlined disequilibrium states as important sources of economic development. Following this line of thought, an innovation can trigger a process of "creative destruction" in which existing technologies are substituted by new ones and new firms compete successfully against old ones. The individual entrepreneur develops and advocates new solutions to existing problems, thereby fighting social inertia. (Schumpeter (2007)) However, Schumpeter's initial line of thought soon experienced a fundamental change. In his later works, based on the discussion of the role of teams in entrepreneurial settings, Schumpeter emphasized the role of entrepreneurial collaboration in large firms. He further suggests that the economy responds to certain events (e.g., the launch of new products) in two distinct ways: (1) adaptive response (predicted *ex ante* by his theory) and (2) creative response (unpredictable *ex ante*). The concept of creative response substituted for creative destruction, emphasizing the role of larger firms Schumpeter (1947). The former idea is often called the creative destruction view or "Schumpeter Mark I". The latter idea is often called cumulative view or "Schumpeter Mark II". (e.g., Vaona and Pianta (2008)) Both views are still encountered in business research, having established two main lines of thought and empirical evidence. With respect to firm characteristics, however, the two perspectives lead to very different and contradictory results. The first implies that young, new and smaller firms in particular can bring about change and innovation because they are more flexible and less dependent on past successes. Young firms therefore have a comparative advantage over old firms that are constricted by organizational inertia. (e.g., Acs and Audretsch (1988); Audretsch and Acs (1991)) The second view states that older and larger firms are more innovative than younger and smaller firms due to their larger resource base, and that they also have better access to the market's resources. The basic assumption inherent in this view is that more resources – more R&D funds, for example – lead to more

year history in power generation and a more than 1000-year tradition in mechanical engineering (Dismukes, Miller, and Bers (2009)).

innovation. (e.g., Galbraith (1997)) Yet more recent research suggests that Schumpeter Mark I and Mark II are complementary and co-exist at the same time. (e.g., Shavinina and Seeratan (2003)) To understand the relationship between these seemingly contradictory views on firms, the arguments presented below draw on the evolutionary framework presented in the previous chapter on technology innovation and diffusion and shown in Figure 3-3.

When a new technology is introduced to the market, it can either be based on an existing one (incremental innovation) or might be completely new (disruptive innovation). As shown, older and larger firms constantly accumulate knowledge and have extensive experience of their own product range. Assuming that the background knowledge needed for innovation is cumulative, older firms would therefore be more likely to innovate than younger firms. This assumption would be reinforced by the fact that older firms already possess the established routines, structures and processes that are needed to develop new technologies and bring them to market. However, at the same time, older firms tend to depend heavily on the paths that have delivered past successes, which diminishes the likelihood of them generating innovations that are completely new. (Soerensen and Stuart (2000)) On the other hand, non-incremental and disruptive innovations may well come from young and small firms outside the industry. These firms are less constrained to follow standard industry procedures and norms and are thus able to develop "out-of-the-box" technologies and solutions. (Schumpeter (2007)) New and radical technologies are therefore more likely to be developed by younger firms. However, these firms tend to be less well endowed with input factors and resources. Seminal early works in the field (e.g., Nelson (1959); Arrow and Rand Corporation. (1959)) show that the inability to appropriate inventions is a key driver for investors' reluctance to fund basic research that may eventually lead to disruptive innovation. It follows that both the limited internal financing capabilities of younger and smaller firms and their limited access to external funding sources can hinder the development of potentially disruptive innovations. According to the innovation model developed by Utterback and Suárez (1993), two types of small start-up firms enter the market: those with growth potential and those without growth potential. While the first type may challenge industry incumbents and set new standards in the long run, the second type remains small. Accordingly, it is small firms with growth potential that introduce disruptive technologies to the market. If such a technology is successful. many firms will follow with imitations and incremental improvements, thereby increasing the dynamics of the business environment. The number of firms will further increase until it reaches a peak. At this point, a dominant design will have become established, locking in a certain technology as the technology of choice. In his formal theoretical model, Klepper (1996) validates these hypotheses and points out that, at least in many industries, as soon as a certain design is locked in, firms start to engage in cost and process optimization and the optimal firm size increases. In other words, unit cost reduction follows increased output with convex adjustment costs. Early entrants therefore do indeed gain a first-mover advantage.

This finding is largely consistent with Schumpeter's concept of creative destruction, in which innovation leads to a short-term monopoly situation for the start-up firm, which can command high product prices, high unit profit margins and increasing sales. (Schumpeter (2007)) A slightly different view is taken by Jovanovic and Lach (1989). In their analytical model, the authors show that early entrants receive higher revenues than later entrants, but that later entrants incur lower costs due to learning effects. Since these effects may cancel each other out, market timing becomes an important success factor. It is therefore not necessarily the first mover who enjoys the greatest commercial success. The observation that the cost of second movers is a lot lower is critical, as the original innovators do not receive the full value of their innovations. This is especially important when looking at potential resource constraints. Investors, for instance, will not provide funds in the first place if the potential return is diminished by second movers. Kline and Rosenberg (1986) critically note that the *"earliest Schumpeterian innovators frequently wind up in the bankruptcy courts, whereas the rapid imitator, or "fast second", who stands back and learns from the mistakes of the pioneer, may experience great commercial success"*. In this context, Agarwal, Sarkar, and Echambadi (2002) point out that the second mover is rarely a single firm, but rather a whole population of firms. Using a firm hazard model, the authors show that population density increases as soon as a technology meets market demand. As the technology becomes more successful, the firm population grows, implying intensified competition. Entrants' liability of newness is therefore lower in industry growth phases but increases over time.

Once a technology choice is taken and the technology grows mature, the market becomes more and more driven by incremental improvements and process innovations that cut production costs. As older and larger firms profit from economies of scale and accumulate more resources, they become more competitive compared to smaller and younger firms (Utterback and Suárez (1993)). In the context of sustainable technologies, Hockerts and Wüstenhagen (2009) also hypothesize that new and radical products are introduced by small firms (which they call "Davids") and at first ignored by larger firms (which they call "Goliaths"). However, when technologies start to grow and new brands evolve, larger firms engage in acquisitions and line extensions. At the same time, competition becomes fiercer and some early entrants exit the market, while the "Goliaths" increasingly try to control industry standards in order to gain further market power. On the other hand, nimbler "Davids" engage in product and process innovation and continue to compete with their older and less agile counterparts. The acquisition of young firms by older and larger firms after a technology is chosen or while it is being chosen can be a successful but also risky strategy. If incumbents wait too long and miss their chance, new technologies can turn disruptive and render incumbents' technologies irreversibly inferior. This effect is called the "disruptive technology trap". (Christensen (2008)) The establishment of a dominant design significantly raises entry barriers for smaller firms because they do not have superior information about standard technology compared to their cost-efficient mature competitors. The marketing of incremental innovation is also more difficult for smaller firms. According to Agarwal, Sarkar, and Echambadi (2002), firm mortality has a U-shaped relationship to the technology innovation and diffusion process until a technology is locked in. Afterward, firm mortality decreases linearly with technology age. This assertion is consistent with the findings of Klepper (1996), who argues that market shares stabilize over time and that process innovation correlates positively (and product innovation negatively) to the size of the firm. Besides increasing competition on cost and process efficiency, firms also start to differentiate products and technologies from those of their competitors. However, if the degree of differentiation exceeds what the market demands, it becomes meaningless. Over-differentiated products do not deliver additional returns but do continue to increase the cost of R&D. The products thus become commodities. (Christensen (2008)) At the same time, differentiation decreases product lifecycles, while the time to market becomes shorter for new generations. This fact further reduces firms' resources and negatively affects their competitiveness. (von Braun (1997)) Eventually, firms become even less flexible, opening the market space up to new opportunities. These may in turn be exploited by young firms with non-incremental innovations that again lead to new discontinuities and disruptive changes. The process described above thus starts over with the same implications for incumbents and new entrants. (Utterback and Abernathy (1975))

Overall, the discussion about whether young and small or old and large firms are more innovative is ongoing. When the different types of innovation are taken into account, it appears that young and small firms posses the flexibility and non-incremental creativeness to develop and market disruptive innovations. These are often technology- and product-oriented innovations that powerfully impact overall market development. However, successful innovation also depends on firms' ability to acquire sufficient financing and resources. Older and larger firms are better endowed and have better access to the necessary sources. These firms therefore tend to bring a larger number of innovations to market. However, these innovations are more likely to be of an incremental nature, because older and larger firms depend heavily on the paths plotted for their existing product base and past successes. It therefore seems that young and small firms have a competitive advantage over old and large firms at the very beginning of the technology lifecycle. However, as soon as a dominant design is determined by the market, larger firms dominate due to their established routines, procedures and advanced process efficiency. This argument is supported by the notion that innovation is based on a process of organizational learning. (O' Sullivan (2007)) To provide a better understanding of the characteristics of innovative firms, the following section reviews the relevant empirical evidence.

3.2.2   Empirical evidence about innovative firms

In his influential work, Utterback (1971) identifies three main factors that determine a firm's ability to generate innovations: (1) characteristics of the firm's environment, (2) internal characteristics of the firm itself, and (3) flows between the firm and its environment. These factors are also called "determinants of innovation". (Souitaris (2003)) The following literature review focuses on the characteristics of the firm only, as the overall research objective is to analyze innovation and financing at the level of the individual firm. It thereby provides some evidence regarding the theoretical concepts discussed above. It also sheds some light on the relationship between innovation, firm size and firm age.

In recent decades, an increasing amount of studies have investigated the main factors driving firms' innovation activity. Table 3-1 shows a sample of studies selected to demonstrate the heterogeneity of the study set-ups, tested variables, countries in scope and methodologies. Empirical relevance remains extraordinarily high, as the determinants differ by industry and region. (Souitaris (2003)) At the same time, certain factors, such as firm size and firm age, are used in most studies. These factors can be extracted and are used, together with their theoretical implications, to formulate the theoretical framework for this dissertation. The publication dates of the selected works, which comprise data sets from all over the world, range from 1965 to 2010. Most studies focus on various industries in the US. Two studies assess German firms. However, no study has yet empirically investigated the innovation activities of German greentech firms. The sample sizes used in the selected studies varies very considerably, ranging from just over 10 to more than 10,000 firm observations. The methodologies used include descriptive statistics, group tests with mean comparisons and a variety of regression models. These methodologies are tailored to each research question and, of course, to the measurement/selection of dependent and independent variables. A comparative discussion of dependent and independent variables is provided in the following section.

Survey	Dependent variables	Selected significant determinants	Sample size	Regions
Correa et al (2010)	Technology adoption (incremental, process innovation)	Firm size, skilled labour, international resources, R&D intensity, industry, technology, GDP	13,575	World
Müller and Zimmermann (2009)	R&D spendings	Firm size, firm age, legal structure, R&D intensity	2,978	Germany
Vaona and Pianta (2008)	Product innovation, process innovation	Firm size, expansion orientation, cost orientation	389	Europe
Savignac (2008)	Innovation	Firm size, financial constraints, patents, industry	1,940	France
Scellato (2007)	Innovation	Total assets, industry, financial constraints	804	Italy
Hyytinen and Toivanen (2005)	R&D spendings	Firm size, sales growth, finance, asset tangibility, patents, profitability, industry, region	500	Finland
Freeman and Soete (2004)	Product innovation, R&D spendings	Firm size, industry technology, path dependency	4,378	UK
Huergo and Jaumandreu (2004)	Product innovation, process innovation	Firm size, firm age, life-cycle phase, industry, GDP	2,356	Spain
Galende and de la Fuente (2003)	Product innovation, process innovation	Firm size, commercial resources, international resources, finance, innovation productivity, expansion orientation, industry	152	Spain
Bougheas et al (2003)	R&D spendings	Firm size, profitability	573	Ireland
Bah and Dumontier (2001)	R&D spendings	Dividends paid, finance	7,004	Europe, Japan and USA
Soerensen and Stuart (2000)	Product innovation	Firm size, firm age, path dependency	237	USA
Scherer and Harhoff (2000)	Returns on product innovation	Firm size	N.A.	N.A.
Love and Ashcroft (1999)	Product innovation	Firm size, skilled labour, innovation productivity, R&D intensity	304	Scotland
Harhoff (1998)	R&D spendings	Employee growth, finance, cash flow	236	Germany
Koberg et al (1996)	Product innovation	Firm size, informational processing, structural formalization, competitive pressure	326	USA
Himmelberg and Petersen (1994)	R&D spendings	Life-cycle phase, finance, market-to-book ratio	3,035	USA
Graves et al (1993)	Product innovation, R&D spendings	Firm size	16	USA
Hansen (1992)	Innovation productivity (innovation sales/total sales)	Firm size, firm age, expansion orientation, industry	598	USA
Damanpour (1992)	Product innovation, process innovation	Firm size, cost orientation	4,000	Various
Khan et al (1989)	Product innovaiton	Firm size, R&D intensity	50	USA
Acs and Audretsch (1988)	Product innovation, process innovation	Firm size, innovation productivity, industry	1,695	USA
Utterback and Abernathy (1975)	Product innovation, process innovation	Firm size, firm age	120	USA
Scherer (1965)	Product innovation	Firm size, innovation productivity, industry	448	USA

Table 3-1: Selected empirical research on the determinants of innovation

3.2.2.1   The measurement problem

One of the fundamental problems in innovation research is how to measure innovation. The OECD has devoted a 90-page manual solely to the question of how technological innovation can be measured: *"the measurement of scientific and technological activities"*. (OECD (2005)) According to Smith (2007), there are three main measures of innovation: (1) R&D data composing basic research, applied research and experimental development, often measured as a monetary unit or ratio; (2) patent applications, grants and citations, often measured as occurrences in certain patent databases; and (3) bibliometric data that can be gathered from databases that specialize in innovation, such as SPRU (UK) and the US Small Business Administration database. A fourth measure should be added. This consists of direct measures that are individually designed to fit certain survey contexts, e.g. by means of questionnaires. This approach is used by Vaona and Pianta (2008) and Savignac (2008), for example. Another disadvantage of the proposed categorization –and also of the nature of each measure – is that no distinctions are drawn between incremental and non-incremental innovations or between product and process innovations. These distinctions are important, however, as they have a bearing on the various theoretical implications summarized in the previous section. Each measure has its own advantages and disadvantages. While R&D (1) is probably a good indicator for innovation, it reflects input more than the actual output. (Hansen (1992)) A firm might have heavy R&D expenditure but may not be able to translate it into successful innovations. Another problem is that small firms do not always have formal R&D budgets and dedicated R&D departments. This leads to the systematic undercounting of R&D at small firms (Kleinknecht and Mohnen (2002)). Patent data (2) is more output-oriented, as it measures the actual number of inventions achieved by a certain research and development team. However, patents can be of low economical and technical significance. They are not a valid indicator of the degree of an invention's commercial success and can thus serve only as a proxy for innovations. (Kleinknecht and Mohnen (2002)) Bibliometric data (3) does trace patents back to the firm and thus builds a bridge to firm-level analysis. However, it too fails to avoid the fundamental problem discussed above. Direct measures (4) can avoid these problems if they are designed properly. Like all direct measures, however, they can be susceptible to individual biases.

3.2.2.2   The determinants of innovation

The different options to measure innovation are rather simple compared to the vast amount of determinants that have been identified. Several authors have tried to consolidate the various factors into clusters, albeit with widely differing results. Vincent, Bharadwaj, and Challagalla (2004), for example, review 134 independent samples from 83 different surveys, all conducted over the past three decades, in order to flesh out the main factors that influence innovation efforts within firms. The authors identify a significant lack of consistency in empirical research and find that different measures appear to be better suited to certain characteristics of firms. They thus distill the following main clustered sets of factors: (1) organizational capacity, (2) organizational demographics, (3) environment, (4) organizational structure and (5) performance. By contrast, Souitaris (2003), reviewing around a hundred articles, identifies four main clusters: (1) contextual variables (e.g., size, sales), (2) external communication (e.g., cooperation, networking), (3) strategic variables (e.g., budget, business strategy) and (4) organizational competencies (e.g., technical competencies, personnel characteristics). Besides the general, seemingly unsuccessful, attempt to find a commonly accepted set of determinants, even basic individual measures such as size and age are the subject of controversy. (e.g., Hansen (1992)) The following analysis thus seeks to synthesize the most important determinants of innovation. Its structure draws on the literature reviews cited above, but is aligned with the analysis of growth provided in the next chapter. The main factors are:

1. R&D (R&D intensity, productivity and spending)
2. Growth (growth of the firm, growth cycle effects)
3. Resources and strategy (skilled labor, strategic orientation, etc.)
4. Financing (equity, debt, financial constraints, etc.)
5. Age (time a firm has been on the market)
6. Size (firm sales, number of employees, etc.)
7. Environmental controls (industry, country, etc.)

3.2.2.2.1 R&D and innovation

R&D intensity has generally been found to increase innovation output of all types. (e.g., Graves and Langowitz (1993); Khan and Manopichetwattana (1989); Hansen (1992)) Although this is a rather perspicuous finding, it still merits discussion. For every unit of R&D, the propensity of a firm to innovate and, hence, the expected amount of generated innovations increases. Efficiency, however, decreases. In other words, the correlation between R&D and innovation output has a concave shape. This finding is important because it implies that smaller and younger firms contribute more efficiently to an economy's overall innovation output. Decreasing returns to scale on R&D expenditures are, for example, found by Love and Ashcroft (1999), Acs and Audretsch (1988), Audretsch and Acs (1991), Hansen (1992) and Galende and de La Fuente (2003). The effect of interaction between corporate size, age and R&D is picked up again in the following.

3.2.2.2.2 Firm growth

It seems that innovation prepares the ground for corporate development. Firm growth and innovation are strongly connected and enhance each other reciprocally. A positive relationship between firm growth, measured as the change in sales, employees or assets, and innovation is claimed by Hyytinen and Toivanen (2005), Harhoff (1998), Himmelberg and

Petersen (1994), for example. However, the direction of this effect is at best unclear. A more detailed analysis of growth and its main drivers is presented in the next chapter.

3.2.2.2.3 Resources and profit

Accepted theory predicts that generating innovations requires the availability of necessary resources. This prediction appears to be reflected in empirical research results. Correa, Fernandes, and Uregian (2010) find that skilled labor significantly affects a firm's innovative output, implying that R&D investment is effective only when it is allocated to capable people. Skilled labor is therefore a necessary prerequisite for a firm to discover inventions. Apart from the actual capability to generate inventions from R&D, however, there are still other hurdles that must be overcome to turn inventions into actual innovations. Galende and de La Fuente (2003), for example, find a positive relationship between innovation and commercial resources. Although commercial resources are a rather abstract measure, the requirement for their availability implies that firms need significant resources to transform inventions into successful products. Innovative firms therefore focus on building an innovation-friendly organization with a strong focus on expansion and product or cost and process orientation. (Vaona and Pianta (2008)) Moreover, effective information processing can improve the overall set-up and thus encourage a positive correlation to innovation. (Koberg, Uhlenbruck, and Sarason (1996)) By contrast, firms with less promising product- or process-innovation-related projects return the funds they generate to shareholders. Accordingly, a negative correlation exists between R&D investment and dividend payouts. However, firms may also achieve superior profitability, which in turn is self-propagating. Empirical evidence seems to confirm a "common sense" finding in business and marketing related literature, namely that innovation is a key factor for long-term competitiveness and profitability. (e.g., Bougheas, Görg, and Strobl (2003); Hyytinen and Toivanen (2005); Del Monte and Papagni (2003); Bah and Dumontier (2001)) Other authors find less tangible factors to be important determinants, such as the number of board members (Hyytinen and Toivanen (2005)), the cognition of managers (Storey (2000)) or the legal structure of the firm (Müller and Zimmermann (2009)). Several studies also provide evidence that the degree of internationality or international exposure of a firm increases innovation due to international learning and increased competitive pressure. (Love and Ashcroft (1999); Correa, Fernandes, and Uregian (2010))

3.2.2.2.4 Financing

Schumpeter dedicated much of his work to the study of resource allocation in the innovation process. Three major implications can be derived from his work with respect to financing: First, innovation depends on investment. Second, innovation is especially strongly embedded in new firms. Third, financing is required in order to release resources from their current usage to achieve new combinations. (Schumpeter (2007)) His change of direction from younger to older firms in later works led to a change in these implications. However, the importance of finance remained. The only difference here was that the emphasis shifted from external to internal financing sources. (Schumpeter (1943)) This is not overly surprising, since innovation is an expensive endeavor. Existing innovation studies regard finance is a necessary resource for innovation and therefore model it as an independent variable. On an abstract level, most studies identify a positive relationship between finance and innovation (e.g., Galende and de La Fuente (2003); Scellato (2007); Harhoff (1998)), and also find that the effect is strongest for smaller firms (e.g., Scellato (2007); Harhoff (1998); Müller and Zimmermann (2009)). On a more detailed level, however, there appear to be certain specific differences with respect to types of financing. Internal finance appears to be the most important factor impacting innovation. Himmelberg and Petersen (1994), for example, identify a substantial effect of internal finance on R&D expenditures. At 0.67, the elasticity of cash flow to R&D is striking. In this context, Scellato (2007) find that stable patenting possible only in the absence of financial constraints. Firms with collateralizable assets are the only exception, since they do not depend on demand-induced cash flow to finance R&D. The existence of constraints leads to reduced R&D and thus to reduced growth in the long run. These results are consistent with those presented by Harhoff (1998) and Bougheas, Görg, and Strobl (2003). Savignac (2008) too arrive at a similar result, although only after correcting for endogeneity problems associated with financial constraints and innovation. Other authors focus on the differences between equity and debt financing. Galende and de La Fuente (2003) find that debt-financed firms seek to reduce risk by introducing innovations that are more incremental than radical. Previously developed technologies and components admit debt financing, because they serve as collateral and a sort of physical demonstration of a firm's track record. The results are complementary to those found by Bah and Dumontier (2001) and Müller and Zimmermann (2009), who claim that smaller and potentially disruptive firms lack collateral and therefore depend on equity financing. For innovative firms, equity enhances R&D investment and, hence, innovation. A very interesting approach is taken by Hyytinen and Toivanen (2005), who model the interaction between innovation and government funding in external capital-dependent industries. Results suggest that government funding can propel innovation and that external capital constraints do the opposite. The key messages are therefore fairly consistent, indicating that an abundance of financial resources has a positive effect and deficits have a negative effect on innovation. Further evidence, drawn from a financial point of view, is presented in section 3.4.

3.2.2.2.5 Firm age

Firm age, mostly measured as the time for which a firm has been on the market, is another common determinant. It represents not only the experience but also the degree of knowledge that a given firm has gathered over time. As such, it embodies a key corporate resource. Empirical findings on the relationship between age and innovation have so far been quite inconsistent. Vincent, Bharadwaj, and Challagalla (2004) point out that there are two main views: The first is that organizations need to be flexible rather than rigid if they are to change. Younger firms do not have the kind of fixed routines, norms and values that older firms have and are therefore more likely to innovate. (e.g., Müller and Zimmermann (2009)) The second is that organizations need a well-defined resource base and a reasonably stable environment in order to allocate resources to innovation activity. Innovation and age are thus positively correlated. (e.g., Galende and de La Fuente (2003)) However, some findings imply that this relationship is not completely monotonic and also depends on the measurement used to represent innovation. In this context, Huergo and Jaumandreu (2004) note that young firms have a greater propensity to innovate while the relationship between innovation and size is increasing monotonically. Whereas this may at first seem contradictory, the authors explain their observation as follows: Small firms that enter the market have different, additional and very unique capabilities compared to small firms that have been on the market for some time. Young firms are not only more flexible than incumbents, but also possess greater potential to do new things in new ways. Irrespective of their size, old firms exhibit the lowest innovation potential and the lowest innovation activity, as they stick closely to proven processes and structures. Contrary to what has been argued with regard to the two persisting views by Vincent, Bharadwaj, and Challagalla (2004), however, it is found that middle-aged firms are the most innovative because they are still sufficiently flexible while at the same time having access to a broader resource base than new entrants. Additionally, the authors note that innovation and age correlate at least in part in that innovation is a means for a firm to achieve age. On the other hand, non-innovative firms are more likely to exit a market and quit operations. Younger firms show more effort toward innovations by setting aside a relatively higher proportion of their resources (as larger firms need only allocate a fraction of their total resources to match the absolute amount spent by small firms). This finding implies a negative correlation between R&D expenditures and firm age, which is found by Hyytinen and Toivanen (2005) and also Müller and Zimmermann (2009), for example. It also correlates to a higher share of sales from new products for younger firms, as claimed by Hansen (1992). Looking at the type of innovation, younger firms are more likely to generate non-incremental and product-oriented innovations, partially confirming the flexibility hypothesis mentioned above and also the propositions made by Utterback and Abernathy (1975). Older firms, on the other hand, generate more incremental innovations, implying a parallel to larger firms as pointed out in the previous section. A generally positive relationship between innovation activity and age is found by Galende and de La Fuente (2003), Soerensen and Stuart (2000), Scellato (2007) and Hansen (1992). The conclusion thus seems to be that age drives overall innovation output. However, further evidence is needed to demonstrate whether this relationship is non-linear and whether the oldest firms are among the least innovative, as well as to flesh out the specific differences between incremental and non-incremental innovators. The heterogeneity of results has previously been found by Souitaris (2003) and Vincent, Bharadwaj, and Challagalla (2004).

3.2.2.2.6 Firm size

On size, it seems that most studies identify a positive relationship between both product and process innovation and firm size (typically measured in terms of (natural logarithmic) sales, employees or assets; e.g., Huergo and Jaumandreu (2004); Vaona and Pianta (2008); Love and Ashcroft (1999); Galende and de La Fuente (2003); Soerensen and Stuart (2000); Savignac (2008); Koberg, Uhlenbruck, and Sarason (1996)). This is consistent with the metastudy on innovation and firm size conducted by Damanpour (1992) and partially consistent with the extensive review of literature on innovation determinants conducted by Vincent, Bharadwaj, and Challagalla (2004), in which the authors state that large firms are quantitatively more innovative but are also associated with significant complexity and inertia. Accordingly, other studies find a negative or at least ambiguous relationship between product innovations and firm size. (e.g., Utterback and Abernathy (1975); Khan and Manopichetwattana (1989); Freeman and Soete (2004)) In this context, Acs and Audretsch (1988) note that the relationship is U-shaped, with small and large firms proving to be more innovative than medium-sized firms. In a later work, however, the same authors find the relationship to be positive but with decreasing returns to scale, claiming that the correlation depends to a large extent on the industry in question (Audretsch and Acs (1991)). Apparently, research into returns to scale seems to be less contradictory. All relevant studies in the given sample find decreasing returns with respect to R&D spending, implying that larger firms with larger R&D budgets are less efficient in their innovation activity than their smaller competitors. (Love and Ashcroft (1999); Scherer (1965); Hansen (1992); Galende and de La Fuente (2003); Graves and Langowitz (1993)) Fewer contradictions also arise when innovation is measured in terms of R&D expenditures, showing a positive correlation in all cases. (Freeman and Soete (2004); Hyytinen and Toivanen (2005); Bougheas, Görg, and Strobl (2003); Müller and Zimmermann (2009)) Likewise, the relationship between firm size and process innovation apparently provokes less discussion. Empirical evidence is largely consistent with the prevalent theory, indicating a positive correlation. In other words, larger firms are more likely to engage in process optimization activities. (e.g., Correa, Fernandes, and Uregian (2010); Galende and de La Fuente (2003); Utterback and Abernathy (1975)) In this context, larger firms also adapt to new technologies faster than smaller ones. (Geroski (2000)) Empirical evidence is largely congruent with the theoretical implications described in the previous chapter. As expected, large firms have more R&D resources and are thus more innovative than their smaller competitors. As an industry matures and firms start to compete on cost, their strategic orientation switches from a product to a process focus. Generally, this happens at a rather later stage of the innovation and diffusion process and, according to theory, when surviving firms already reached a large scale. The implication here is that process innovations are essentially the preserve of larger firms.

3.2.2.2.7 Environment

It appears that there is broad acceptance that some industries are more innovative than others. Most studies therefore control for industry and technology specifics by introducing relevant dummy variables into the analysis. (e.g., Freeman and Soete (2004); Correa, Fernandes, and Uregian (2010)) Horizontal studies also use GDP and the change in GDP as control variables. (e.g., Huergo and Jaumandreu (2004); Correa, Fernandes, and Uregian (2010)) Country differences are taken into account by Hyytinen and Toivanen (2005), although most studies focus on one specific country.

3.2.3 Conclusion – Characterizing innovative firms

The one-firm and one-technology case presented in section 3.1.5 was a rather simplistic portrayal of the characteristics and growth of innovative firms. It was therefore enriched with findings from innovation theory and empirical research to identify the actual determinants of innovation at firm level. A wide range of factors have been identified as significant drivers of innovation. The most relevant determinants with respect to the context of this dissertation have been outlined. It was shown that firm size and age are still the subject of controversy and discussion. For example, Gompers (1996) derive five factors from literature that imply a positive relationship between innovation and firm size, as well as three factors implying the opposite relationship. A synthesis of the studies presented was proposed by differentiating between innovation types and measurements. It follows that young and small firms have extraordinary capabilities that can lead to new products and new procedures that may in turn lead to successful non-incremental innovations. Larger firms have stronger resource bases compared to new entrants. If they remain flexible enough, the former will be able to engage in both incremental and non-incremental innovation. Older and larger firms thus tend to be more innovative in absolute terms. However, the relationship is not monotonic and does not necessarily include the oldest firms, which may be less innovative than their younger and apparently larger competitors. It thereby seems that it is not the development of inventions but the commercialization of inventions that establishes a critical path. Large and mature firms are able to gather significant resources and allocate a large absolute amount to R&D. For example, firms with more than 1,000 employees accounted for 2/3 of innovations in the US between 1945 and 1983 (Freeman and Soete (2004)). If we also take R&D productivity into account, it appears that larger firms are relatively less efficient. Larger R&D investments thus yield larger absolute returns but smaller marginal returns. Further factors, such as internationality, resource endowment and strategy are mentioned as examples of innovation drivers. Another important factor is the availability of finance. Yet, this has only attracted limited attention from researchers in the field. Existing studies suggest a broad consensus that financing is a necessity for innovation, and that the effect is strongest for small firms and in the case of equity and internal financing instruments. Further assessment is nevertheless needed and further links to financial research must be established. The role of innovation in finance literature is picked up again and analyzed in section 3.4.4.

The results are summarized in Figure 3-8. Four main clusters lend structure to the factors identified: R&D, growth, firm characteristics and controls. A firm's strategy moderates the factors. A product strategy will, for example, increase the likelihood of a firm using its resources to bring new technologies to market. This will then increase innovation activity in respect of products. Innovation breaks down into the categories described in previous chapters. R&D intensity too is modeled as a dependent variable to account for its common use in literature. Product and non-incremental innovation on the one hand and process and incremental innovation on the other are clustered, as the relationships did not differ from each other. The plus and minus signs indicate the vertical direction of the relationship. For example, R&D expenditures correlate positively to product innovations and negatively to R&D intensity.

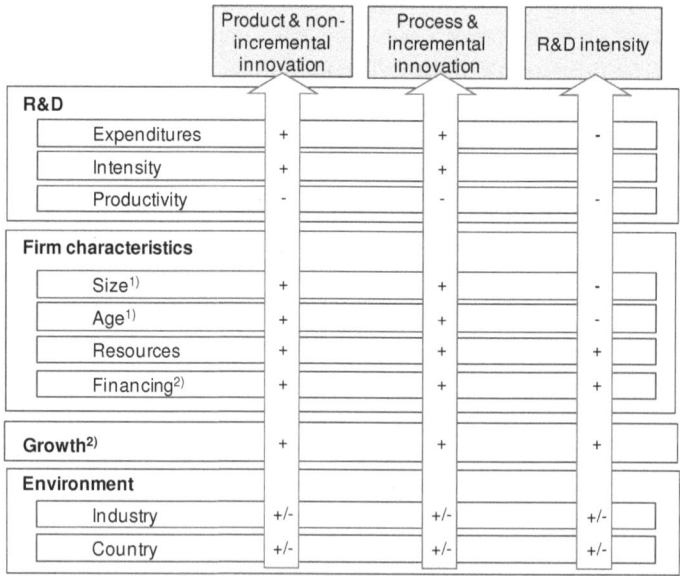

	Product & non-incremental innovation	Process & incremental innovation	R&D intensity
R&D			
Expenditures	+	+	-
Intensity	+	+	
Productivity	-	-	-
Firm characteristics			
Size[1]	+	+	-
Age[1]	+	+	-
Resources	+	+	+
Financing[2]	+	+	+
Growth[2]	+	+	+
Environment			
Industry	+/-	+/-	+/-
Country	+/-	+/-	+/-

1) Relationship not monotonic   2) Potentially simultaneous relationship

Figure 3-8: Determinants of innovation – A synthesis

R&D investment reflects a firm's commitment to innovation, driving innovation output for products as well as processes. The increase in output plots a concave curve, meaning that innovation becomes less productive as innovation output increases. R&D expenditures are expected to correlate negatively with R&D intensity, because although larger firms spend more on R&D than smaller firms in absolute terms, they only have to devote a much smaller fraction of their total resource base to R&D than is the case for small firms. Theoretical considerations fuel the argument that small and young firms have a higher propensity to develop disruptive innovations because they are more flexible, do not face the problem of organizational inertia and are less prone to basing decisions and procedures on past routines and successes. This assumption is supported by the fact that the greentech market space is fairly new and driven by young firms such as Q-Cells, Solarworld, etc. These innovations have considerable potential to change the entire market space, posing a serious threat to the incumbents and current market leaders. However, significant hurdles stand in the way of this development path, as young firms in particular often do not have sufficient resources or financing to achieve successful commercialization. As pointed out by Kline and Rosenberg (1986), early innovators may fail because "fast seconds" often experience greater commercial success due to R&D spill-over effects and organizational learning. If this is truly the case, it would imply that young and small firms with disruptive inventions may not stay in the market until their inventions turn into innovations, and hence that they may simply not figure in empirical studies. Large firms would then appear to be more innovative as they are more likely to be the "fast seconds". More generally, there is strong evidence that larger firms are indeed more innovative in terms of total innovation output than smaller firms, as they possess and have better access to resources. However, the relationship between size, age and innovation is not monotonic. Empirical literature suggests that the overall effect is dominated by larger firms and total innovation output, although small and young firms may be very innovative as well. The overall relationship between size, age and innovation seems to plot an inverted U- or N-shape, whereas the positive correlation between size and innovation dominates the overall effect. "Large and old" are therefore not prerequisites for successful firms. In fact, some studies show that young/middle-aged and medium-sized firms are the most innovative ones, combining the flexibility of newness with the resource endowment of largeness. Additionally, old firms that do not innovate are at least empirically shown to have a high propensity to exit the market. Innovations by larger firms are also more likely to be based on existing products with lower returns per unit of R&D, due to decreasing returns to scale. In this case, the theoretical implications are mostly consistent with empirical findings. The theory also predicts that larger firms will dominate a market space when a dominant design emerges. Empirical research does not provide sufficient support to falsify this hypothesis, as it is scarcely possible to measure the rise of a dominant design. Notwithstanding, the higher degree of innovation activity at larger firms could be interpreted as an indicator that large firms have superior technology marketing capabilities. An incremental innovation could, for example, be the new generation of an existing dominant design. This would indirectly support the stated theoretical proposition. Firm profitability and growth have a positive effect on innovation. However, the effect might also be concurrent, as existing research has modeled both variables as dependent and independent. For example, although studies of innovation model growth as a driver of innovation, it may well be that innovation is also a driver of growth. Internationality and legal structure are also thought to impact innovation. The more international exposure a firm has, the greater will be its commitment to innovate, in order to cope with global competition and fulfill the need for local differentiation. Industry and country controls are included as standard in literature.

To summarize: Not all theoretical propositions seem to find support from empirical studies. Three main issues remain very clear in contemporary empirical research: First, small firms' R&D budgets are undercounted. Second, empirical studies only cover survivors. And lastly, none of the studies capture the actual impact of an innovation. Accordingly, the question of whether smaller and younger firms are more likely to generate disruptive innovations than larger firms could neither be fully supported nor falsified. Furthermore, and with respect to the research questions tackled by this dissertation, it appears that the relationship between innovation and finance has only attracted limited attention and deserves further investigation. Finally, it must be noted that the synthesis presented above reflects – at least to some extent – an unavoidable selection and, hence, the view of the author. However, in the absence of a commonly accepted construct, it should still serve its purpose: to prepare the ground for empirical analysis of this dissertation to the best of the author's knowledge. The next section examines the process of technology diffusion, i.e. firm growth. It picks up where this section ends, i.e. at the point where an invention has been brought to market and gained initial acceptance, thereby impacting a firm's growth trajectory.

3.3   Growth – A measure of corporate development

Firm growth is commonly accepted as one of the main forces that drives economic development (e.g., Acs and Storey (2004)) and has thus become one of the most researched fields in literature (Bahadir, Bharadwaj, and Parzen (2009)). Growth is a desirable goal not only from a macroeconomic point of view, but also from a business point of view. For instance, growth is an important measure of shareholder performance, representing the prospect of wealth generation. As such, it is widely celebrated in business media – one example being Fortune's 100 fastest-growing firms.[10] It is therefore commonly accepted that growth is one of the key objectives of young firms in particular, as it increases their size and leads to a more optimal scale of operations (Barrow (2006)). There are plenty of studies on how to grow a business by a variety of means, including geographic expansion, mergers and acquisitions, adding service provision and, of course, innovation. This is especially true of publications that primarily target practitioners. In academic literature, two main fields of research have emerged: (1) corporate growth cycles and (2) determinants of growth. The first field is more conceptually driven, while the second is driven mostly by empirical research. Although both fields should be closely related to each other, a lack of mutual integration persists. Most studies can therefore be allocated exclusively to one or the other of the areas referred to. (Garnsey, Stam, and Heffernan (2006)) Moreover, both fields seem to deliver differing results with respect to apparently similar problems. Both issues have contributed to the absence of an overall theoretical concept of firm growth and development. This has, of course, not gone unnoticed by researchers over the past few decades. (e.g., Garnsey, Stam, and Heffernan (2006); Evans (1987); Kazanjian and Drazin (1990); Lichtenstein, Levie, and Hay (2007)) The following analysis therefore summarizes the status quo in research by (1) reviewing theoretical concepts and lifecycle-oriented frameworks, (2) analyzing empirical literature and synthesizing its main findings, and (3) deriving conclusions for the theoretical framework of this dissertation.

3.3.1   Patterns of growth

Before looking at the theoretical concepts that underpin growth, this section builds on the findings presented above with regard to technology innovation and diffusion. As previously demonstrated in section 3.1.5 and shown in Figure 3-6, a firm's growth would, in a one-firm and one-technology world, entirely match the sales growth from the technology in question. Firm growth would then be driven by the number of product or technology adopters in a given setting and could be described by models used in technology innovation diffusion research.

---

[10] http://money.cnn.com/magazines/fortune/fortunefastestgrowing/2010/full_list/, accessed in March 1st, 2011

(e.g., Bass (1969)) However, in the real world, firms engage in various activities that span not only a broad range of technology types but also a series of services and internal projects. To understand the process of technology diffusion, one needs to understand the process of firm development. Growing greentech firms are thus the drivers that push the diffusion of sustainable technologies. Let us now assume that more technologies are added to the framework (Figure 3-6). For the sake of simplicity, let us only take into account a firm's sales. The firm would possess a product portfolio in which several lifecycles reflect the adoption of each individual technology by the firm's customers over time. The firm's overall sales volume will then be the sum of sales for each technology. Sales for each technology will, at constant prices, be driven by the number of adopters. Now let us assume that more firms are competing with regard to the given technologies and their substitutes. The existence of competition will influence not only the degree of adoption, but also the price, which will affect a firm's sales in two ways. Figure 3-9 shows the hypothetical example. At point t_1, the firm introduces it's first innovation (P1). The increase in adoption leads to sales growth and thus the growth of the firm. At point t_2, sales from technology P1 stabilize and the firm introduces innovation P2. As sales from P2 pick up, cumulative sales increase too. The market as a whole could grow faster, allowing either new firms to enter the market or causing the firm in question to lose market share to its competitors, or vice versa. Relative development is not reflected in the diagram. At point t_3, the firm introduces innovation P3. P3 is not successful and is ultimately taken off the market. There are several reasons that could precipitate such a development: price erosion due to fierce competition, for example, a better substitute at a lower price available on the market, or information asymmetries with regard to adopters. The overall sales volume and growth of the firm would thus, in the given example, be reflected by the outer line, i.e. the cumulative sales volume for each product. Overall sales growth would then determine the firm's lifecycle in the example.

Again, the real world is naturally more complex, and circumstances are influenced by the firm's characteristics and capabilities. (Penrose (1997)) Moreover, firms may choose to grow in different ways, i.e. organic growth, growth through acquisition or hybrid growth. (McKelvie and Wiklund (2010)) The growth trajectory will thus neither necessarily follow the path shown in the example nor reflect the typical S-curve, as we saw in the previous section. Delmar, Davidsson, and Gartner (2003) analyze the growth of 1,501 Swedish fast-growing firms over a period of ten years and argue that there is an insufficient inclusion and acceptance of the heterogeneous nature of this growth in literature. Based on their empirical analysis, the authors identify seven clusters of firms with distinctive growth patterns: (1) super absolute growers (knowledge-intensive and manufacturing SMEs); (2) steady sales growers (large firms in traditional industries); (3) acquisition growers (large and old firms in heavy and mature industries); (4) super relative growers (high knowledge intensive and independent SMEs); (5) erratic one-shot growers (mature SMEs in low-tech industries); (6)

employment growers (SMEs in low-tech industries); and (7) steady overall growers (large and affiliated firms in manufacturing industries).

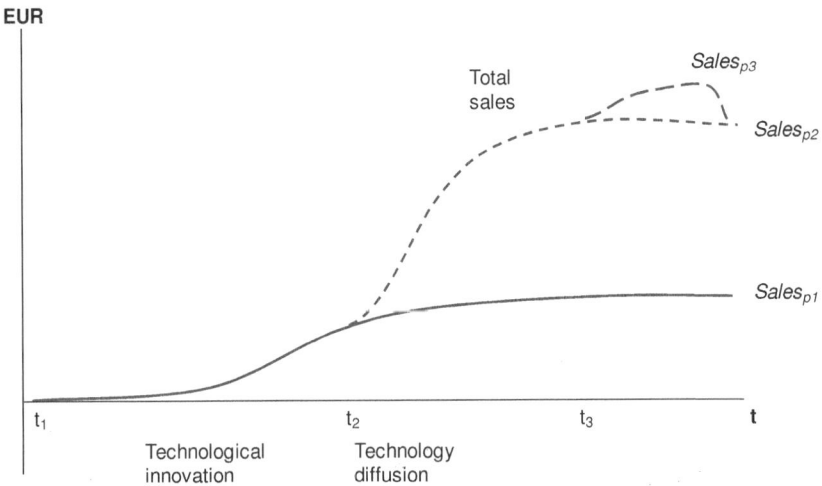

Figure 3-9: The multiple-firm and multiple-technology innovation diffusion case

According to Brush, Ceru, and Blackburn (2009), growth is not even normally distributed among all firms on the market. Based on several case studies, the authors identify four model growth patterns: (1) rapid, (2) incremental, (3) episodic and (4) plateau. These potential growth patterns are shown in Figure 3-10. Rapid growth patterns are associated with firms that face entirely favorable market conditions that do not harm prices and ensure sufficient market demand. Competition is either low or is circumvented by serving niche markets or preferred regions. Incremental growth, on the other hand, is the strategy adopted by firms that depend on a limited customer base, for example. For them, each new customer acquisition increases sales by a relatively high amount. Episodic growth trajectories force firms to operate in unstable environments with volatile internal resources and capabilities. The final pattern, plateau growth, is often associated with declining margins as firms operate in a stable or, sometimes, declining market space.

According to the authors, each pattern is established by a combination of three relevant factors: management, market and money. Management must have the objective and ambition to grow if they are to do so. A management strategy that merely targets steady income generation, for example, will hardly generate additional sales and probably lead to a plateau growth pattern. Lastly, firms must have the financial resources they need to invest in product development and commercialization. Although the proposed patterns of growth lack the backing of rigorous empirical evidence, they still highlight two important facts. First, growth trajectories can take on various forms. And second, they are determined by the resources and capabilities a firm possesses. Both assumptions are consistent with fundamental theory (Penrose (1997)) and partially consistent with lifecycle concepts (e.g., Churchill and Lewis (1983); Greiner (1998)). The latter are reviewed in the section that follows.

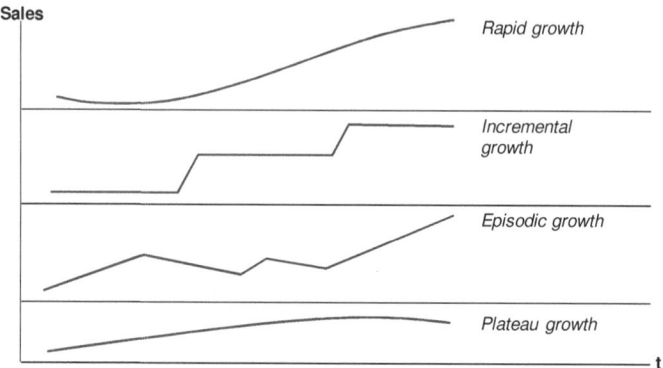

Figure 3-10: Exploratory study results on growth patterns
Source: Adapted from Brush, Ceru, and Blackburn (2009)

3.3.2 Theory of firm growth

Despite the importance of and increasing interest in firm growth, past attempts to develop a consistent and exhaustive theory have been rather fruitless. (e.g., McKelvie and Wiklund (2010); Levie and Lichtenstein (2008); Phelps, Adams, and Bessant (2007); Kazanjian and Drazin (1989); Miller and Friesen (1984); Poutziouris (2003)) The development of relevant theories is thus progressing slowly and is subject to severe limitations. One of the earliest, most complete and still very popular theories on growth was first published in 1959 by Penrose (1997). Even recent articles in peer-reviewed journals still refer to it (e.g., Macpherson and Holt (2007); Garnsey, Stam, and Heffernan (2006)). The basic idea of Penrose's theory is that firms are administrative constructs or units that possess certain sets of resources. Moreover, a firm is composed of two main functions: the entrepreneurial and the managerial function, where growth is achieved by matching business opportunities with appropriate resource combinations. Entrepreneurial creativity drives the ways in which resources are combined. At the same time, growth is only achieved if all relevant tasks and procedures are managed effectively. The managerial function is therefore a sort of instrument to bring creative resource combinations into action and exploit certain business opportunities. This has one important implication: The older a firm becomes, the more it will rely on implemented procedures and thus limit the entrepreneurial function, which will in turn limit growth potential despite the greater availability of resources. On the other hand, young firms face a trade-off between a forceful entrepreneurial spirit and limited managerial capacity and capability. Both implications reveal close parallels to innovation theory, in which young firms can threaten incumbents with disruptive inventions but are limited by their scant resource base. (Utterback (1971); Utterback and Abernathy (1975)) In her work, Penrose points out that growth can be achieved either organically or through acquisitions. In the current context, an acquisition would broaden the resource base and, in so doing, increase the potential resource combinations. However, 50 years on, McKelvie and Wiklund (2010) argue, in an extensive literature review on firm growth, that only few studies even consider these distinct growth modes. This appears to be one of the problems with respect to theory building.

3.3.2.1 The corporate lifecycle concept

Most of the later studies that assess the nature and characteristics of firm growth have adopted an evolutionary perspective. (e.g., Chandler (2001); Steinmetz (1969); Greiner (1998); Churchill and Lewis (1983)) Like the technology diffusion models presented at the beginning of chapter 3, evolutionary models of business growth assume that firms move through certain stages over time. This idea originates in the view that firms develop, like organisms, from birth through growth until they reach maturity and eventually exit the market. A large number of such models have been developed in the past 50 years. In their literature review on "models of entrepreneurial growth and change", Levie and Lichtenstein (2008) present 104 different models from management research. The majority of these models assume a certain number of development stages that involve distinct characteristics of the firm. The overall process is thought to be of a predictive – i.e. deterministic – nature. This means that every firm has a predefined growth path with a certain number of stages, i.e. a corporate lifecycle. Adizes (1988) define corporate lifecycles as follows: *"Organizations have lifecycles just like living organisms do; they go through the normal struggles and difficulties, accompanying each stage of the Organizational Lifecycle and are faced with the transitional problems moving to the next phase of development."* This definition adds another attribute to firm development, namely that firms pass through phases by solving phase-specific problems. A successful growth path thus implies that a firm must overcome internal crises that will then lead to new stages of development. These crises may be rooted in the market, organizational issues, financial problems, etc. However, literature has identified a large variety of supposedly stage-specific problems. Inconsistencies in literature not only with regard to such problems, but also with regard to the actual number and characteristics of stages, makes the problem of insufficient theory building even worse (Levie and Lichtenstein (2008)). The paragraph below therefore summarizes some of the most important models in literature.

One of the first lifecycle-oriented concepts was formulated by Rostow (1959), albeit from a macroeconomic perspective. The framework compares the constitutive differences between geographical regions and their impact on economic development. The author's fundamental belief is that industries are characterized by lifecycle phases or development stages. This assumption sparked the idea that firms too are characterized by certain stages of development, such that subsequent research focused strongly on the firm. An influential model of the corporate lifecycle was formulated by Chandler (2001), who argued that organizations are dynamic in nature. In this context, lifecycle phases have a powerful impact on how firms act and behave on the market. The work analyzes early institutional development during the 1960s in the US. One of its main propositions – that "structure follows strategy" – has since become rather famous. The proposition implies that a firm's growth also changes its general characteristics. Although the work is more a call to heed the needs of administration and an analysis of which corporate structure best supports different sorts of growth, it also supports one fundamental assumption of lifecycle theory, which is that small, young firms and large, old firms differ in terms of both growth and structure. A direct link between size and stage is proposed by Steinmetz (1969), who claims that firms grow in a roughly S-shaped manner. Steinmetz believes that, as in most lifecycle models, stage transition occurs when stage-specific problems are overcome. Four main stages are formulated: (1) direct supervision (25-30 employees); (2) supervised supervisor (250-300 employees); (3) indirect control (750-1000 employees); and (4) divisional organization (more than 1,000 employees, plateau phase, new entrants who challenge how things are done). The first stage is dominated by the firm's efforts to survive while managing increasing complexity and competition. Work and responsibilities must be delegated more effectively to move the firm successfully to the next stage. The supervised supervisor stage is characterized by principal-agent issues as the owner becomes a manager and focuses on financial and organizational efficiency. Growth in overheads and the consumption of resources pose a serious problem to the firm as it does not yet enjoy economies of scale. When a firm does make the transition, however, it significantly improves efficiency. However, at the same time growth rates and profitability slow down as the firm becomes large enough to suffer from organizational inertia. The last phase evidences a plateau growth path. New firms enter the market and challenge how things have been done up to now. Mature firms have to reinvent themselves to cope with competition. Otherwise, they risk losing their dominant position.

A later but equally influential model was developed by Churchill and Lewis (1983). This model claims that firms vary widely in terms of size, growth and several other characteristics, all of which are moderated by a firm's development stage. At each stage of development, firms seem to encounter similar problems that, once resolved, begin the transition to the next stage. The concept comprises six distinct stages: (1) existence, (2) survival, (3) success, (4) take-off, (5) resource maturity and (6) ossification. At the existence stage, firms have only just begun their operations. They are dominated by the founder, who usually owns the firm entirely or to a large extent. The main problem at this stage is to attract enough customers to generate revenues and enough external funds to finance operations. When this is achieved, the firm moves on into the survival stage. Although is the criticism is often leveled that lifecycle models are deterministic (e.g., Miller and Friesen (1984)), assuming that every firm must pass through all development stages, Churchill and Lewis (1983) note that firms can either pass through the survival stage or stay in this category. If a firm successfully completes the transition, it will improve its profit and sales situation. If not, it will generate only marginal returns but may serve as a source of income until the owner finally sells up or retires. The third stage is called the success stage. The firm generates plenty cash, employs management staff and provides wealth to the owner. The owner may then either turn to other activities or actively work on further growth. The firm may thus either stay as it is or attract further resources to foster transition. The fourth stage is called take-off, because work is delegated, resources are attracted and the firm develops at fast pace. Increased size brings with it greater complexity, causing the firm to start to implement standard procedures and structures. However, it will only continue to grow if it can preserve its culture and uphold both the spirit and the level of innovation. At some point, it enters the resource maturity stage, at which growth rates stabilize and products may eventually become commodities. If the firm does not leverage new market opportunities, it may then fall into the ossification stage, at which the firm's products and technologies have been rendered obsolete and the firm's structure is dominated by inertia and inflexible structures and procedures. Ultimately, it may fail to survive on the market. The overall model serves as good basis for most other lifecycle models as it predicts certain characteristics for each stage. It is based on exploratory research into 83 firms and was thus formulated on an empirical basis. A fairly similar concept was proposed by Greiner (1998). However, Greiner argues that the relationship between age and size is also driven by industry. In other words, low-growth industries are populated by smaller firms of a more advanced age, while high-growth industries are populated by larger firms of a younger age. The model also assumes stage transition as a response to internal crisis. Five phases are defined: (1) creativity, (2) direction, (3) delegation, (4) coordination and (5) collaboration. The first two stages are comparable to the existence and survival stages in the model formulated by Churchill and Lewis (1983) and the direct supervision stage in the model proposed by Steinmetz (1969). The third and fourth stages compare to the supervised supervisor and indirect control stages in the Steinmetz model and the delegation and coordination stage in the Churchill and Lewis models, respectively. The last stage is rather different as it focuses on the role of teams and collaboration within the firm. Moreover, Greiner adds a more detailed conception of organizational structures. In early stages, the structure manages to stay informal and flexible. In later stages, however, a functional orientation dominates as profit centers and formal communication lines are introduced. At the last stage, the firm assumes a divisional or matrix organization, while formal incentive-based compensation systems emerge. More detailed analysis would naturally reveal more differences between the models cited. However, a detailed review of the characteristics is neither the purpose nor part of the scope of this dissertation. And while the examples cited do not claim to be a representative selection, they unquestionably rank among the most influential works, featuring over 300 citation links in the EBSO database. (This statement applies to the three journal papers by Greiner (1998), Steinmetz (1969) and Churchill and Lewis (1983).)

The aim of this section was to introduce different models for development stages and describe some of their main characteristics. First, lifecycle models share the view that firms can only pass a certain stage when they acquire the resources they need to grow and stay on the market. Second, lifecycle stages moderate firms' characteristics and their relationship with the environment. Third, lifecycle models prescribe a deterministic growth path that firms either follow or may eventually fail to follow. Finally, lifecycle models are composed of a specific number of stages covering a certain time frame. However, the number of stages and the time frames covered differ widely across the various studies. The fact that these claims also hold true for a larger sample of developed models is shown in the next section, which discusses literature reviews and contemporary criticism.

3.3.2.2 Critical appraisal of the corporate lifecycle concept

As indicated above, lifecycle models have been the subject of intensive discussion in literature. Several literature reviews conducted to assess the overall validity of concepts (e.g., Phelps, Adams, and Bessant (2007); Levie and Lichtenstein (2008)) concluded that the concepts concerned have major shortcomings and are in need of modification. Critical appraisal is therefore vital to the formulation of a proper theoretical framework.

Weber and Antal (2001) note that although organizational lifecycles seem to be widely accepted in literature, organizations do not, like organisms, follow a predetermined path of development. Indeed, firms may skip stages, fail to pass through stages or accelerate organizational learning. The general acceptance seems to come from the appealing notion that stages stand for the role of time in the corporate development process, moderating the age and size of firms. Miller and Friesen (1984) also argue that stages are not connected in a deterministic way and that firms may move through stages in either direction. Levie and Lichtenstein (2008) and Lichtenstein, Levie, and Hay (2007) review 104 academic articles on the organizational lifecycle concept published from 1962 to 2006, of which 20 were published after 1994 – an indication that the subject remains highly topical. Essentially, the authors find large differences among the main conceptual pillars. The number of stages varies between three and as many as eleven. The attributes examined likewise vary considerably, whereas the most common attributes are (in descending order of frequency): outcomes (size, age and growth), management characteristics, organizational structure, strategy, systems, stage-related problems, process and product characteristics, staff, market factors, innovation, external factors, profitability, geography, culture and risks. Furthermore, the authors are unable to identify any commonly used definition of what a development stage actually is. The conclusion is that there is no consensus on the main characteristics of a stage, the number of stages or even the actual definition of what a stage is. With respect to transition processes and mechanisms, the authors draw a similar picture, claiming that the increasing number of publications did not lead to a decreasing number of baseline models. in other words, only 32 of the newer models are linked to one of the main sources, while 44 models have no cited connection whatsoever. The authors claim that the lack of consensus is also reflected in *"theorists' increasing aversion for claiming universality for their particular model"* (Levie and Lichtenstein (2008)). Phelps, Adams, and Bessant (2007) also review literature on lifecycle models. The authors find that most models deal with the growth of particularly small firms, and that earlier models in particular tend to be rather conceptual, involving huge variations in assumptions. Their conclusion that existing literature lacks convergence is consistent with Levie and Lichtenstein (2008): *"Taken as a whole, the results of these studies are inconclusive and provide limited support for the thesis that structural, functional or problem-type patterns are congruent with stages models."* (Phelps, Adams, and Bessant (2007)). The problem is also rooted in the heterogeneous nature of small firms. In particular, the authors point out that the assumption of the organism metaphor (that growth is linear, deterministic and sequential) is false. To deal with these problems, two approaches have emerged in literature. The first approach aims to identify commonly accepted principles for deriving a more valid, aggregated lifecycle model. (Hanks et al. (1993); Dodge and Robbins (1992)) The second approach completely ignores the basic assumptions of lifecycle theories and proposes new context models. (Levie and Lichtenstein (2008); Phelps, Adams, and Bessant (2007))

Moreover, as we saw above, many studies – and particularly the early ones – lack a rigorous empirical foundation. These studies are either entirely conceptual or use only exploratory research methods. A different approach is taken by Hanks et al. (1993) in an effort to overcome the heterogeneity of existing models. The authors use cluster analysis based on various factors such as corporate structure, size, growth, etc. to identify internally homogeneous but nevertheless distinctive clusters of firms. Two interesting results emerge. First, all study participants were able to assign their firm to a certain category. Second, the authors were able to derive common categories from ten major lifecycle models that fit their data:

1. Stage 1 (start-up): Young and small firms, R&D focus, informal organization
2. Stage 2 (expansion stage): Older and larger firms compared to stage 1, highest growth rates, functional organization, commercialization and product focus

3. Stage 3 (expansion, early maturity stage): Firms are slightly younger than those in stage 2 but almost double their size.; focus on product and process orientation; growth slightly lower than in stage 2; slightly lower growth compared to stage 2

4. Stage 4 (maturity, diversification): Large and old firms, emergence of divisional structures, product portfolio and process orientation

5. Stage 5 (non-fitting firms): Old firms of similar sizes to stages 1 and 2, lowest growth rates, not really participating in growth cycle but relatively stable

Although not entirely consistent with the above taxonomy, Dodge and Robbins (1992) presents fairly comparable results. His analysis is also based on an effort to integrate models from literature and optimize their validity using empirical data. He notes that all models have different implications and a varying number of stages, but that four main stages appear to be commonly accepted: formation, early growth, later growth and stability. Kazanjian and Drazin (1989) and Kazanjian and Drazin (1990) also use a mixed-method approach to identify common stages. Like Dodge, the authors identify four stages that they refer to as (1) conception and development, (2) commercialization, (3) growth and (4) stability. In addition to Dodge and Robbins (1992) and Hanks et al. (1993), the authors also derive an aggregate of stage-specific problems. In the first phase, the dominant problem is prototype development; in the second it is phase product improvement and the acquisition of facilities and resources; in the third it is phase production and the need to balance profit and growth; in the fourth it is the need to fight inertia. Apart from attempts to find a general taxonomy that integrates the miscellany of corporate growth-cycle models, several researchers have found it useful to use aggregated lifecycle models to assess a whole series of problems in different disciplines. (e.g., Elsayed and Paton (2009); Hoy (2006); Berger and Udell (1998); Rutherford, Buller, and McMullen (2003)) Nevertheless, as appealing as the idea of finding a generally accepted framework for organizational lifecycles undoubtedly is, the three metastudies discussed above do not solve all of the problems raised. First, none of them seems to apply to all firms. For example, Kazanjian and Drazin (1989) explicitly reduce the scope of their aggregated model to young technology firms. Second, the number and naming of stages still varies, even though it seems that most models now propose either 4 or 5 stages, and that time may ultimately lead to more convergence. Lastly, even aggregate models are based on the assumption of organismic development.

These problems are addressed in works that aim to develop new or substantially modified models. While the Hanks et al. (1993) model reveals support for the existence of stages, Phelps, Adams, and Bessant (2007) criticize that it lacks stability. In fact, the model explains around 40% of the firms studied. Hite and Hesterly (2001) points out that stages "function as a proxy for many strategic issues" Hite and Hesterly (2001), but that the lifecycle framework has reached its limits with respect to its predictive qualities. Levie and Lichtenstein (2008) conclude that only one aspect remains true, namely that *"growing business displays distinguishable stage or configurations at different times"* (Levie and Lichtenstein (2008)) and that *"businesses tend to operate in some definable state for some period of time"* (Levie and Lichtenstein (2008)). The authors propose that stage models should no longer be used by entrepreneurship scholars, before going on to develop a more dynamic alternative model. In this model a state, rather than a stage, is derived from the business model and from configuration of firm attributes in accordance with a dynamic logic. It therefore assumes neither consecutive stages nor a certain number of stages. The sequence and order of states, representing management's attempts to best match internal organization capacity to environmental needs, is predicted by the individual context. The model thereby takes into account market change and the creation of opportunities, focusing on the effectiveness of a firm's business model in a given environmental context. Phelps, Adams, and Bessant (2007) also propose an alternative, state-oriented model. In the authors' view, firms encounter problems that establish a "tipping point" that needs to be overcome to continue growth. Although this idea is shared with traditional models (e.g., Greiner (1998)), the authors make no assumptions about linearity, time frame or sequence. They identify six areas for potential tipping points: operational improvement, people management, obtaining finance, formal systems, strategy and market entry. According to the authors, a firm's absorptive capacity is the means to acquire external knowledge, which is then used to navigate these "tipping points". Absorptive capacity is focused on the specified tipping points and is composed of four cognitive steps: ignorance, awareness, knowledge and implementation. Since assumptions are made neither about types of tipping points nor about the means to overcome them, the overall concept remains rather abstract. This criticism also applies to the model proposed by Levie and Lichtenstein (2008). Although they solve the problems associated with stage models, the proposed state concepts have one major shortcoming. Due to the degree of abstraction, their exclusive focus on a certain business model and the dynamic logic argument, neither model admits interfirm comparison. Both are thus of virtually no use in empirical research. Although useful when analyzing a single firm, they provide no insights from which to derive policy implications in a broader institution-driven setting.

As discussed above, there is widespread skepticism regarding lifecycle models in literature. Two approaches to resolve this skepticism have been summarized, and both have different implications. The aggregation approach, which tries to find a common pattern and common ground for an integrated theory, fails to overcome the fundamental assumption that firms do not grow in a deterministic way in several stages. At the same time, it appears undeniable that the propensity and potential to grow depends on the characteristics of a firm's current stage. (Poutziouris (2003)) Another problem is that aggregation approaches have not yet managed to consolidate either the number of stages or the nomenclature used. Only time will tell if these problems can be resolved in the future. The second approach, proposing substantially modified models, may turn out to be a promising path. However, these models suffer from a high degree of abstraction and thus make it hard to draw general conclusions for more than one firm. At present, it therefore seems that the general idea of the organizational lifecycle is not quite "dead" (Levie and Lichtenstein (2008)).

3.3.3   Empirical evidence about growing firms and the corporate lifecycle

Empirical research has focused strongly on identifying the factors that impact a firm's growth. Some work has also focused on providing evidence for the corporate lifecycle concept. Three of these studies (Hanks et al. (1993); Dodge and Robbins (1992); Kazanjian and Drazin (1990)) were briefly discussed in the previous section as their focus is on theory development. According to McKelvie and Wiklund (2010), distinctions can be drawn between three areas of research: (1) growth as an outcome, (2) the outcome of growth and (3) the growth process. The first area treats growth as the dependent variable in order to find factors of influence. The second area models growth as the independent variable. These models assess lifecycle stages. The third area assesses the internal level of the firm. The analysis that follows summarizes the main findings of empirical literature for the first two areas only, because the firm-immanent process of growth is of no interest for the purposes of this dissertation. It thus provides evidence for the determinants of growth and for at least some characteristics of the organizational lifecycle models. The wealth of work that has been done is evidenced by the selected surveys shown in Figure 3-11. As can be seen, several metastudies have been conducted. These provide a highly representative view of past achievements. The following discussion accordingly attaches greater importance to their propositions than to the findings of individual studies. The discussion is organized as follows: First, the measures of growth used in literature are discussed. Second, evidence for the organizational growth models is presented. Third, determinants of growth are synthesized. Finally, conclusions are drawn with respect to the above theoretical considerations and the theoretical framework of this dissertation.

Survey	Dependent variables	Selected significant determinants	Sample size	Regions
Lee (2010)	Growth rate	Size, age, innovation, resources, finance	1500	Various
Bahadir et al (2009)	Growth rate	Size, age, strategy, innovation, resources, finance, environement	n.a.	Various
Levie and Lichtenstein (2008)	Life cycle	Size, age, strategy, innovation, resources, environement	n.a.	Various
Musso and Shiavo (2008)	Growth rate, firm hazard rate	Size, resources, finance, industry, time	15,000	France
Macpherson and Hold (2007)	Growth rate	Resources, other	n.a.	Various
Phelps et al (2007)	Life cycle	Age, size, resources	n.a.	Various
Pasanen (2007)	Growth rate (organic and acquisitive)	Age, size	110	Finnland
Scellato (2007)	Investment/capital stock	Resources, finance, industry	800	Italy
Gilbert et al (2006)	Growth rate	Size, age, strategy, innovation, resources, environment	n.a.	Various
Hyytinen and Toivanen (2005)	Growth rate	Size, age, resources, finance, industry	450-500	Finland
Del Monte and Papagni (2003)	Growth rate	Size, innovation	500	Italy
Poutziouris (2003)	Growth orientation	Age, resources	922	USA
Audretsch and Elston (2002)	Investment/capital stock	Size, resources, finance	100	Germany
Carpenter and Petersen (2002)	Growth rate	Resources, finance	> 1,600	World
Plötscher and Rottmann (2002)	Investment	Resources, country	867	Germany
Lee et al (2001)	Growth rate	Size, resources, innovation	137	Korea
Wijewardena and Shiran (1995), (1999)	Growth rate	Size, age, resources, industry	> 50	Japan and Australia
Weinzimmer (1998)	Growth rate	Size, age, strategy, innovation, resources, environement	n.a.	Various
Harhoff (1998)	Investment/total assets	Resources, finance	236	Germany
Sutton (1997)	Growth rate	Size, (Gibrat's law falsification)	1,501 out of 11748	Sweden
Himmelberg and Petersen (1994)	Investment	Resources, finance	3,035	USA
Dodge (1992)	Growth cycle	Problem types	364	USA
Kazanjian and Drazin (1989)	Growth cyce	Age, size, resources, growth	105	USA
Miller and Friesen (1984)	Growth cycle	Age, size, resources	36	USA

Table 3-2: Selected empirical research into the determinants of growth

3.3.3.1 The measurement of growth

Growth is a development whereby a certain measure increases from at least one point to another. (Penrose (1997)) Growth can be measured using various demographic variables for a firm, such as assets, sales, employees, number of branches, countries of operation, etc. To avoid comparing apples with oranges, so to speak, it is important to be aware of which measure is being used. According to Gilbert, McDougall, and Audretsch (2006), the most common measures of growth in literature are changes in sales, employment and market share.

How growth is measured has a significant impact on the implications drawn. Bahadir, Bharadwaj, and Parzen (2009), for example, note that, when growth is measured in terms of market share, competition has a powerful effect, which would not be the case if growth were measured in terms of sales. Besides the different concepts of growth, Weinzimmer, Nystrom, and Freeman (1998) argue that the problem of varying results also stems from the different formulas (e.g., averages, absolute changes) used to quantify growth, resulting in significantly different relationships for dependent and independent variables. Moreover, the authors point to an excessively weak correlation between employees, sales and asset growth. Independent variable relationships therefore also vary depending on the measure that is chosen. Independent variables that are used in literature to assess firm growth have significantly greater explanatory value with respect to sales growth (42.8%) than they do for employees (29.2%) or asset growth (28.3%). (Weinzimmer, Nystrom, and Freeman (1998)) Shepherd and Wiklund (2009) investigate the specifics of growth regarding knowledge accumulation in firms. The authors' literature review reveals a huge variation in the measures used and time frames considered, examples including sales, employment, assets, profit and equity for time spans from one to five years. On the other hand, some 60% of all studies used the sales growth measure, confirming the assessment of Gilbert, McDougall, and Audretsch (2006). The study further shows that relative and absolute measures of growth are comparable only to a limited extent. Relative measures should therefore only be compared to relative measures and absolute measures only with absolute measures. Sales and employment growth correlate very closely and are therefore interchangeable. In other words, studies that use sales measures may also refer to employment growth studies and achieve high degrees of validity. However, sales and employment measures differ significantly from the other growth measures used in literature. Another specific problem arising from horizontal analysis is mentioned by McKelvie and Wiklund (2010): *"Nevertheless, while it may be possible to use the same distinguishing feature (i.e., firm name, contact individual, organizational number) to follow "the firm" over time, the actual existing firm itself may not be something that is comparable with the firm that was first studied in the previous temporal period."* Firms can change their name, legal structure, strategy, financing relationship and/or product offers, for example. Many of these factors cannot be assessed by empirical studies, as they measure changes in certain variables on an aggregate level. As in empirical innovation studies, further problems arise from the selective inclusion of firms in empirical data sets, with too little attention being paid to firm exits, for example. Overall, there appears to be a growing consensus in literature that, if only one growth measure is to be applied at firm level, the best one is sales, as it applies to most industries and conceptualizations and is readily accessible. Even so, measurement still depends on the specific research interest. As Delmar, Davidsson, and Gartner (2003) put it: *"A review of prior academic scholarship on firm growth suggests substantial heterogeneity in a number of factors that characterize this phenomenon. Failure to recognize this heterogeneity appears to have led to some confusion and conflict in current theory and research findings. [...] All high-growth firms do not grow in the same way. This implies that researchers should measure different forms of growth with different growth measures."* Accordingly, it may also be the inconsistency in growth measurement that is leading to slow progress in theory building.

3.3.3.2 Empirical evidence about the corporate lifecycle concept

As discussed earlier, conceptual corporate lifecycle models appear to lack empirical validation. This assertion is consistent with the findings of Kazanjian and Drazin (1989), who note that lifecycle research has been conducted in an empirical vacuum. Based on a data set of 105 young and small technology-based ventures, the authors provide one of the first rigorous empirical analyses. They find significant evidence for stage transition after firms have resolved dominant stage-immanent problems. The occurrence of dominant problems is also confirmed by Dodge and Robbins (1992), who tests a consolidated model with four main stages using a data sample from 364 firms. Findings suggest that specific problems exist in each stage. While problems of an external nature prevail in early stages, later stages are characterized by internal problems. The most important problems, measured in terms of their frequency of occurrence, are in the field of marketing, management and finance and are found to be predictable throughout all stages. In a second study, Kazanjian and Drazin (1990) use correlation analysis to show that the degree of organizational centralization decreases along stages while formalization increases. Additionally, findings suggest that, over time, firms focus combines with an increasing functional specialization. These findings are in line with those presented by Hanks et al. (1993). Both studies thus seem to confirm the implications drawn by Greiner (1998) and Churchill and Lewis (1983). Furthermore, both studies suggest an S-shaped development of the firm's sales in which growth rates increase during some phases and decrease in the last phase. This is consistent with the propositions made by Steinmetz (1969). By contrast, Miller and Friesen (1984) test a consolidated model consisting of four stages derived from literature. Although the authors find some patterns that are consistent with theoretical predictions, they conclude that more than 20% of firms move backward from "later" stages to "earlier" stages. Accordingly, the evidence does not seem to support the consecutive phase transition hypothesis. Lester et al. (2008) focus on products and innovation. The authors of this study discover evidence about strategic choice differences within the model they adopt from Churchill and Lewis (1983). Their findings suggest that the strategic orientation of the best-performing firms is, at least partially, consistent with prevailing theory. In early stages, outperforming firms focus on first-mover strategies instead of copy strategies. In later stages, high-performers actively differentiate their products to achieve a unique selling proposition. The very last stage is driven by reinventing products and pushing innovation. Efficiency-oriented firms are less successful than their innovation-oriented counterparts. These findings are partially supported by Koberg, Uhlenbruck, and

Sarason (1996). However, their explanation is somewhat broader as it also takes organizational and environmental attributes into account. The authors essentially conclude that innovation is moderated by the firm's lifecycle. Contrary to the proposed theory, their findings suggest that formalization has a negative impact on firms' innovation activity in the early stages. It seems that formal structures suppress early-stage creativity. Scanning, formal analysis and centralization of the environment help later-stage firms in particular to foster innovation. The authors thus claim that lifecycle stages should be incorporated into innovation research as they have an important moderating effect with respect to resources, structures and organizations. According to Pasanen (2007), a distinction should be drawn depending on whether growth through new products is achieved organically or through acquisition. With respect to corporate lifecycle stages, Pasanen argues that older and larger firms in particular are more likely to engage in acquisitive growth due to the strength of existing products and fragmented customer bases. Younger and smaller firms have greater growth potential, but also have a higher risk profile due to their more consolidated customer structure. Another factor that should be considered is the characteristics of the industry and sample. Some authors devote their model to the small business sector only. (e.g., Churchill and Lewis (1983), Headd and Kirchhoff (2009)) Other scholars focus exclusively on high-tech sectors (Kazanjian and Drazin (1989), Kazanjian and Drazin (1990), Hanks et al. (1993)). One unique study with a focus on industry differences and the high-tech sector in particular was conducted by Agarwal and Gort (2002). In his model, the author estimates the probability of firm survival as a function of learning by doing, differences in resource endowments and changes in resource endowments. The probability of survival increases with firm size in the course of the corporate lifecycle. Contrary to most hazard rate models, however, the function of age is not monotonic. The basic idea is that resource endowments increase over time, but that initial endowments become obsolete at the same time. This is especially true for high-tech industries, which are shown to have higher hazard rates due to faster endowment obsolescence than in other industries. The industry in question thus matters and should be taken into account.

Overall, it seems that, on the one hand, lifecycle models suffer from major shortcomings. As Macpherson and Holt (2007) put it: *"The growth process is significantly more challenging and complex than stage models portray."* On the other hand, these models are still popular in many fields of business and economic research. Empirical findings reveal what has already been suggested in the review of theoretical concepts and summarized by Weber and Antal (2001): *"Even if one does not concur with this narrow idea of a predetermined, endogenous developmental logic of organizations, certain phases related to their age and size are discernible."* The following conclusions thus are drawn as the basis on which to develop the theoretical framework: First, stages exist. Stages moderate firm characteristics such as funding, organizational structure, strategy and innovation, and are thus a very useful tool for structuring and comparison. Second, only future research will show whether stages can be consolidated to a certain number and a harmonized nomenclature. This is likely to be a challenging endeavor, however, and the outcome is anything but clear. For now, it is fair to say that stages are context-specific, and that several different models may exist at the same time. Researchers can thus develop a model based on the presented findings and adopt it to their specific research context. This has already been a common approach in literature. (e.g., Elsayed and Paton (2009); Hoy (2006); Berger and Udell (1998); Rutherford, Buller, and McMullen (2003)) Of interest in this context is a study by Chan, Bhargava, and Street (2006), which discovers fast-growing firm cohorts to be internally homogenous. Third, the industry matters. Researchers must take into account the sectoral context when applying lifecycle-oriented frameworks. Fourth, there is clear evidence that stage transition is not deterministic. A firm may not pass through all stages, may jump back and forth, or may not even survive the first stage. The predictive value of growth cycle models cannot be sustained. Relaxing this assumption would yield a sort of status-quo model. Stages would then not predict further development, but would imply a potential development path, i.e. growth potential for a certain firm at a certain stage. Finally, modified lifecycle stage or state models overcome some of the associated problems. However, although these models are appropriate in one-firm contexts, they are of little use as tools to derive recommendations in wider institutional contexts. Some of these conclusions are consistent with those made by Levie and Lichtenstein (2008). Summarizing efforts that have been made in existing literature, the authors point out the following shortcomings: First, with reference to Eggers and Leahy (1994), differences between individual firms cause predicted lifecycle characteristics to deteriorate. Second, models cannot be generalized for all industries and geographic regions. Third, stages are not necessarily consecutive. The latter point refers to the study by Miller and Friesen, among others. One point that should be noted, however, is that the overall study *"results seemed to support the prevalence of complementarities among variables within each stage and the predicted inter-stage differences"* (Miller and Friesen (1984)). Finally, stage-related problems as proposed by Greiner (1998) can hardly be predicted empirically. Nevertheless, the overall conclusion is in sharp contrast to the one drawn here. While this dissertation accepts some of the properties of lifecycle models to be true and applicable, Levie and Lichtenstein (2008) in particular believe that stage theory is "dead" and should not be used at all.

3.3.3.3 The determinants of growth

There is a huge amount of empirical literature assessing the determinants of growth. One explanation is probably that existing theories, especially the organizational lifecycle concept, have arrived at results that are contradictory to some extent and, hence, not satisfactory. The wealth of available literature is demonstrated by several review articles in established journals that attempt to define patterns and consolidate past results. Despite the problem of the segregation of theory and empirical research on firm growth, it seems that empirical reviews too have shown limited convergence. Four literature reviews thus revealed four different sets of factors. One of the first reviews, albeit not an entirely representative one, was carried out by Storey (1996). Three main determinants or sets of factors affecting growth are identified as: (1) the entrepreneur's background (e.g., motivation, skills, education), (2) firm characteristics (e.g., size, age, legal structure, ownership); and (3) strategic planning (e.g., R&D, financing, human resources). At least the importance of entrepreneurial characteristics is supported by Macpherson and Holt (2007), who analyze 113 peer-reviewed articles, half of which were quantitative empirical studies on growth. The authors identify three sets of impact factors: (1) entrepreneurial and human capital (i.e. skills, knowledge, education, creativity); (2) structuring the firm ready for growth, a factor that relates to the entrepreneur and management (i.e. innovation orientation, the systems and structures used to improve the firm's absorptive capacity); and (3) networks (network depth, network diversity). Entrepreneurial characteristics are also mentioned by Gilbert, McDougall, and Audretsch (2006) as a major category. Indeed, these authors' literature review is somewhat broader and covers the factors mentioned by Storey (1996) and Macpherson and Holt (2007). The following determinants are proposed: (1) entrepreneurial characteristics, (2) resources, (3) geographic location, (4) strategy, (5) industry context and (6) structures and systems. A different approach was taken by Bahadir, Bharadwaj, and Parzen (2009) in a recent article. The authors not only review the largest volume of works, but also use a quantitative regression approach to condense their broad array of findings. The study identifies eleven factors that were frequently used in studies between 1960 and 2008: (1) innovation, (2) market orientation, (3) advertising, (4) interorganizational networks, (5) entrepreneurial orientation, (6) management capacity, (7) firm age, (8) firm size, (9) competition, (10) munificence and (11) dynamism. Wherever possible, the following analysis of factors refers to significant findings from this study first, as its reliability is comparatively strong due to the large sample and advanced methodology used. At the same time, the structuring of factors in this dissertation involves a mix of all reviews. With respect to Bahadir, Bharadwaj, and Parzen (2009), factors (2) and (5) are clustered in just a single category, namely strategy. Factors (3), (4) and (6) are clustered in a resource category that also comprises internationality and financing, neither of which factors is considered. Factors (9), (10) and (11) are clustered in an environment category that is enriched by industry variables. The three clustered categories are consistent with Macpherson and Holt (2007) factors (4), (2) and (5), respectively. The following factors are derived and analyzed:

1. Innovation (innovation, R&D, etc.)
2. Willingness to grow (growth ambition, entrepreneurial orientation, etc.)
3. Resources and profit (advertising, skilled labor, etc.)
4. Financing (equity, debt, financial constraints, etc.)
5. Firm age (time a firm has been on the market)

6. Firm size (sales, number of employees, etc.)

7. Environment (industry, competition, etc.)

3.3.3.3.1 Innovation

According to Bahadir, Bharadwaj, and Parzen (2009), innovation correlates positively to the growth of a firm. The relationship is significant and found in 66 models. 114 models do not find significant relationships, whereas 14 find significant negative relationships and 55 find non-significant negative relationships. The authors argue that findings are highly sensitive to the inclusion of other variables, particularly size and marketing activities. Also, geographic region has an effect, although most evidence is drawn from US samples. The positive impact of innovation seems to be very consistent across different disciplines and various measures, delivering the largest relative impact compared to all other measures. In contrast to the moderating impact of firm size, Gil (2010) find, in their literature review, that R&D intensity is independent of size. However, they also note that a noticeable proportion of smaller firms in particular do not report any R&D expenditures. Nor do the authors take into consideration the fact that small firms may not have formalized R&D budgets and may thus actually invest more than they report. Lee (2010) points out that knowledge accumulation and technology competence drive firm growth. Essentially, knowledge accumulation and R&D intensity are related, while technology competence is a mediating factor. This implies that R&D and innovation do not correlate directly as claimed above, but are rather latent constructs reflecting a firm's accumulation and utilization of knowledge. An interesting contribution is made by Dasgupta (1986), who develops a model of technological competition in which firms choose R&D investment and output. The implication is that growth and R&D are, to some extent, trade-off substitutes. In the long run, however, investing in R&D has a strong effect on a firm's production costs as it competes in the arena of process innovation. *Ceteris paribus*, an R&D-intensive firm will therefore outperform its less R&D-intensive competitors. The authors support findings suggesting that the relationship is moderated by industry characteristics, such as the degree of concentration, the existence of technological opportunities, etc. Del Monte and Papagni (2003) use horizontal econometric analysis to assess the impact of R&D on growth and specify the impact of industry variables. The authors find that R&D levels correlate systematically and positively to firm growth. This impact increases in traditional industries compared to research-intensive industries. They conclude that R&D does not seem to raise any large barriers to entry. Overall, it is fair to say that R&D and innovation drive firms' development and growth. However, it is important to account for moderating effects and potential R&D measurement biases.

3.3.3.3.2 Willingness to grow

Bahadir, Bharadwaj, and Parzen (2009) mention two strategy-related factors that affect a firm's growth path, i.e. its market and entrepreneurial orientation. This meta-analysis suggests a positive relationship between both factors and growth. However, market orientation is not significant. The positive relationship between entrepreneurial orientation and growth is underlined by the fact that several studies (Gilbert, McDougall, and Audretsch (2006); Storey (1996); Macpherson and Holt (2007)) explore a significant relationship between the characteristics of the entrepreneur and firm development. In their literature review, Gilbert, McDougall, and Audretsch (2006) explicitly cover strategy as a main determinant of growth. Likewise, their findings seem to support the non-significant relationship for the other strategy variables, hence the mixed results of individual studies. The authors conclude that the industrial context and a firm's positioning are of great importance and that many models do not properly account for this fact. They propose to use market share growth as solution to overcome these problems and achieve more consistent results with respect to strategic orientation. Pasanen (2007) points to growth orientation as an additional strategic factor in this context. This is consistent with many lifecycle-oriented analyses, suggesting that only a limited number of firms achieve high growth rates. (e.g., Dodge and Robbins (1992); Kazanjian and Drazin (1989)) This is a crucial but unspectacular finding. Firms that grow must be willing to grow. It is therefore likely that firms that do not choose to grow distort the empirical findings of lifecycle literature. (e.g., Hanks et al. (1993))

3.3.3.3.3 Resources and profit

According to Bahadir, Bharadwaj, and Parzen (2009), advertising, interorganizational networks and management capacity correlate positively to firm growth. These three factors are grouped together in the resource category due to their shared positive sign and their common character. The metastudy identifies significant relationships for advertising and management capacity, but not for interorganizational networks due to inconsistencies in measurements. The positive relationship of advertising is not accepted by all authors. For example, Wijewardena and Cooray (1995) see no significant relationship in a less consumer-oriented industrial context. On the other hand, management capacity has also been found to correlate positively to growth in the literature review conducted by Storey (1996). More detailed measures are also used in literature. For example, Wijewardena and Cooray (1995) suggest using 'skilled labor share' as the regressor to account for the relative importance of well-trained labor in quality- and technology-driven industries. Lee (2010) and Wijewardena and Cooray (1995) correct their analysis for internationality. Moreover, Storey (1996) points out that legal form and ownership also play a crucial role and are factored into many empirical surveys. In his literature review, he finds an unambiguous positive relationship between firm growth and a limited legal structure due to the advantages of limited liability. However, he also critically notes that the relationship may be concurrent, meaning that legal structure may be the result of rapid growth. On the subject of ownership, it seems that affiliated firms profit from better access to resources and knowledge without losing their structural flexibility. Another resource that may have a positive effect on firm development and growth is profitability. The basic idea is that profitable firms have more funds to invest and can focus on growth activities rather than efficiency issues. (Kang, Heshmati, and Choi (2008)) This idea is closely related to the role of internal funds with respect to financial constraints (see the discussion in section 3.4).

3.3.3.3.4 Financing

The perception of financing in growth research is comparable to that found in innovation research. Together with the effect of other resources, and according to Pasanen (2007), finance can be raised and set in a broader context. Essentially, most research postulates that adequate firm resources will generally correlate positively to firm growth. Consequently, many of the studies that identify determinants of innovation also analyze the determinants of growth. Scholars thus stress the substantial role played by sufficient financial resources. (e.g., Harhoff (1998); Hyytinen and Toivanen (2005)) Like innovation projects, however, growth projects are highly intangible and hard to assess. This makes it difficult to estimate the actual value of a project, not only for the management, but particularly for the investor. It therefore seems that internal capital in general and cash flow in particular are important means of financing growth. (Audretsch and Elston (2002); Carpenter and Petersen (2002); Himmelberg and Petersen (1994)) With respect to external capital, studies suggest that equity Lee, Lee, and Pennings (2001) and government support programs Hyytinen and Toivanen (2005) have a positive effect. On the downside, it seems that long-term debt Scellato (2007), leverage Becchetti and Trovato (2002) and financial constraints harm growth Musso and Schiavo (2008), thereby negatively affect a firm's development cycle. Only Honjo and Harada (2006) find ambiguous results and reports that debt has a negative effect on growth in assets and employees but a positive effect on sales. These effects are most pronounced for small firms. (Harhoff (1998) Further evidence is presented, together with financial considerations and reasoning from a financial point of view, in section 3.4.4. The findings presented here have been culled from growth studies that model finance as an independent variable only.

3.3.3.3.5 Firm age

The literature review by Bahadir, Bharadwaj, and Parzen (2009) reveals a significant negative relationship between firm age and growth, with a strong sensitivity to marketing. This finding suggests that younger firms need to act more aggressively while older firms may rely on their reputation. It is also consistent with the literature review by Storey (1996), which suggests an almost unanimously negative relationship between firm age and growth. Only one study in the sample showed a positive relationship, while one other study indicated no significant relationship. In addition, a study by Evans (1987) shows that firm growth decreases with firm age when size remains constant, and that this holds true for firms of all age categories. The latter finding contradicts the theoretical belief that learning is a driver of growth. (Jovanovic (1982a)) Literature thus seems relatively clear on the view that age is negatively correlated to growth. This view also supports lifecycle concepts in which firm growth is assumed to increase rapidly at the very beginning of firm development and then to decrease over time, approximately reflecting the S-curve phenomenon. A more recent study focusing on the role of age and knowledge accumulation by Lee (2010) strongly supports this view. However, the author argues that this pattern is not necessarily always the case. It may well be that a firm's technology capability and learning potential may lead to a sustained growth path in older age categories. For these firms, the relationship between age and growth may thus turn out to be U-shaped. An asymmetrical growth structure is also found by Garnsey, Stam, and Heffernan (2006), using case study analysis to account for the dynamic structure of growth paths. The authors' results suggest that young and small but growing firms have a relatively high risk of failure. More generally, it seems that turning points exist in the growth path and that historic growth also affects future growth. Accordingly, organizational learning may still play a role, but not in a monotonic way.

3.3.3.3.6   Firm size

Unlike firm age, and according to Bahadir, Bharadwaj, and Parzen (2009), size correlates positively with firm growth. The authors explain that firm size may be an asset to help overcome the problem of slack resources. However, the relationship is not unambiguous and has been the subject of disagreement in literature. This disagreement traces back to a theoretical consideration modeled and tested by Gibrat (1931), who generated a large amount of empirical research. His most famous hypothesis, known as Gibrat's law, states that growth is proportional to size and that the factor of proportionality follows a random walk. In other words, the proposition implies that size and growth are independent of each other. Although the meta-analysis by Bahadir, Bharadwaj, and Parzen (2009) reveals a positive correlation, many other studies (Storey (1996); Sutton (1997); Evans (1987)) arrive at the opposite conclusion, i.e. that growth rates decrease with firm size. This is also consistent with a literature review conducted by Gil (2010), who finds that smaller firms tend to have higher growth rates due to their greater flexibility and market potential. In this context, Barron, West, and Hannan (1994) notes that, despite the increasing dominance of large and global firms, economists find young and small firms to have the highest growth rates. However, the effect is jeopardized by the fact that larger firms are more likely to survive. Moreover, the author points out that larger firms' growth is less volatile. This view also has a certain practical appeal, as larger firms have more diversified risk structures. A lifecycle-oriented perspective would probably support a negative correlation, because most models imply a positive correlation between age and size. (e.g., Hanks et al. (1993)) However, as was noted earlier, some firms may choose not to grow and may thus age at a constant size. Since a large number of firms fall into this category (e.g., Storey (1996)), the overall relationship will be rather biased. Although results differ from study to study, it is fair to say that firm size has an important impact on growth. To properly assess the actual relationship in a given context, it is important to consider other factors as well and account for the fact that some firms may not want to grow.

3.3.3.3.7 Environment

Ceteris paribus, firms in fast-growing industries obviously experience higher growth rates. A firm's environment must thus be assumed to have a constitutive effect on its growth. Bahadir, Bharadwaj, and Parzen (2009) include three environmental factors in their study: competition, munificence and dynamism. Their findings suggest that competition significantly harms growth. Additionally, industry dynamism, which refers to the volatility of an industry, correlates negatively to firm growth. Only ten out of 16 significant models support this hypothesis, however, To put that another way: six out of 16 models find the opposite relationship to be significant. Munificence is closely related to resources, as it measures the availability of resources in the economy. The relationship is thus expected to be positive. However, the metastudy result is not significant. In addition to the three factors proposed above, many studies (e.g., Evans (1987); Gilbert, McDougall, and Audretsch (2006); Lee (2010); Wijewardena and Cooray (1995)) include industry, sector and market as major determinants too. This is an intuitive finding, since some industries generally grow faster than others. With respect to existing empirical research, Macpherson and Holt (2007) find a general sectoral bias toward high-tech and manufacturing. This implies bounded comparability on the one hand but a good match to the target industry of this dissertation on the other. Finally, whatever environmental factors drive firm growth, one fundamental necessity is that the market space in question must provide sufficient growth opportunities in the first place. (Pasanen (2007))

3.3.4 Conclusion – Characterizing growth firms

Growth is not only a major force in economic development but also a key objective for firms. It has therefore become one of the most important topics in business and economic literature. The two major research streams focus on corporate growth cycles and determinants of growth. This chapter has reviewed theoretical concepts and empirical evidence and summarized the most important findings. It has shown that a firm's sales would, in a "one-technology, one-firm world", follow the S-curve seen in technology innovation diffusion. The previous section showed that more firms and technologies can affect this development by altering prices and volumes sold. Moreover, the growth path is characterized by a firm's overall sales development, which in turn is the accumulation of individual technology and product sales. The growth pattern has been the topic of several studies with some consistency, depending on the business model, industry, growth mode, etc. Nevertheless, the S-curve is not entirely unlikely and would occur if the firm in question were to develop at the same rate as the market at a constant product portfolio price and with a constant market share. It may be for this reason that the S-curve is assumed in many corporate lifecycle models. The basic idea behind these models is that firms, like organisms, move through certain characteristic stages from birth to death. However, although still commonly used in various contexts, the models are exposed to substantial skepticism. The above analysis showed that several assumptions cannot be maintained in light of empirical evidence. First, a firm's growth path is not deterministic, nor does it hold for all firms. Second, firms do not necessarily enter all stages. Third, neither an agreed number of stages neither a shared nomenclature seem possible given the current state of the art of research. Literature has coped with these problems in two ways: first, by using aggregate models; and second, by proposing completely new approaches. Yet it seems that neither alternative has delivered a satisfactory solution. Empirical evidence has not added insights that resolve the problems relating to stage-oriented concepts. Indeed, two main streams of literature on the subject co-exist but are scarcely interlinked: One deals with stage models and that the other focuses on the determinants of growth. On the whole, it seems that empirical results regarding determinants of growth are not as mixed as the propositions arising from lifecycle models. Although growth measures are not used consistently (Gilbert, McDougall, and Audretsch (2006)) and a broad variety of determinants are identified, there is recent evidence of convergence for most important factors (Bahadir, Bharadwaj, and Parzen (2009). We have seen that one fundamental requirement for achieving high growth rates is a firm's willingness to do so. Empirical evidence from various studies paints a fairly consistent picture, indicating that innovation and R&D correlate positively to growth. Firm age correlates negatively, indicating that flexibility and relative market potential are more important than a large resource base. The effect of size is not as clear and is still treated differently. The effect may be positive, as most studies claim, but also negative given the assumption that size and age are correlated. Measurements appear to be biased, since many studies take no account of a firm's intention to grow or otherwise. Firms' resources, such as labor, finance, internationality, firm capability, etc., have extensively been analyzed. It seems very likely that resources are general enablers for growth. Abundance will thus correlate positively and scarceness negatively to growth. Lastly, industry and environmental characteristics play a crucial role. They not only drive industry growth, but also moderate the firm characteristics that determine actual growth at Firm level. However, empirical research also suffers from certain problems that are summarized by McKelvie and Wiklund (2010). First, empirical studies do not account for time variations, as they assume linear relationships and development paths. Second, growth modes are not sufficiently integrated into empirical models. Additionally, hybrid growth forms such as franchising, licensing and alliances are barely discussed in academic literature. Finally, it is noted that several literature reviews were unable to identify the same patterns and thus derive consistent determinants of growth to provide common ground for theory development. Nevertheless, the authors ignored the work of Bahadir, Bharadwaj, and Parzen (2009), cited herein, which makes a remarkable contribution to the field. Accordingly, the above analysis identifies three major determinants of growth: (1) innovation, (2) firm characteristics and the (3) environment. The willingness to grow seems to be a prerequisite if firms are to grow at all. Figure 3-11 summarizes the findings and postulated relationships.

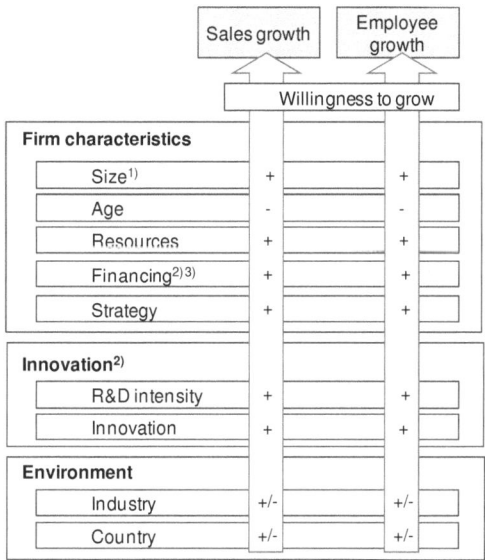

1) Relationship not monotonic   2) Potentially simultaneous relationship
3) Relationship for debt capital reversed

Figure 3-11: Determinants of growth – A synthesis

By matching the overall empirical evidence to the characteristics of aggregate stage models, this dissertation makes the following four conceptual propositions: (1) Stages exist and moderate firm characteristics. (2) Unlike in the case of organisms, these stages are not deterministic. However, they are very useful to structure firms and assess certain problems via stage-related sampling. The model conceptualized herein follows the recommendations of Dodge and Robbins (1992) and Kazanjian and Drazin (1990) and defines the following four states of development: formation, conceptual design and development; early growth and commercialization; later growth; and growth and maturity. (3) Lifecycle models are not "generally valid" but may reflect a certain research context. (4) The growth path shown in Figure 3-6 reflects a sort of "best-practice development" and thus reflects the S-curve for innovation and diffusion. However, several firms and technologies alter the curve as outlined above. Firms may thus move forward, backward, enter a market, exit it or stay at certain point of development. The three premises lead to the proposition of a new aggregate model summarized in Figure 3-12.

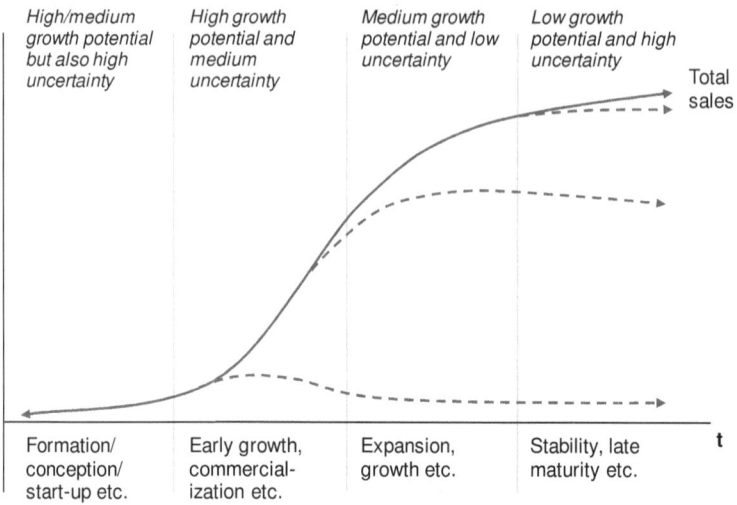

| High/medium growth potential but also high uncertainty | High growth potential and medium uncertainty | Medium growth potential and low uncertainty | Low growth potential and high uncertainty | Total sales |

| Formation/ conception/ start-up etc. | Early growth, commercial-ization etc. | Expansion, growth etc. | Stability, late maturity etc. | t |

Figure 3-12: The corporate lifecycle – A modified growth model

The model accepts that firms tend to fall into one of these four states, determining not only a firm's growth potential – represented by the slope of the curve – but also its demographics. However, contrary to most models (e.g., Greiner (1998); Hanks et al. (1993)), it assumes neither universal transition mechanisms nor a forward-oriented organismic development path. Firms may actually move forward or backward. For example, a mature firm may come up with a new innovation that could follow the S-curved innovation and diffusion model. As noted above, the overall sales curve is the sum of the individual technology and product sales curves. Given the assumption that the innovation will quickly dominate the overall product portfolio, the firm may once again pass through all four states until the innovation reaches maturity. Conversely, firms may fall into stability at any state. This could be one possible explanation for the large number of small firms with low growth rates (Storey (1996). Finally, it is suggested that firms may enter a certain market space at any point, whereas the overall product portfolio determines the actual development state of the firm. For example, a large conglomerate may enter a new market segment but still be a firm in the expansion state. Similarly, firms may exit a certain market space at any time. The proposed model deals with a specific state in an unknown process. It therefore assesses a firm's growth potential rather than predicting its next development steps. The basic idea is partially consistent with models that cluster firms into growth type categories, e.g. growth firms, survival firms, exit-oriented firms, etc. (Poutziouris (2003).

3.4   Financial economics – A necessity for corporate development

The previous sections examined in detail the determinants of firms' innovation and growth. Recapitulating the one–firm, one-technology case (Figure 3-6) allows us to derive initial implications regarding the firm's financial requirements. Figure 3-13 shows the one-firm, one-technology case and is complemented by the phases proposed in the previous section.

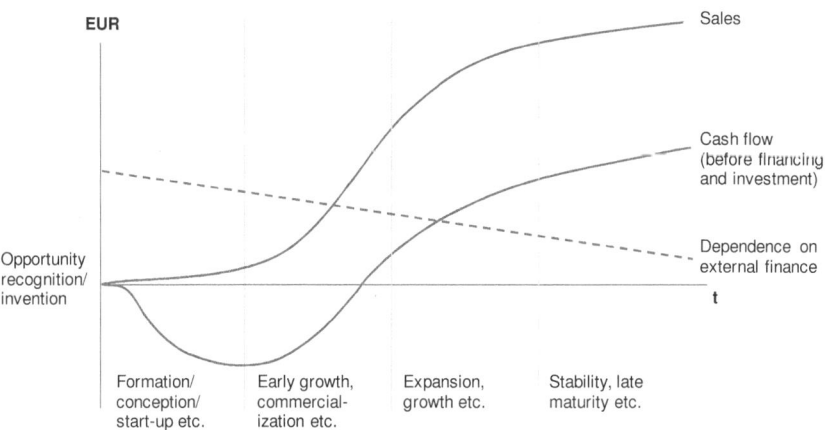

Figure 3-13: The one-firm, one-technology case – A financial perspective

As can be seen, the firm's external financial requirements are determined by its ability to generate cash. At the very beginning of the lifecycle, when a technology is first developed, the firm does not generate any sales. At the same time, it incurs significant expenditures to pay for the process of creating the technology. Although the graph begins where a technology is introduced on the market, the actual development phase may last for some time, resulting in a negative cash flow. If the technology is successful and gains market share, the firm starts to receive revenues to cover at least part of its expenditures. External finance is still required as long as operating expenditures are not fully covered by sales.[11] The early growth and expansion states are probably the ones in which the largest amounts of capital are invested and capital needs are the greatest. As sales increase, a firm's dependence will decrease due to

---

[11] For the sake of simplicity, the graph assumes that expenditures and operating costs represent the actual cash outflows from the firm. Nor is any distinction drawn between fixed and variable costs, although the cost curve implies the existence of both forms. Moreover, the distinction between internal and external funds is, for the purposes of the example, not consistent with the definition provided in chapter 5. Specifically, it is assumed that funds provided by the entrepreneur are considered external to the firm.

its improved ability to use internals funds such as retained earnings to finance not only operations but also potential investments. The need for external financing is represented by the shaded area. Again, the real world is more complex and the assumptions made here are very simplistic. However, at least one very important finding can be obtained: The firm's innovation activity and growth in a one-technology, one-firm world essentially depend on the availability of sufficient funds. Moreover, ignoring the availability of capital, the assumptions made here are consistent with several empirical studies (e.g., Bruno and Tyebjee (1985); Cassar (2004); Fluck, Holtz-Eakin, and Rosen (1998)). According to Bruno and Tyebjee (1985), financing must cover the main activities of the firm in each lifecycle state, product development in the seed and start-up phase, commercial manufacturing and sales in the early growth state, and working capital, plant expansion, marketing and product development in the expansion phase. Following the logic developed in the previous sections, financial funds are necessary to finance the activities of a firm at a given state in order to foster its growth, its innovation and, ultimately, its development.

The following analysis therefore focuses on financing firms' development. Besides the financial requirements identified in the example above, it primarily considers the availability of capital and the characteristics of capital types. In this context, two main bodies of literature have emerged. The first includes literature on the structure of capital, e.g. capital forms, the optimal capital structure and financing instruments. The second includes literature on the availability of capital, e.g. financial constraints, bank relationships and collateral. Both fields are important to the theoretical framework of this dissertation, which attempts to assess not only how greentech firms are financed, but also whether they can find sufficient capital. Financing is thus seen not only as an enabler for innovation and growth, but also as a potential barrier to it. The section is organized as follows: First, the main terms will be introduced. Second, a brief overview of fundamental financing theories is provided. Third, empirical evidence about the structure of capital and, lastly, about financial constraints is presented.

3.4.1 Definitions and terminology

3.4.1.1 Capital structure and financing instruments

Firms can use a wide range of different financing instruments. The overall financing basket, i.e. the use of different financial instruments and the precise mix, determines a firm's capital structure. The capital structure can thus be defined as the configuration of financing instruments for a certain firm to enable that firm to fund its assets and constituting the liabilities side of its balance sheet. The many different funding sources can be assigned to three classes: equity, debt and hybrid instruments. Equity instruments are composed of retained earnings, the principal owner's equity, angel finance and venture capital. Debt instruments are comprised of commercial bank instruments, borrowings from other financial institutions, trade credit, government funding, credit cards and borrowings from individuals. Hybrid instruments combine aspects of equity and debt capital forms, e.g. government grants with subsidized loan elements (Berger and Udell (1998)). According to Fluck, Holtz-Eakin, and Rosen (1998), financing instruments belong to two major categories: inside or outside financing. Inside finance embraces all capital sources that are either directly connected to the firm or are closely related to it, e.g. friends and family. Outside financing is received from banks, private investors, government institutions, institutional investors and stock- and bondholders. The properties and typical patterns of utilization are discussed individually for each instrument in the sections that follow.

3.4.1.2   Financial constraints

In this dissertation, financial constraints are defined in accordance with Lamont, Polk, and Saá-Requejo (2001), for whom they are *"frictions that prevent the firm from funding all desired investments. The inability to fund investment might be due to credit constraints or inability to borrow, inability to issue equity, dependence on bank loans, or illiquidity of assets. [Financial constraints do] not mean financial distress, economic distress, or bankruptcy risk, although these things are undoubtedly correlated with financial constraints."* Financial constraints are thus a measure representing the degree to which a firm is restrained in its pursuit of investments. This dissertation focuses particularly on investment in R&D and in firms' growth. Measurement options and determinants are discussed in the subsequent analysis.

3.4.2   Corporate finance and capital structure theory

Essentially, corporate finance theory is a theory of investment and firm value as well as a theory of how investments and firms are financed. While most fundamental work in the field focuses on corporations (e.g., Modigliani and Miller (1958); Myers (1984)), the past few decades have seen an increase in research into the financing situation of younger, smaller and medium-sized firms. (e.g., Berger and Udell (1998); Bruno and Tyebjee (1985); Cassar (2004)) The second stream in particular is of importance to the theoretical framework of this dissertation, as there are only a small number of large public firms in the greentech sector.

3.4.2.1   The capital structure irrelevance theorem

Corporate finance theory often assumes perfect capital markets, whereas SME finance explicitly takes account of market imperfections that are most severe for younger and smaller firms. Perfect capital markets imply perfect information, perfect competition, no transaction costs and, hence, a perfect match between risk and return. In this context, Modigliani and

Miller (1958) were the first to raise the question of the importance of capital structures. In their model and brief empirical evidence, they argued that, in a world without taxes and with perfect capital markets, a firm's capital structure does not have any impact on its cost of capital. Consequently, the capital structure is irrelevant to the market value of the firm or, as Stewart Myers puts it: There is *"no magic in leverage"* (Myers (1984)). This is in line with Grabowski and Mueller (1975), who argues that a firm earns a similar return on capital whether money comes from equity, debt or plowback. However, in a somewhat later work, Modigliani and Miller (1963) introduced taxes to their model. In a world with taxes, debt has one important advantage over equity financing: the tax shield. The model implies that, when firms increase leverage, they also increase their corporate valuations due to the avoidance of tax payments. However, contrary to the common belief that debt is generally "cheaper" and should be maximized, the authors critically note that the *"tax advantage for debt financing [...] does not necessarily mean that corporations should at all times seek to use the maximum possible amount of debt in their capital structures"* (Modigliani and Miller (1963)). By implication, innovative growth firms thus have similar capital structures to their less innovative and slower-growing competitors. The Modigliani Miller theorem (MM theorem) has triggered a large variety of studies, especially with respect to market imperfections such as taxes.

3.4.2.2   Asymmetric information and the principal-agent problem

An important concept to account for market imperfections is asymmetric information, a concept originating from the considerations of Akerlof (1970). In essence, asymmetric information is the result of adverse selection, for instance because insurance firms know less about their insured clients than the clients themselves, or moral hazard, which occurs when an agent is employed by a principal to fulfill a certain task. With regard to adverse selection, entrepreneurs and managers differ in their ability, honesty and capability. The problem of asymmetric information arises from the fact that entrepreneurs know more about their own and their firms' characteristics than potential investors do. Firms with a focus on innovation and growth must, by implication, be associated with a greater degree of information asymmetries due to their intangible asset structure and the uncertain outcome of their projects. Moreover, it is very likely that the extent of information asymmetries decreases with firm size. In other words, the smaller the firm, the greater its information opacity (Berger and Udell (1998)). On the other hand, Jovanovic (1982b) points out that very new and young firms often know less about their business than banks may actually do. This is the case especially because banks have broad experience of the market space, while entrepreneurs only have a very imperfect understanding of how their business may work and develop over time.

Moral hazard is a problem attributed primarily to the principal (the investor), who cannot perfectly observe the actions of her agent (the entrepreneur or manager). An important contribution to this line of thought was made by Jensen and Meckling (1976), who argue that the acquisition of external capital is associated with substantial agency costs. In their model, the authors show that the entrepreneur or manager acts as an agent for the principal lender. More specifically, the bank represents the principal whereas the entrepreneur or manager is the agent who has an obligation to act in the principal's best interests, i.e. by repaying debt and paying interest. Since the interests of the principal do not necessarily match those of the entrepreneur, the bank must conclude a contractual agreement that must then be monitored. Specifically, the manager is given incentives to increase project risk after receiving funds at a relatively lower market price by way of debt finance for the original project. In other words, while the manager gains from increased project volatility, the bank gains only from actual repayment of the lent funds. When managers invest in overly risky projects, the overall value of the firm decreases, because the pursued investments may yield negative net present values (NPVs). This problem is therefore also often called the overinvestment problem. In this context, Leland and Pyle (1977) show that firms can signal superior quality if the management invests in ownership. The agent's participation in the firm's risk exposure thus aligns the interests of principal and agent. Considerations by Ross (1977) arrive at a contradictory result. The authors argue that bankruptcy costs reduce a firm's ability to take on debt. Following this line of thought, "good" firms are better able to increase leverage than "bad" firms that cannot shoulder the associated additional risk. Accordingly, it is not a higher share of equity but a higher share of debt that signals firm quality. Apart from this contradiction, both models share the conviction that an optimal capital structure can be derived by assessing the costs and benefits of debt – an argument which, of course, contradicts the MM theorem.

3.4.2.3 The "capital structure puzzle" and the quest for a universal theory

The lack of a single accepted model has been noticed by many scholars and was summarized by Myers (1984) in an influential article entitled "The capital structure puzzle". In this article, Myers distinguishes between two major theoretical constructs: (1) the static trade-off theory and (2) the financial pecking order. The static trade-off theory basically refers to the works referred to above. On a very abstract and simple level, it predicts that firms will choose their optimal capital structure by weighing up the advantages (e.g., the tax shield) and disadvantages (e.g., the bankruptcy cost) of capital forms. On the other hand, the notion of the financial pecking order refers to a hermeneutic concept implied by past anecdotal evidence. Myers was the first to formalize the idea, claiming that static trade-off theory shows an *"unacceptably low R^2"* Myers (1984). His model predicts a hierarchical order of financing instruments where internal funds are preferred over debt and debt is preferred over equity.

The concept integrates asymmetric information, e.g. firms use internal funds to avoid overpriced capital or the loss of positive-NPV projects and the cost of financial distress. In other words, firms limit their debt to manageable levels. Another important implication is that the debt levels also depend on a firm's absolute financing requirements. The Myers (1984) model states that, when a firm is undervalued and is thus offered financing at a relatively high rate, it will rather forego investment opportunities than take on expensive financing. In doing so, it decreases the actual value of the firm and thus fails to pursue all positive-NPV projects. This problem is also known as the underinvestment problem. In a partially related work, Myers (1977) propose to distinguish between in-place assets and growth options. Growth options are often associated with innovation. They are exposed to greater risks than in-place assets and should therefore be financed differently. More specifically, the model predicts a negative relationship between debt financing and growth options (with options priced as a share of total assets). This implies that young firms with substantial growth potential must rely on capital forms other than debt. In other words, the proportion of inside capital increases with firm size but decreases with the number of growth projects and innovativeness of the firm. This is an important finding, because it supports the belief that young, small and innovative growth firms use different means of capital to old, large and less innovative firms.

3.4.3   SME finance and the importance of firm demographics

Corporate finance and capital structure theory have traditionally focused on large publicly traded corporations. However, as already indicated, it seems that small, young and innovative growth firms in particular suffer from information asymmetries. To understand the financial specifics of greentech firms, most of which are undoubtedly neither large nor publicly traded, it is important to look at financial research that focuses on the role and importance of firm demographics. This aspect too builds a logical bridge to the theoretical concepts developed in the previous sections.

3.4.3.1   The financial growth cycle and firms' capital structure

Berger and Udell (1998) propose a model of SME finance that accounts for the different development states of firms and varying characteristics of financial instruments. Essentially, the authors point out that the key characteristic of small and medium-sized firms is their information opacity compared to larger firms. For example, contracts to which these firms commit are kept private and are not publicly visible. Moreover, small firms do not have audited financial statements and thus have difficulties communicating their quality and establishing a good reputation. Contrary to the assumption of perfect information inherent in the MM theorem, the authors claim that information opacity is a key factor determining the state of a firm's lifecycle. Demonstrating some similarities to the state models for firm growth, they assume that firm size, firm age and information availability moderate the availability of capital and the use of certain instruments. The overall theory consists of theoretical elements borrowed from trade-off, principal-agent and pecking order concepts. The financial pecking order (Myers (1984)) predicts that a firm will first exhaust its internal capital before it turns to debt, which in turn is preferred over equity. This is in line with the Berger and Udell (1998) model for older and larger firms. Although the model does not consider internally generated funds, i.e. retained earnings, it shows that younger and smaller firms may not be able to follow this path. Essentially, younger firms' capital structures contradict the pecking order principle because they are simply not free to choose from all the available financing instruments. Bank loans, for example, may be strictly rationed. These firms may either not find sufficient capital or have to rely on other, more expensive capital forms. This theoretical concept is also consistent with propositions made by Jensen and Meckling (1976), stating that moral hazard is more acute in early-state firms. These firms thus have to rely on business angel and venture capital, making use of the alignment of ownership and management interests through partial ownership. Figure 3-14 summarizes the main propositions:

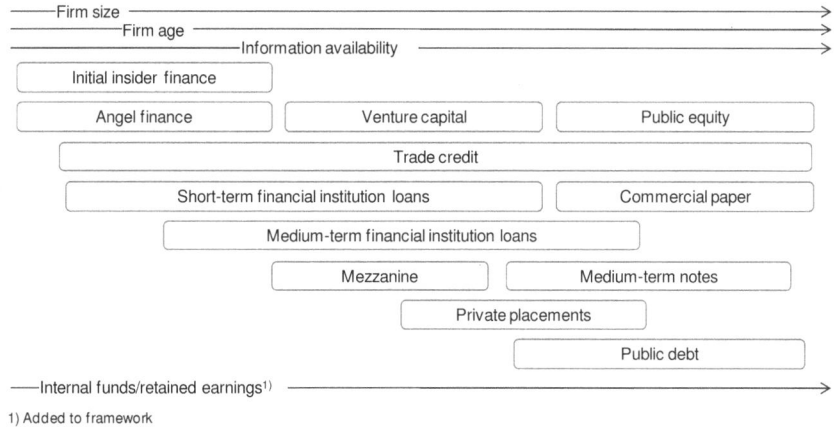

Figure 3-14: Firm continuum and sources of finance
Source: Adopted from Berger and Udell (1998)

When a firm is first launched, it has a limited track record and can thus hardly signal creditworthiness. As the firm ages, it builds up relationships with various stakeholders and develops its business activities accordingly. Information about the firm becomes more detailed and decreases the firm's opacity over time. The characteristics of the firm, particularly the degree of information opacity, therefore determine a firm's propensity to acquire certain financing instruments. While some investors have a higher risk and return affinity, others are keen to minimize the volatility of repayment. According to the model, initial inside finance is the most important source for start-up firms and remains important for some time. Business angel funding and venture capital are financing instruments suited to the requirements and characteristics of small and young firms that evidence significant growth potential. The two capital forms are connected. According to Berger and Udell (1998), around 15% of all US IPOs in the 1980s were backed by VC. Angel investments are made directly, whereas VC is intermediated, representing 3.59% and 1.85% of total finance respectively. It should also be noted that government support funds can also be important to younger and smaller firms, although Berger and Udell (1998) find only very limited relevance in the US. Another financing instrument that seems to be available throughout the entire lifecycle is trade credit. Trade credit is granted by suppliers in the form of accounts payable for which a certain invoice due date is set. However, trade credit can be very expensive when firms are offered trade discounts. A trade discount of 3% would, for instance, translate into an annual interest rate of 36%, ignoring compound effects. It is for this reason that Ploetscher and Rottmann (2002) propose that the use made of trade credit should serve as a measure of a firm's financial constraints.

As already indicated, smaller firms are often managed by the owner and financed by private funds. This reduces the principal-agent problem with respect to equity, but does increase the agency cost of debt. However, issuing debt is desirable from an ownership perspective, because it does not affect control rights over the firm. The younger and smaller the firm, though, the higher the agency cost of debt. This is especially the case for growth-oriented firms. Growth basically implies a greater need for funds. At the same time, an increase in the number of growth projects may lead to conflicts between owners and capital providers due to the resultant increase in agency-related project risks. (Jensen and Meckling (1976)) Young growth firms are therefore offered short-term debt contracts that partially mitigate these effects. As briefly mentioned above, the most common action that banks take to reduce agency costs is to ask for collateral. Collateral can be split into inside and outside collateral. Inside collateral is provided by the assets of the firm (such as accounts receivable) while outside collateral is usually provided by the owner (such as personal wealth). Inside collateral leads to a reorganization of security providers' claims. Outside collateral favors a single party, because it provides additional resources to secure repayment to a specific lender in the event of bankruptcy. It thus improves the firm's potential to acquire debt by reducing the risks to the lender. This is consistent with theoretical observations made by Stiglitz and Weiss (1981) in which collateral reduces information asymmetries and, hence, credit rationing. Other forms of bank monitoring include covenants and loan maturity. Covenants introduce standardized monitoring measures and KPIs. They make the firm's financial situation and prospects more transparent, but also make monitoring less complex and less expensive. Loan maturity, on the other hand, reflects considerations raised by Myers (1977) in which banks aim to match maturity agreements with asset lifecycles. Shorter maturities may, for example, reduce the principal-agent problem by reducing renegotiation cycles. However, firms can also improve their own access to financing. One way is to build long-term relationships with investors. In literature, relationship lending has been shown to reduce information opacity of small firms, thereby improving not only the availability of debt capital but also the contractual terms. Additionally, Berger and Udell (1998) point out that dependence on trade credit correlates negatively to bank relationships. Accordingly, firms that have established relationships with banks do not need to draw on unfavorable terms afforded by suppliers. In the data set presented, more than 15% of all firms use trade credit as a source of financing. Moreover, the model accepts the importance of signaling as proposed by Leland and Pyle (1977). With respect to debt financing, the authors point out that announcing loan commitments may boost the stock price and reduce the cost of equity finance. Also, borrowers from banks who do not pledge collateral may send a favorable signal about quality that also lowers the cost of other instruments. Loan commitments and credit lines are forward contracts that ensure liquidity and capital availability on a very short-term basis. On a more general level, it seems that acquiring capital is not an easy endeavor. Smaller and younger firms in particular have limited access to financing as they do not always fulfill all the requirements. Some financing instruments, such as public securities, also remain the exclusive preserve of older and larger firms due to high fixed costs and economies of scale. Additionally, financing instruments become less complex during the firm's lifecycle. Large firms can, for example, access more standardized and publicly available instruments on generic terms. By contrast, smaller firms often obtain capital via financial intermediaries, who play an important role in SME finance. They screen, evaluate and monitor firms and offer highly structured and customized constructs.

It seems that Berger and Udell (1998) are very aware of the general skepticism with regard to organizational lifecycle models. Accordingly, they stress that "*firm size, age, and information availability are far from perfectly correlated*" Berger and Udell (1998). They also avoid defining consecutive states. On the other hand, they structure firms into four main categories: (1) infants (very small firms, possibly with no collateral and no track record); (2) adolescents (small firms, possibly with considerable growth potential but often with a limited track record); (3) middle-aged (medium-sized firms with some track record and collateral available); and (4) old (large firms whose risk and track record are known quantities). These categories correspond to the typical set of categories found in practitioner-oriented literature, namely seed, start-up and later states (e.g., Burtis (2004); Burtis (2006); Murphy and Edwards (2003)). They are also largely consistent with the ideas presented in the growth section. The model thus faces comparable problems, which will not be restated at this point. The model further lacks any treatment of internally generated funds, quite apart from their significant impact on a firm's external finance needs. (see Figure 3-13, for example) From a theoretical perspective, however, the effect of liquidity on financing seems somewhat unclear. Trade-off theory implies that profitable firms have greater leverage due to their ability to carry higher levels of debt. Pecking order theory would predict the opposite, since firms normally exhaust their internal funds before accessing external ones. Overall, the model seems a promising basis on which to incorporate financial aspects into innovation and growth literature.

3.4.3.2   The availability of funds and the concept of financial constraints

Building on corporate finance and capital structure theory, a large body of literature has emerged in recent decades that focuses on the specific problems of small and medium-sized firms with respect to capital acquisition. Storey (1996) summarizes eight assumptions that characterize the wide variety of theoretical works in the field: (1) asymmetric information, (2) agency issues, (3) systematic risk and firm size, (4) costly monitoring, (5) competing banks, (6) characteristics of the entrepreneur, (7) entrepreneurial benefits from increased project valuation and (8) bank benefits limited to repayment and interest. These assumptions are not mutually exclusive. Some are interrelated and may occur at the same time. Asymmetric information (1) refers to differences in what a firm's management, owners and investors know and how they do their reckoning. Essentially, in light of the comments made above, it seems that asymmetries favor small firms while the bank does not have sufficient monitoring mechanisms to cope with the high investment risk. Its ability to control the firm is thus limited and gives rise to substantial agency issues (2). Agency issues are closely connected to the objectives, ability and honesty of the management or entrepreneur (6). Moreover, the actual business environment determines the potential for cheating. In other words: The greater the benefit to the entrepreneur (7) of deviating from contractual constraints, the greater the likelihood that the bank will incur agency costs. With respect to a firm's systematic risk (3), Storey (1996) notes that there is *"little doubt that, by objective measures, bank lending to smaller businesses is more risky than to otherwise comparable large firms."* In order to keep their actions economically feasible even so, banks must to engage in close controlling activities (4) to deal with these issues. These activities lead to higher transaction costs. Some studies are specifically concerned with competition (5) among banks, assuming that the owner or manager may approach several banks for funding. In a functioning market, prices would reflect the associated risk of a certain contract. In practice, one key mechanism to reduce risk, from a banker's perspective, is to ask for collateral. Additionally, the bank may look for quality signals or define additional terms, e.g. covenants and/or personal liability, in individual contracts. However, one fundamental problem remains: The bank only gains from repayment at a fixed rate, while the business owner gains from the project returns (8). If the above risks are too high for the bank or cannot be mitigated, e.g. through collateral or monitoring, the bank may ultimately ration its credit supply. The theory thus predicts that economies of scale lead to a lower degree of information asymmetries, lower relative transaction costs, better access to resources and less volatile cash flows. Consequently, smaller firms either incur higher cost of financing or receive less financing than needed. With respect to debt capital, Stiglitz and Weiss (1981) argue that competition leads neither to efficient security prices nor to market equilibrium. In essence, assuming the bank would accept higher interest, its expected return would decrease due to information asymmetries. The authors' seminal work states that higher interest rates would, on the one hand side, attract high-risk borrowers, while low-risk borrowers would not accept the offer (adverse selection). On the other hand, firms that are granted funds would always prefer high-risk projects to low-risk projects (moral hazard). Both considerations imply that investors and intermediaries must tighten their monitoring and controlling activities which, in turn, will further reduce their expected return. It follows that banks prefer to reduce the supply of capital than to increase interest rates to avoid firms whose information is opaque (Storey (1996)). Small, young and innovative growth firms with limited collateral in particular may thus not be able to attract sufficient capital from banks and may thus face financial constraints. These firms may still be able to attract capital from other sources (such as venture capital or personal funds) to cope with these constraints. Otherwise, they will simply have to progress at a much slower rate. The implications are consistent with the lifecycle concept proposed by Berger and Udell (1998) and seem to be particularly harsh for the innovation- and growth-driven greentech industry.

3.4.4 Empirical evidence about SME finance, capital structure and firm demographics

Many empirical surveys on capital structure have been conducted in recent years. There is still no commonly accepted theory, however, and financing patterns seem to differ depending on the country, industry and other factors. Since a firm's financial growth cycle (Berger and Udell (1998)) integrates several aspects of all fundamental finance theories and also provides a solid basis for integrating innovation, growth and finance theory, this issue will be discussed first.

3.4.4.1 Measurement of capital structure, financial constraints and the diversity of capital instruments

There are many different ways to measure capital structures. Indeed, the sheer diversity of studies in this field makes it difficult to consolidate the results. Empirical research on capital structures typically assesses capital structure as a dependent variable. Most rigorous studies thus focus on only one or few aggregated measures. (e.g., Cassar (2004); Cassar and Holmes (2003); Börner and Grichnik (2010); Fluck, Holtz-Eakin, and Rosen (1998)) A few studies consider a larger variety of instruments. (e.g., Gregory et al. (2005); Hogan and Hutson (2005); Gottschalk et al. (2007); Bozkaya and de van Potterie (2008)) However, the latter are often descriptive in nature or also use aggregated categories for prediction. The diversity of potential measures becomes obvious when one considers the large number of potential financing sources. For example, Berger and Udell (1998) discuss 12 different financing instruments in their model. And even this model is probably not exhaustive, excluding cash flow, for instance. It follows that there are 12 over x possible combinations (where x=1, 2,..., 12), i.e. a total of 4,096 distinct potential dependent variables. Not all combinations make sense, of course. Even this small example nevertheless demonstrates the complexity of the subject. The literature review underlines this consideration. The variety of measures is vast, comprising alternatives such as long-term debt, short-term debt, total debt, leverage, owner's funds, liquidity, retained earnings, outside finance and equity. A further aggregation of measures seems to yield two major topics of discussion, i.e. internal versus external funding and equity versus debt capital. However, the following review of existing empirical knowledge does not pursue the aggregation approach, nor does it focus on specific instruments. Instead, it seeks to flesh out the most common and most important determinants of capital structure and discuss the implications for selected capital sources and financial instruments. It thereby tries to avoid drawing wrong conclusions due to aggregation effects. For example, theory teaches us that internal capital may be composed of negatively correlated single variables such as owner's equity and retained earnings. It also accounts for the interconnection of instruments, as argued by Berger and Udell (1998), and for the importance of financing baskets.

The variety of measures of financial constraints has been equally diverse. One seminal study was conducted by Fazzari, Hubbard, and Petersen (1988), who claims that investment/cash flow sensitivities are a good proxy for measuring financial constraints. The reasoning behind this approach is that firms facing financial constraints use all possible internally generated capital to finance investments. If cash flow decreases in one year, the firm will not be able to maintain its level of investment, as it can generally not acquire the necessary equity externally. Since Fazzari's study, a large body of literature has assessed financial constraints using sensitivities, for example. The original work is cited over 40 times in the EBSCO database alone. However, an essay by Kaplan and Zingales (1997) questions whether financial constraints can indeed be measured in this way. The authors use qualitative data derived from financial statements to show that less constrained firms, according to the prevailing definition used in state-of-the-art literature, show greater cash flow investment sensitivities than more constrained firms. The authors conclude that cash flow sensitivities are of no use when assessing financial constraints. The discussion of whether sensitivities are a useful measure has yet to be resolved. Both alternatives are still put forward as valid arguments. (Fazzari, Hubbard, and Petersen (2000); Kaplan and Zingales (2000)) According to the literature review conducted by Carreira and Silva (2010), *"the best one can do [...] is to construct indexes that allow one to measure the degree of constraints"* directly. The authors summarize the following proxies: (1) dividend payout ratio, (2) firm self-evaluation, (3) cash stocks, (4) degree of leverage, (5) age, size, (6) institutional affiliation and (7) credit ratings. It is fair to say that literature has not yet come to a conclusion about which measure is best and should primarily be used. Empirical findings therefore remain questionable and the type of measurement should be considered carefully when analyzing results.

3.4.4.2 Empirical evidence about the financial growth cycle of the firm

To date, empirical evidence about the financial growth cycle model has been rather scarce. The authors themselves note that "*until recently, data constraints have made it difficult to examine this paradigm empirically*" (Berger and Udell (1998)). Although the growth cycle paradigm is itself well established in literature, only very few studies test the model explicitly. In this context, Bozkaya and de van Potterie (2008) use a mixed-methods approach to test for financial growth cycle patterns in small Belgian technology-based firms. Their findings suggest that firms have distinct financing patterns in different states of development. At the outset, firms are more dependent on internal sources, especially funding provided by the owner, while in later states external sources become more important until a peak is reached at the expansion state. The authors conclude that their "*results mostly support the financial development cycle paradigm*" (Bozkaya and de van Potterie (2008)). Bhaird (2010) tests the financial growth cycle theory using a sample of around 300 small and medium-sized firms located in Europe. His findings also suggest a pattern that is broadly consistent with the growth cycle framework. Firm size and age are thus strong moderators of a firm's financing basket. Older firms are more likely to use external debt, while younger firms depend on funds provided by the owner. However, contrary to the Berger and Udell (1998) model, Bhaird finds that younger and smaller firms in particular use a wide range of different financing instruments. The explanation seems to be that a single source does not provide sufficient amounts of capital. As a result, these firms have to accept greater transaction complexity and, ultimately, less attractive offers from non-bank investors. However, this view contrasts not only with the financial growth cycle, but also with several other empirical study results. Examples include Börner and Grichnik (2010), who use ordinal regression on a sample of German SME, a study by Fluck, Holtz-Eakin, and Rosen (1998), who use a Tobit model on smaller firms, and a study by Hogan and Hutson (2005), who analyze 117 Irish manufacturing firms. All three studies indicate that the use of external capital normally correlates positively with firm size and firm age. Gregory et al. (2005) use a slightly different approach, modeling financial baskets as shown in Figure 3-14. The implications are tested based on a US data set consisting of 954 SMEs and a multinomial logistical model. The authors find only partial support for the financial growth cycle hypothesis. Specifically, size is a good predictor of capital structure, but age is not, as it yields contradictory results. Another sharp contrast to the growth cycle concept is the finding that younger growth firms are most likely to attract long-term debt (LTD) and private equity (PE). Nevertheless, the authors support the hypothesis that there is no "one-size-fits-all" model. They therefore conclude that SME practices are too diverse to be reflected in a single pattern. A comparable analysis is conducted by La Rocca,

La Rocca, and Cariola (2009), who use cluster analysis to define distinct lifecycle states and regression analysis to explore the lifecycle concept. In line with the research referred to above, it seems that firms' information opacity is a key driver of financing decisions. Additionally, the study reveals some evidence with regard to the pecking order hypothesis, in which greater profitability allows firms to deviate from the model, as more internal funds can reduce dependence on external finance in all phases. Essentially, the Berger and Udell (1998) model suffers largely from not integrating cash flow into the model. In empirical studies too, retained earnings seem to play a major role in firm development and financing. (e.g., Bhaird (2010); Börner and Grichnik (2010)) As cash flow increases, firms become less dependent on external capital and increasingly use their own funds. This finding is in line with the pecking order concept. However, it undermines the aggregate internal equity profile, according to which younger firms are more reliant on internal equity but older firms may have greater internally generated financing capacity. To further assess the financial growth cycle and other theoretical predictions, it appears to be necessary to broaden the scope of analysis. The following discussion thus reflects the results of empirical research into the determinants of capital structure and into financial constraints.

3.4.4.3   The determinants of capital structure

In 2003, Cassar and Holmes (2003) noted that empirical research into the capital structure at SMEs has been sparse. Since then, several studies have attempted to close this research gap. Table 3-3 provides a brief overview of selected empirical surveys on the determinants of capital structure and dependent variables in scope. At first glance, it seems that factors affecting capital structure are more diverse than in the case of innovation and growth, where research appears to reflect greater convergence. This could also be one reason why it was not possible to find an exhaustive literature review. However, key demographic characteristics of firms, such as size and age, are also covered in most studies on capital structure. A total of eight key factors were identified:

1. Firm size (Sales, employees etc.)
2. Firm age (Time on the market)
3. Innovation (R&D, innovation output etc.)
4. Growth (Sales growth rate, employment growth etc.)
5. Asset structure, tangibility and collateral (Long-term assets, collateral etc.)
6. Profitability and retained earnings (Margin, profit etc.)
7. Ownership and firm structure (Shareholder structure, legal form etc.)
8. Environment (Industry, technology etc.)

Two main distinctions can be drawn in order to condense the results: (1) the effect on the use of internal versus external capital; and (2) the effect on the use of equity versus debt capital. Exceptions and special cases, such as short-term debt, are mentioned wherever necessary.

3.4.4.3.1 Firm size

Regarding the use of internal or external capital (1), Hogan and Hutson (2005) argue that internal finance dominates in early states of development, while external capital gains ground in later phases. A positive relationship between external capital, defined as the sum of all external financial instruments, is found by Bruno and Tyebjee (1985); Cassar (2004); Cassar and Holmes (2003); Colombo and Grilli (2007); Egeln and Licht (1997) and Vos et al. (2007). Only the Fluck, Holtz-Eakin, and Rosen (1998) study cited earlier posits an unexpected negative relationship. The authors reason that smaller firms do not receive sufficient funds from single capital sources and therefore need to approach various investors. A corresponding negative relationship between firm size and internal capital, provided by the owner, is found by Bhaird (2010); Gregory et al. (2005) and Colombo and Grilli (2007). Only one study, by de Haan and Hinloopen (2003), finds a non-significant relationship, while one other study, by Romano, Tanewski, and Smyrnios (2001), identifies a positive relationship. However, the latter uses a family-owned business sample in which firms' ability to acquire capital is overcompensated by their determination to stay independent. This argument is in line with findings that support the hypothesis that owner's preferences (for independence and growth, say) are an important factor to be considered. (Hogan and Hutson (2005); Bhaird (2010)) The use of internal capital contrasts rather sharply with the availability of internal funds, revealing the conflict between the financial pecking order concept and the financial growth cycle concept. The results are at least mixed, whereas Bhaird (2010) finds a positive relationship, Magri (2007) a non-significant relationship and Chittenden, Hall, and Hutchinson (1996) a negative relationship between liquidity and size. However, the latter study uses a sample of UK SMEs that may be too large to reflect the expected positive sign. This would also explain why it finds a positive relationship between liquidity and age. Overall, it seems that evidence tends to support implications from the financial growth cycle hypothesis, although it has not yet been possible to resolve the question of the internal cash flow effect. If internal funds are sufficiently large, it may well be that predictions derived from the pecking order hypothesis are more appropriate.

With respect to the debt and equity choice (2), study results are less contradictory. Most studies report a positive correlation between size and equity (de Haan and Hinloopen (2003); Romano, Tanewski, and Smyrnios (2001); Egeln and Licht (1997); Casson, Martin, and Nisar (2008)). Only a few studies either report a negative correlation Bruno and Tyebjee (1985) or a non-significant correlation (Bhaird (2010); Magri (2007)). This is consistent with the financial growth cycle concept, because studies mainly assess public equity. Only Bruno and Tyebjee

(1985) use ownership fragmentation as a dependent variable. Apparently, regression models cannot properly account for venture and business angel capital, since these capital forms are available to a very small number of firms. For example, La Rocca, La Rocca, and Cariola (2009) find a high proportion of debt in small firms compared to external equity, which is generally sparse in Italy. Gottschalk et al. (2007), who analyze German high-tech firms, also note that only 5% of all firms receive financing from external equity sources. Furthermore, as proposed by the financial growth cycle model, evidence indicates a strong link between business angel and venture capital (Bozkaya and de van Potterie (2008)). Additionally, Bruno and Tyebjee (1985) find that firms actually receiving VC have access to significantly more capital than firms using other sources. In their horizontal study, the authors provide evidence that VC-backed firms on average raise 75% of the capital they originally anticipate, whereas non-VC-backed firms only raise 25% of their targeted capital needs.

Survey	Dependent variables	Selected significant determinants	Sample size	Regions
Bhaird (2010)	STD, LTD, total debt and liquidity	Age, asset tangibility, profitability	275	France
Börner and Grichnik (2010)	Internal capital, external capital	Age	10,692	Germany
La Rocca et al (2009)	Leverage, STD, LTD	Growth, total assets, long-term assets, tax shield, tax rate, profit	69,694	Italy
Heyman et al (2008)	Leverage, STD, LTD	Size, growth, asset tangibility, cash flow, profitability, industry	1,132	Belgium
Hyytinen and Pajarinen (2008)	Internal capital, MTD, external capital	Size	3,825	Finland
López-Gracia and Sogorb-Mira (2008)	Leverage	Growth, technology complexity, industry	3,569	Spain
Casson et al (2008)	Owner funds, retained earnings, external capital, STD, LTD, total debt	Age, size, collateral, owner collateral, ownership, industry	820	UK
Colombo and Luca (2007)	Rating	Age	386	Italy
Vos et al (2007)	Total debt, LTD, equity	Size, growth, R&D, profit	15,750	UK
Magri (2007)	Total debt	Size, age, asset tangibility, governement support, ownership structure, R&D, profit, industry	4,500	Italy
Gregory et al (2005)	External capital	Collateral, seed capital obtained, industry	4,637	USA
Sogorb-Mira (2005)	Multiple sources of capital	Size, age, growth, profit, industry	6,482	Spain
Thornhill et al (2005)	Leverage, equity, cash flow	Size, asset tangibility, R&D, firm structure, technological capability, profit, industry	30,712	Canada
Cassar (2004)	Total debt, family loans, retained earnings, equity	Size, age, growth, ownership, industry	292	Australia
Voulgaris et al (2004)	LTD, total debt	Size, growth, long-term assets, asset tangibility, rating, profit	143	Greece
Bollingtoft et al (2003)	Equity, outside capital	Size, age, industry	654	Germany
Cassar and Holmes (2003)	LTD	Size, R&D, profit	1,555	Australia
Fama and French (2002)	Total debt	Size	0	USA
Haan and Hinloopen (2002)	Total debt, STD, LTD, external capital	Growth, total assets, long-term assets, profit, industry	150	Netherlands
Romano et al (2001)	Total capital, leverage, personal capital, bank loans	Size, collateral, firm structure, owner and management characteristics, industry	1,490	World
Bah and Dumontier (2001)	Leverage	Age, growth, asset tangibility, ownership structure, profitability, industry	7,004	Europe, UK, Japan and USA
Hall et al (2002)	Total debt, external capital, LTD	Size, growth, total assets, long-term assets, industy	3,500	UK
Fluck et al (1998)	Internal, bank loans, public debt, public equity	Size, cash flow, profit	197	USA
Egeln and Licht (1997)	Leverage	Size, age, growth, cash flow, tax shield, profit	n.a.	n.a.
Chittenden et al (1996)	External capital, equity	Size, growth, industry	3,480	UK
Bruno and Tyebjee (1985)	STD, LTD	Size, age, long-term assets, profit, industry	458	USA

Table 3-3: Selected empirical research on the determinants capital structure

Overall, it seems that external equity is more important to large firms, but that at least some small firms use it as primary source of financing. Statistical results therefore depend heavily on the methodology and sample construction. The impact of size on debt financing is rather more homogeneous, with the exception of short-term debt. Most studies use either LTD, the LTD proportion or public debt as the dependent variable and report a positive correlation with firm size. (Chittenden, Hall, and Hutchinson (1996); Cassar (2004); Cassar and Holmes (2003); de Haan and Hinloopen (2003); Hall (2002); Voulgaris, Asteriou, and Agiomirgianakis (2004); Bah and Dumontier (2001)) These findings are consistent with studies that use leverage (Börner 2010 #165}; La Rocca, La Rocca, and Cariola (2009); López-Gracia and Sogorb-Mira (2008)) or medium-term debt (Gregory et al. (2005)) as the dependent variable. Only Heyman, Deloof, and Ooghe (2008) report a negative relationship; and only Casson, Martin, and Nisar (2008) and Bhaird (2010) report a non-significant relationship. However, the identified relationship does not hold for all maturities. Principal-agent theories propose to shorten repayment time frames for smaller firms to reduce the risk of deviant behavior. The implied negative correlation to size is confirmed by the studies of Chittenden, Hall, and Hutchinson (1996) and Hall (2002). A contradictory result is reported by Voulgaris, Asteriou, and Agiomirgianakis (2004), although it must be noted that the sample comprises only 143 Greek firms and may therefore not be fully representative.

3.4.4.3.2 Firm age

The impact of a firm's age on capital structure should, according to the theory, be comparable to the impact of the firm's size. This assertion is largely reflected in empirical research. Besides this fundamental alignment, two further observations can be made. First, it seems that firm size is a slightly better predictor of capital structure than firm age. Second, some studies report a positive relationship between capital and size but still find a negative relationship between capital and age, indicating that young firms may be large and small firms may be old. (e.g., López-Gracia and Sogorb-Mira (2008); Vos et al. (2007)) By implication, it follows that young firms can compensate for their limited track record by achieving a sufficient size.

3.4.4.3.3 Innovation

Innovation implies greater cash flow volatility and, hence, greater risks to the firm, for three main reasons. First, inherent uncertainty is rooted in particular in the problem that knowledge generated by R&D suffers from substantial spill-over effects. In other words, R&D returns are appropriated to the originator only to a limited extent (Peneder (2008)). Investors may therefore be reluctant to invest in R&D. Second, compared to investment in fixed assets, R&D has a low liquidation value because preliminary or even final research outcomes are hard to market. According to Myers (1977), an asset with a low liquidation value cannot be used as a specific security for a debt claim and is thus heavily dependent on equity sources. Third, information asymmetries appear to be relatively high in innovation projects because the actions of the agent are difficult to observe and assess. By implication, innovation-intensive firms therefore have a relatively high degree of information opacity. The financial growth cycle hypothesis would predict limited use of long-term debt and greater usage of equity. This is in line with Thornhill, Gellatly, and Riding (2004), who argues that knowledge-intensive firms, working in uncertain areas, have lower levels of debt and make greater use of internal funds. All three implications seem to be reflected in empirical research, albeit subject to certain limitations. According to Bhaird (2010) and Casson, Martin, and Nisar (2008), equity is the financing instrument of choice for innovative firms, indicating a positive relationship between R&D spending and the use of equity. Additionally, many studies point to the important role of venture capital in innovation. Gottschalk et al. (2007), for example, find that VC-backed firms are more innovative and invest in R&D more constantly than their non-VC-backed competitors. There nevertheless seems to be some sort of bias associated with the measurement used. Contrary to expectations, equity does not appear to be a clear substitute for debt, implying a negative correlation between R&D and LTD. In some cases (Bhaird (2010); Magri (2007), the relationship is not significant. In others (Bah and Dumontier (2001); Börner and Grichnik (2010)) it is negative, as expected. In still others (Casson, Martin, and Nisar (2008)), it is positive. However, the latter study reports a negative relationship between R&D intensity and debt. R&D intensity may thus be the better measure due to potential size distortions with respect to the absolute amount of spending. A large firm may, for example, spend heavily on R&D in absolute terms but still not be considered innovative, as it invests only a tiny fraction of total assets. The size effect is thus dominant, implying a positive effect on LTD.

3.4.4.3.4 Growth

It seems obvious that growing firms have higher capital needs than stagnating ones. However, it is less obvious which financing sources are most suitable. Many factors exert an influence. For example, Berger and Udell (1998) find that small firms in high-growth industries depend on equity, while small firms in low-growth industries tend to depend on debt. Theoretical considerations on innovation could also be adapted to growth. Myers (1977), for example, distinguishes between growth options and tangible assets. This would imply a high proportion of equity financing and a low proportion of debt financing for growth firms. Like theoretical concepts, empirical research is rather mixed and has not delivered unambiguous results. Several studies (e.g., Cassar and Holmes (2003); La Rocca, La Rocca, and Cariola (2009); Sogorb-Mira (2005); Casson, Martin, and Nisar (2008); Thornhill, Gellatly, and Riding (2004); Voulgaris, Asteriou, and Agiomirgianakis (2004)) report a positive relationship between firm growth and debt on the one hand. On the other hand, many studies also report exactly the opposite (e.g., López-Gracia and Sogorb-Mira (2008); Romano, Tanewski, and Smyrnios (2001); Heyman, Deloof, and Ooghe (2008)), or do not find any significant relationships at all (e.g., Cassar and Holmes (2003); Hall (2002)). This seems to hold true for all types and maturities of debt. The impact of firm development on the use of equity and internal funds is also mixed, although slightly less and with a tendency toward a positive relationship. Romano, Tanewski, and Smyrnios (2001) Venture capital and business angel capital are often not included in large, rigorously empirical studies of capital structure. However, they do seem to be commonly accepted as a means to positively affect a firm's growth path. (e.g., Davila, Foster, and Gupta (2003); Gompers and Lerner (2001)) On a more general level, there seems to be some support for the assertion that growth requires capital. Not only do surveys of firm growth find evidence for finance as a critical resource, but empirical studies on finance and capital structure too show less contradictory results with respect to general external capital sources. From a financial perspective, it thus seems that growth positively affects the use of external capital sources. (e.g., Bruno and Tyebjee (1985); Cassar and Holmes (2003); Cassar (2004); Vos et al. (2007); Watson and Wilson (2002))

3.4.4.3.5 Asset structure, tangibility and collateral

Asset structure and tangibility are further important determinants of capital structure. The more tangible and generic a firm's assets are, the better is the likelihood of a potential sale and the higher the price will be. The liquidation value therefore increases in line with the firm's degree of tangibility. (Harris and Chaplinsky (2006)) This is particularly important in the context of innovation and growth, since such projects are often characterized by the opposite attributes, i.e. intangible and highly specific assets and resources. From an investor's perspective, tangible and generic assets reduce the magnitude of financial loss if a firm goes bankrupt. A firm's assets thus serve as a sort of indirect collateral. The degree of asset tangibility therefore correlates positively to the use of debt capital. (Börner and Grichnik (2010); La Rocca, La Rocca, and Cariola (2009); Voulgaris, Asteriou, and Agiomirgianakis (2004); Chittenden, Hall, and Hutchinson (1996); Heyman, Deloof, and Ooghe (2008)) Moreover, the theory implies that loan maturity should be aligned with asset utilization periods to prevent entrepreneurs and managers from allocating external funds to other purposes. Firms with a high proportion of short-term and intangible assets thus use short-term debt (STD), whereas larger firms with a high proportion of capital investment will prefer to use long-term debt (LTD). In other words, the proportion of long-term assets correlates positively to LTD and negatively to STD. (Cassar and Holmes (2003); Cassar (2004); Hall (2002); Sogorb-Mira (2005)) In this context, Storey (1996) argues that banks do not necessarily provide capital based on firms' profitability, but rather on collateral. Where firms lack collateral, banks would rather ration credit than adjust their terms. From a theoretical perspective, collateral sends a favorable signal about firm quality and overcomes persisting principal-agent problems. For both reasons, it seems that collateral is a very common tool in bank lending, particularly with respect to SME-driven markets. There are two types of collateral: (1) internal collateral, where the firm assigns certain assets to certain financing contracts. thereby altering the principal payback structure for capital providers; and (2) external collateral, where the scope of liability is extended to the personal wealth of the owner. Empirical evidence seems to support these theoretical considerations. For example, Chittenden, Hall, and Hutchinson (1996) find that banks base lending decisions on collateral rather than on profitability. Hogan and Hutson (2005) report that banks require collateral in the form of fixed assets. Evidence about internal collateral is thus closely related to that relating to tangibility. Firms with intangible assets will therefore show lower levels of debt. Also, external collateral is common in loans to SMEs. Bhaird (2010), for example, show that more than 60% of young French firms have to provide external collateral in order to acquire capital. LTD in particular correlates positively to the provision of external collateral. (Bollingtoft et al. (2003)) However, since the pledging of collateral is also a signal, it too may have a positive effect on other applications for financing. By implication, the existence of collateral leads to a larger share of external capital. This is supported by the study by Colombo and Grilli (2007).

3.4.4.3.6 Profitability and retained earnings

A firm's profitability and its capital structure are closely but, from a theoretical perspective, not unambiguously related. According to the pecking order hypothesis, and as indicated above, profitable firms tend to use less external capital as they prefer to finance their activities out of cash flow. On the other hand, it is difficult to argue that less profitable firms would thus use more external funds, because they are exposed to a higher risk of bankruptcy and are therefore less likely to successfully acquire external capital. Empirical evidence tends to support the pecking order hypothesis. Profitability drives cash flow and thus internal financing. Internal financing in turn reduces both the level of debt and the firm's equity. Profitability therefore correlates positively to internal financing (Chittenden, Hall, and Hutchinson (1996); de Haan and Hinloopen (2003); Magri (2007)) and negatively to leverage (Heyman, Deloof, and Ooghe (2008); La Rocca and La Rocca (2007); López-Gracia and Sogorb-Mira (2008); Sogorb-Mira (2005); Magri (2007); Bah and Dumontier (2001)), total bank debt (Börner and Grichnik (2010); Casson, Martin, and Nisar (2008)), long-term debt (Chittenden, Hall, and Hutchinson (1996); Voulgaris, Asteriou, and Agiomirgianakis (2004)) and short-term debt (Chittenden, Hall, and Hutchinson (1996); Hall (2007)). Only its effect on equity financing seems to be somewhat inconsistent. de Haan and Hinloopen (2003), for example, report a negative relationship and Casson, Martin, and Nisar (2008) a positive one. This may go back to the problem of circularity. On the one hand, profitability can increase equity through the accumulation of retained earnings. On the other hand, profitability may decrease dependence on the acquisition of external equity.

3.4.4.3.7 Ownership and firm structure

Corporations (such as German firms with the legal forms AG, GmbH or GmbH & Co. KG) are bound by stricter publication rules and must provide accessible information to the market.

In line with the financial growth cycle, these firms' information should be less opaque, giving them better access to financing. Börner and Grichnik (2010) thus propose to control for the legal structure of the firm. Another factor that may play a role is the ownership structure. Essentially, if a firm has many shareholders, it is harder for banks to monitor their actions. Information asymmetries are thus associated with fragmented ownership structures, which in turn correlate negatively to debt capital. (e.g., Bhaird (2010); La Rocca, La Rocca, and Cariola (2009); Romano, Tanewski, and Smyrnios (2001))

3.4.4.3.8 Environment

Hall, Hutchinson, and Michaelas (2004) point out that country differences are major determinants of capital structure. This largely explains the mixed results found by empirical research. For example, firms operating in countries with well developed equity markets such as the US will, *ceteris paribus*, have better access to equity than firms of comparable quality in countries with less well developed equity markets. Most international studies (e.g., Hall (2002); Bah and Dumontier (2001)) therefore control for country effects. Means of financing means may also differ from industry to industry. For example, biotech firms rely heavily on equity financing, because they do not generate cash flow to pay back loans for several years. The majority of studies (e.g., Bruno and Tyebjee (1985); Vos et al. (2007)) thus control for industry too.

3.4.4.4   The determinants of financial constraints

The availability of capital and the existence of financial constraints were briefly touched on in the last section on capital instruments. However, literature on capital structures is more interested in the use of certain financing means than in their availability. It has also been noted that research results on capital structures may be biased due to the thwarting effects of strategy and possibility. Moreover, even the causality of strategy is at least equivocal. La Rocca and La Rocca (2007), for example, point out that, while financing is part of a firm's decision space, the relationship is simultaneous in that the capital structure also influences a firm's strategy. Research into financial constraints has thus become an important field in finance literature. Table 3-4 provides an overview of selected studies that primarily use direct measures.[12]

A recent empirical literature review on financial constraints conducted by Carreira and Silva (2010) provides a summary of recent works and serves as a basis for discussing the main

---

[12] It should be noted that research into capital structure already and implicitly contains a wealth of information about potential constraints, and that this more focused literature stream should be seen as an extension and validation. Results that are consistent with those presented in the section on capital structure are therefore be discussed only briefly. Additional insights naturally receive more attention.

determinants. The authors identify ten summarized facts derived from literature: (1) Financial constraints are more severe for younger and smaller firms. (2) The size distribution of firms is highly skewed. (3) Start-up firms and entrepreneurs are subject to particularly tight constraints. (4) Financial constraints are crucial to a firm's survival. (5) R&D investment is more constrained than other activities. (6) Government support may reduce financial constraints. (7) Close relationships with banks reduce the degree of constraints. (8) Developed financial markets reduce the number of constrained firms. (9) Financial liberalization improves the situation of constrained firms. And (10) foreign-owned firms are less financially constrained than domestic ones.

Survey	Dependent variables	Selected significant determinants	Sample size	Regions
Carrcira and Silva (2009)	Financial constraints (literature review)	Size, age, banking relationship, internationality, R&D, firm survival, industry	n.a.	World
Canepa and Stoneman (2008)	Financial constraints (direct measure)	Size, industry	1,016	UK
Savignac (2008)	Financial constraints (direct measure)	Size, growth, innovation, industry	1,940	France
Piga and Atzeni (2007)	Intention to adopt more credit/debt, application denied	Size, STD, R&D, profit, industry	2,144	Italy
Vos et al (2007)	Loan application (denied and approved)	Age, ownership, rating	3,239	US
Beck and Demirguc-Kunt (2006)	Financial constraints (direct measure)	Size, growth	n.a.	n.a.
Wolf (2006)	Fiancial constraints (direct measure)	Size, R&D, management capacity	554	Germany
Plötscher and Rottmann (2002)	Trade credit	Size, growth, past investment, profit	867	Germany
Giudici and Paleari (2000)	Financial constraints (direct measure)	Size, age, growth	46	Italy
Freel (1999)	Bank debt application	R&D	238	UK
Guiso (1998)	Credit rationing	Size, growth, total assets, LTD, STD, internationalisation	608	Italy
Egeln and Licht (1997)	Financial constraints (direct measure), rating	Size, age	1,919	Germany
Westhead and Storey (1997)	Financial constraints (direct measure)	Age, collateral, R&D, industry	188	UK

Table 3-4: Selected empirical research into the determinants of financial constraints

These summarized facts can be condensed into seven main determinants of financial constraints: firm size, firm age and development state, innovation and R&D, government support systems, bank relationship, firm structure and environment. However, with respect to the research questions posed by this dissertation and the literature under consideration, the following review of determinants only briefly touch on firm structures. It also adds firm growth (e.g., Beck and Demirguc-Kunt (2006)) and profit (e.g., Piga and Atzeni (2007)) to the analysis. This yields the following seven determinants of financial constraints at the level of the firm:

1. Firm size (Sales, employees etc.)
2. Firm age (Time on the market)
3. Profitability (Margin, profit etc.)
4. Innovation (R&D, innovation output etc.)
5. Firm growth (Sales growth rate, employment growth etc.)
6. Banking relationship (Number of banks, relationship satisfaction etc.)
7. Environment (Industry, technology etc.)

3.4.4.4.1 Firm size

In the previous chapter it was shown that smaller firms primarily use internal funds due to the opacity of their information. Empirical results on capital structure have a direct link to financial constraints, because they suggest that smaller firms do not have access to all capital sources. This is not only consistent with the implications of the financial growth cycle hypothesis, but also with findings from literature on financial constraints. Most studies thus seem to report a negative correlation between firm size and the degree to which a certain firm does not receive the desired financing. (e.g., Beck and Demirguc-Kunt (2006); Canepa and Stoneman (2008); Egeln and Licht (1997); Guiso (1998); Piga and Atzeni (2007); Ploetscher and Rottmann (2002); Vos et al. (2007); Wolf (2006); Giudici and Paleari (2000)) This view is also supported by the literature review cited above.

3.4.4.4.2 Firm age

The theoretical prediction of a firm' age and constraints relationship is similar to the one for size: Younger firms have a shorter history and therefore have difficulty communicating their quality to investors. Consequently, the younger a firm is, the lower is its propensity to acquire external capital. A limited ability to generate cash flow adds to the problem. Accordingly, Carreira and Silva (2010) postulates that start-up firms in particular find themselves financially constrained. The proposed negative correlation is supported by several empirical studies. (e.g., Hyytinen and Pajarinen (2008); Vos et al. (2007); Westhead and Storey (1997); Giudici and Paleari (2000)) Only Egeln and Licht (1997) point out that the relationship may also have a reverse U-shape in which younger and older firms are less heavily constrained than middle aged ones.

3.4.4.4.3 Profitability

As we have seen, profitability measures a firm's ability to finance investment projects internally out of cash flow. From a theoretical perspective, profitable firms should therefore have lower external capital needs, implying a negative relationship between profitability and financial constraints. This view is supported by empirical studies predicting financial constraints via regression analysis. (e.g., Piga and Atzeni (2007); Ploetscher and Rottmann (2002)) The results are comparable to those presented in capital structure research, in which profitability correlates negatively to the amount of external debt, equity and, in particular, to more expensive capital sources such as STD.

3.4.4.4.4 *Innovation and R&D*

From a financial perspective, innovative firms constitute a greater investment risk – a situation exacerbated by highly intangible assets and correspondingly lower liquidation values. As we saw above, innovative and R&D-intensive firms have limited access to debt capital. Younger and smaller firms in particular must rely either on internal funds or on comparatively small private equity markets. Public equity, on the other hand, is associated with high fixed costs, making this option suitable for large financing volumes only. Innovation and R&D intensity therefore correlate positively to financial constraints. In other words, the more innovative a firm is, the more difficult it will be to acquire sufficient capital (Carreira and Silva (2010); Piga and Atzeni (2007); Westhead and Storey (1997); Wolf (2006); Freel (1999)). This assertion is in line with capital structure research, which proposes that innovative firms use more internal and less external capital sources in general, and less debt in particular.

3.4.4.4.5 *Firm growth*

Surprisingly, it seems that firm growth correlates negatively to financial constraints. (Guiso (1998); Piga and Atzeni (2007); Ploetscher and Rottmann (2002); Giudici and Paleari (2000)) The observation seems puzzling, because growth projects, like innovation projects, can also be considered risky and intangible. The effect of firm growth, again like innovation and R&D intensity, on the degree of financial constraints would thus be expected to be positive. In other words, the faster a firm grows, the greater the financial constraints that would be expected. However, the results are consistent with the (reversed relationship) implications derived from capital structure research, suggesting a positive relationship between firm growth and both internal and external capital sources. One solution might be that strong growth relaxes financial constraints because historic growth improves investors' mood. Growth, unlike innovation, may therefore not be a proxy for uncertainty, but rather an indications of a firm's capability and growth potential. This proposition is supported by the findings of Piga and Atzeni (2007), who report a negative correlation between growth and financial constraints, but a positive correlation between growth variance and financial constraints.

3.4.4.4.6 *Banking relationship*

Berger and Udell (1998) argue that a close banking relationship may overcome small firms' opacity problem and resolve information asymmetries. It may also reduce the need to secure credit lines with collateral. (Chakraborty and Hu (2006)) The relationship is characterized by the financial institution's efforts to gather information about the firm, and is assumed to improve over time, thereby adding to the negative size and constraint correlation.

Occasionally, however, increased market power on the part of a bank may lead to unfavorable credit terms for the firm. Firms may therefore engage in relationships with multiple banks. There is also a second implication that underlines the weaker relationship between age and constraints. A small but older firm may have a good banking relationship and, for this reason, be free from financial constraints. Empirical evidence seems generally to support the view that banking relationships can reduce potential financial barriers for firms (Carreira and Silva (2010)). Consistent with the credit rationing assumption, it thus seems that interest rates are impacted less by relationship lending than by credit lines. (Berger and Udell (1992)) A recent literature review on relationship lending conducted by Elyasiani and Goldberg (2004) summarizes empirical research. Existing evidence supports the view that relationship lending reduces financial constraints. This seems to be the case for small firms in particular. The overall number of relationships has the contrary effect, albeit with a smaller magnitude. Additionally, Casson, Martin, and Nisar (2008) point out that Germany's system in particular is dominated by banks and credit. As a result, good relationships with banks have a greater impact on financial constraints here than in other countries.

3.4.4.4.7 Environment

Following standard practice in empirical literature, most studies on financial constraints correct for industry (e.g., Canepa and Stoneman (2008); Piga and Atzeni (2007)) and geographical region (e.g., Guiso (1998); Ploetscher and Rottmann (2002)). On the whole, it seems that, while the magnitude of constraints may differ from industry to industry and region to region, the overall relationships remain relatively robust. Westhead and Storey (1997) find that firms with considerable technological complexity are much more likely to report financial constraints.

3.4.5   Conclusion – Characterizing financial structures and constraints

The first section of this chapter discussed theoretical works in corporate finance literature and, more generally, in financial economics literature. The existing concepts are as diverse as the empirical evidence. Some of the implications derived actually contradict each other. For example, trade-off theory implies that larger firms gain advantages from leverage, while pecking order theory predicts that larger firms possess more internally generated funds and will thus make less use of debt. However, there are also some similarities between theories. Apparently, most theories share one important characteristic, namely that the risk to a firm, measured as the variance of cash flows and information availability, determines its choice of financial instruments and the availability of these instruments from investors. Only the MM theorem does not account for differences between firms and information availability. By implication, it follows that these issues are irrelevant to the financial structure of the firm. This finding would also be consistent with Gibrat's law, discussed earlier, according to which firm size should have no impact on firm growth. Nonetheless, empirical studies show that both theories do not hold. Both firm size and age are therefore significant factors influencing capital structure. (e.g., Chittenden, Hall, and Hutchinson (1996)) More recent research has thus strongly focused on comparing trade-off theory with pecking order theory. Support is provided for both: Some authors (e.g., Fama and French (2002); López-Gracia and Sogorb-Mira (2008)) claim that the trade-off theory has a better fit, while others (e.g., de Haan and Hinloopen (2003); Heyman, Deloof, and Ooghe (2008)) find the pecking order hypothesis more appropriate. However, none of the studies discussed in this chapter presents either striking or unambiguous results in one direction or the other. In fact, most of them point out that evidence supports elements of both theoretical concepts. For example, Bozkaya and de van Potterie (2008) note that there is some support for the financial pecking order theory with respect to the capital structure of firms in later states. Unlike trade-off theory, however, this theory cannot explain the capital structure decisions taken by younger firms that lack cash flow and retained earnings. A more integrated framework, accounting for both concepts, is characterized by the information opacity hypothesis proposed by Berger and Udell (1998). Information opacity has become an important concept in literature but has barely been underpinned by empirical studies (although many studies include it in the form of proxies such as age, size, etc.). Moreover, studies focusing on the financial growth cycles in particular have often been descriptive in nature and have refrained from presenting testable hypotheses. While it is thus fair to say that the overall evidence is mixed, three main implications are still hard to deny. First, firm size and age reflect the degree of information opacity and serve as moderators of a firm's financial structure. Second, financing instruments that suffer acutely from moral hazard and information asymmetries in general, such as long-term debt, are more likely to be used by larger and older firms. Conversely, younger and smaller firms are more reliant on internal financing sources and external equity. Lastly, younger and smaller firms are more likely to face financial constraints due to the limited availability and rationing of external funds as a result of a firm's information opacity. In this context, Hyytinen and Pajarinen (2008) are the first to present empirical evidence based on an innovative approach that analyzes rating disagreements among firms. The results suggest that, despite certain limitations, size and age are indeed proxies for information opacity. The financial growth cycle concept is therefore a good basis for understanding the financial characteristics and problems of greentech firms with respect to their lifecycle state.

The second part of the analysis focused on the determinants of capital structure and financial constraints. Regarding capital structure, most of the studies reviewed control for firm demographics, such as size and age, but also incorporate innovation, growth and the firm's environmental factors. Overall, it seems that, despite the vast variety of potential dependent variables, there are some at least some areas in which the results converge, examples being size, age, profitability and, more importantly, innovation and growth. The review provided a synthesis of contemporary literature and proposes two major distinctions: (1) internal versus external capital and (2) equity versus debt capital, treating short-term debt as an exception. In essence, it was found that internal capital is very important to younger and smaller firms that lack access to external capital due to information opacity. Internal capital is distorted by a firm's ability to generate cash flow, which allows it to finance investment internally without accessing external funds. However, cash flow correlates positively to firm size and age, which thwarts small firms' dependence on internal funds. It seems that firm size and age positively affect the use of both equity and debt financing. This is due to the fact that most studies focus on public equity only. Innovative firms use less debt and less external capital. Investors seem reluctant to invest in highly uncertain assets with low liquidation values. Interestingly, this does not hold for growth firms that seem to have access to the entire spectrum of financing instruments. Although, the results discussed are important to understand firms' capital structures, they should nevertheless be treated with caution. Capital structures reflect two different things: a firm's ability to raise capital; and a firm's strategy and decisions on how to use certain financing instruments. Both may be aligned but still move in different directions. Empirical results may thus be biased. It is therefore important to explicitly assess a firm's ability to access capital, i.e. the likelihood that a firm will experience financial constraints. Research into financial constraints largely supports the concept of information opacity proposed by the financial growth cycle model. Smaller and younger firms are therefore limited in their ability to raise capital. This correlates closely to the results suggested by empirical research into capital structures. The correlation is especially strong for debt-related instruments. In this context, it seems that debt rationing cannot be offset by private equity investors. This constraint is most severe in Germany, where equity markets are small and provide capital only to a very small number of firms. (e.g., Müller and Zimmermann (2009); Metzger et al. (2010)) However, those firms that do acquire venture capital reduce their barriers to growth and attain sufficient capital in the process. This is consistent with Bruno and Tyebjee (1985), who report that only VC-backed firms actually manage to acquire financing that comes closely to their real needs. Schulte (2006) summarize empirical literature mostly from Germany and draw the following conclusions: (1) Young and small firms have significantly more problems acquiring funding than older and larger firms. (2) The characteristics of firms and their owners moderate the degree of constraints. (3) Internal finance is the most important source to overcome constraints. (4) Bank loans and government support systems play a very crucial role in Germany. (5) German SMEs prefer to stick to existing equity capital sources. (6) Alternative financing instruments are used inconsistently. And (7) only larger firms have specialized financing departments. The overall results from the literature review are synthesized in Figure 3-15.

	Internal capital	External capital	Public equity	Debt	STD	Financial constraints
Firm characteristics						
Size[1]	-	+	+	+	-	-
Age	-	+	+	+	-	-
Tangibility	+/-	+	+/-	+	-	
Profitability	+	-	-	-	-	-
Legal form	+/-	+/-	+/-	+/-	+/-	
Relationship						-
Innovation	+	-	+	-	-	+
Growth	+	+	+	+/-	+/-	-
Environment						
Industry	+/-	+/-	+/-	+/-	+/-	+/-
Country	+/-	+/-	+/-	+/-	+/-	+/-

1) Relationship distorted by the ability to generate cash flow

Figure 3-15: Determinants of capital structure and financial constraints – A synthesis

A total of five key insights can be derived. First, younger and smaller firms have potentially less access to capital sources than larger firms. This is particularly true for debt, a finding that supports the credit rationing hypothesis (e.g., Storey (1996)). The tighter financial constraints for these firms not only supports this hypothesis, but also indicates that equity cannot fully offset the rationing effect, as the market for dedicated instruments, i.e. business angel and venture capital, is rather small. Second, the availability of debt capital depends on the degree of tangibility of a firm's assets, as tangible assets serve as sort of collateral for banks. The only other way to acquire debt capital is to build close relationships with banks, which can then also significantly reduce a firm's financial constraints. Third, in line with the pecking order hypothesis, profitability (as a proxy for cash flow) significantly reduces dependence on external financing and is also preferred by firms to other instruments. The effect seems to overcompensate for better access to external funds due to greater credibility. Fourth, innovation and growth have different effects on capital structure and financial constraints. This is surprising when one considers that both factors are characterized by considerable uncertainty and information asymmetries, as well as by the existence of intangible assets. Apparently, innovation activities must rely on internal capital and equity. Contrary to the findings of innovation literature, innovation correlates positively to financial constraints given the assumption of reverse causality. In other words, the more innovative a firm is, the more it will experience financial constraints. Growth, on the other hand, correlates positively to all financing sources. It seems that track record is more important than the limited tangibility of growth-related activities, thereby improving capital access and reducing financial constraints to the firm. Lastly, several studies have shown that the specific industry is also important, and that both financing and the predictive value of theory vary significantly from one industry to another. Even so, many studies suffer from three deficiencies: (1) survivorship bias, (2) a lack of empirical testing of capital structure theories, and (3) a limited geographic and industry focus. (Cassar (2004)) Research into financial constraints is also short of reliable results due to measurement problems, such as cash flow sensitivities and the self-serving biases that arise from questionnaire-based research. (Kohn (2009)) Further research is also needed to assess reasons for potential constraints and ways to overcome them. One way could be through government intervention, which is the subject of the following section.

3.5   Public policy – Bypassing financial constraints

This chapter examines the importance and impact of public policy. It draws on the previous chapters and thus presupposes a proper understanding of how firms innovate and grow. As argued by contemporary policy research into greentech, negative external effects are not fully allocated to the originator and are thus incurred by all market participants. The so called "free rider problem" causes inferior technologies to be locked in and raises substantial barriers to innovation and to the growth of small innovative technology firms, because the expected returns remain below investors' requirements. An associated problem is that returns on innovation suffer from spill-over effects relating to the knowledge generated by R&D. Both effects lead to underinvestment by the firm. Accordingly, the theory predicts that, in unregulated markets, private firms' R&D investment levels will not be sufficient to generate technological breakthroughs, because investors' profitability considerations do not incorporate the related social benefits. (Spence (1984)) Limited access to financing due to capital market imperfections confronts smaller and younger firms with yet another problem. For this reason, governments support SMEs in particular in order to alleviate potential financing constraints. In this context, Pissarides (1999) summarizes attempts by the European Bank for Reconstruction and Development to support the SME sector. Her analysis shows that, although many obstacles persist, most activities aim at reducing the financial constraints for firms. To ultimately drive sustainable technologies, governments should thus take action to offset market imperfections and improve the economics of greentech firms and new technologies. However, in order *to justify intervention in a market economy it is necessary to identify precisely where the market failure exists, and whether it is possible to rectify that market failure through intervention"* (Storey (1996)).

This section differs from the previous ones in that it focuses specifically on greentech. Although this appears to break out of the logical pattern of the dissertation, it is justified by three reasons. First, the review of literature on greentech research demonstrates that only very few studies have adopted a firm-level perspective of how greentech innovation and growth is financed. It is therefore necessary also to draw on findings in other fields. Second, greentech policy is well covered as a research topic due to the high specificity of the relevant policy measures. Third, findings from policy-oriented research can be adapted to the context of this dissertation by assuming a one-firm, one-technology world as derived in the previous chapters. However, most studies of greentech have analyzed the impact on technology innovation and diffusion, distinguishing between push and pull policies. The same categorization can be adapted to the firm. At this point, however, the discussion focuses specifically on the implications of policy on innovative activity, growth and finance. A broad literature review was provided in the first part of the dissertation and is not recapitulated here.

3.5.1 The gap theory and the one-firm, one-technology case

In her explorative and innovative work, Chertow (2000) assesses the factors that have impacted the emergence and development of green technology markets. On the basis of various expert interviews, contemporary policy research and existing governmental support systems in the US, the author formulates a theory of financing for greentech innovation and diffusion. However, the work has two limitations. First, it does not base findings on rigorous empirical analysis. Second, it adopts a technology perspective and therefore lacks consistency with financial economics literature. However, Chertow's main implications provide fertile soil for this dissertation and can be directly linked to a one-firm, one-technology world in which technology and firm financing are congruent. Figure 3-16 presents the main findings of her work. As supported by financial economics and the growth cycle hypothesis, smaller and younger firms have difficulty accessing external capital sources. In essence, firms are heavily dependent on internally generated funds to invest in innovation and growth. However, these funds are particularly rare for early-state firms. This is shown by the industry/private funding graph, which depicts the relative availability of external funds to firms. The dashed line shows the impact of policy programs. Apparently, policy support is realized through the provision of funds during the various phases. To identify the total availability of capital to firms, private and public funding must simply be added together. The resulting curve is demonstrated by the first line, "relative money available".[13] As can be seen for a firm in the very early state, private funding is limited due to information opacity, although policy actions step in to foster

---

[13] In this context, it must be noted that the graph does not include pull activities, because the focus of Chertow's work was R&D rather than large-scale diffusion. Pull activities will most likely affect later-stage firms and may therefore not alter the propositions made.

the development of new technologies. Government's reckoning is that, at this state, firm and technology risk exceeds the maximum acceptance level for private investors.

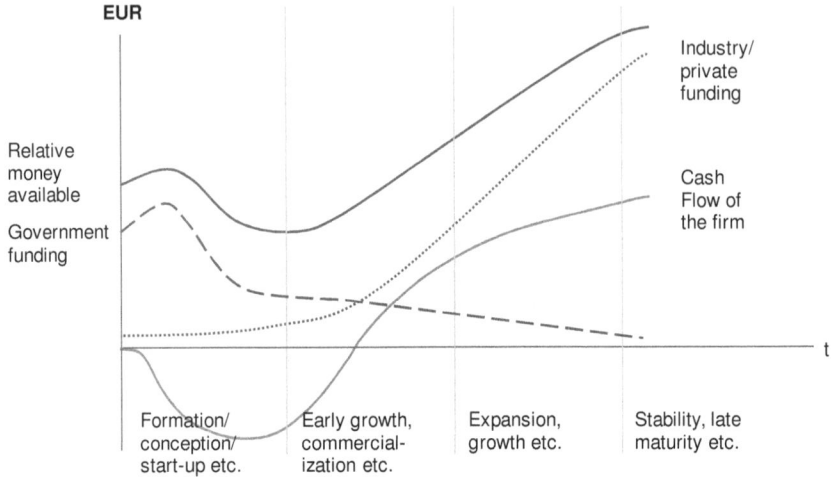

Figure 3-16: The need for and impact of policy intervention on firm development – The one-firm, one-technology case

Policy instruments should therefore significantly reduce financial constraints for early-state firms, although such initiatives are assumed to be scaled back over time as private financing options become more appropriate. However, it appears that these options are not available early enough. Accordingly, the relative availability of capital is worst when firms begin to commercialize their technologies. At this point, it seems that private investors still perceive the risk to be unbearable while, at the same time, governments regard the risk as too low to justify support mechanisms. In other words, there seems to be a funding gap for early growth firms that are aiming to expand their product sales. If this is really the case, policy activities may be ineffective because they do not remove the market barriers created by market imperfections. The goal of the following paragraph is to validate Chertow's findings with respect to the German greentech market space.

3.5.2   The status quo for greentech policy measures in Germany

When examining existing research on greentech, there seems to be no single study that summarizes policy efforts in Germany in a detailed, exhaustive manner. As was shown in the literature review, most studies assess policy effectiveness on a macroeconomic basis, primarily by comparing fed-in systems and other pull activities in different countries. The emerging consensus is that feed-in tariffs yield the best technology diffusion results.

However, they may not be suitable to foster innovation and avoid technology lock-in. In order to resolve these issues, the first step is to gain a transparent overview of existing programs. The following analysis therefore attempts to review the support mechanisms that are available in the German market space. Fortunately, assistance was provided by the German "Förderdatenbank", a database of national, regional and EU-backed support programs conducted by the German Ministry of Economics and Technology.[14] A total of 86 different greentech- and environment-related programs were identified using the database's own categorization option and relevant search arguments. Each support program was screened and described using the following categories: name, regional scope, policy type, instrument type, impact on greentech, target group according to lifecycle state and industry characteristics, issuing institute, policy goals, details, amount granted and duration. To fill all categories, additional information was gathered from the issuing institution. The complete list and sources are attached in the appendix (A). Five types of instruments are currently available on the market: (1) grants/allowances, (2) subsidized loans, (3) equity participations, (4) hybrid instruments and (5) feed-in tariffs. The instruments examined here are often called "direct funding mechanisms" (e.g., Peneder (2008)). They do not include the fiscal incentives used to encourage investors or regulatory action designed to influence capital markets. The most common instruments are grants and subsidized loans, although most money is allocated to feed-in tariffs. Figure 3-17 provides an overview of the greentech-related government support programs available in Germany, segmented by the targeted firm development state and instrument type. The majority of support mechanisms are offered by the federal states, with an average of three programs per state. 28 programs are operated on a national level and five on a European level. With respect to the financial growth cycle presented in the previous section, it seems that most programs aim to support SMEs in general and therefore focus on established and older firms in the expansion phase. Although this contradicts the propositions made by Chertow (2000), there appears to be some congruence with the proposed pattern at first glance. A large share of support programs thus targets firms in the technology development state. 31 programs specify the development of new products and prototypes as a selection requirement. Also consistent with Chertow's theory is the fact that early growth firms have access to only eight different programs. Apart from the rationale of decreasing government support when a firm becomes self-sufficient, programs are in shortest supply in a phase in which firms are still bound by substantial constraints due to credit rationing. Most instruments are managed either by the ministry directly or by the state-owned KfW bank. Subsidized loans are also accessible through commercial banks, which act as intermediaries. The most common goal of each instrument appears to be to improve efficiency and reduce emissions either within the firm or by developing technologies for the market. The volumes granted vary considerably, but often involve covenant-like terms, such as a stipulation that

---

[14] The database is accessible via internet under http://www.foerderdatenbank.de/, accessed 2010, March-August

50% of the investment must be carried out using private funds. Furthermore, it seems that policymakers have addressed the lack of green venture capital (e.g., Diefendorf (2000); O'Rourke (2009); Wüstenhagen and Teppo (2006)) by introducing equity substitutes in the form of government-backed VC, such as the "Technologie Gründerfonds". There are three such instruments on the market.

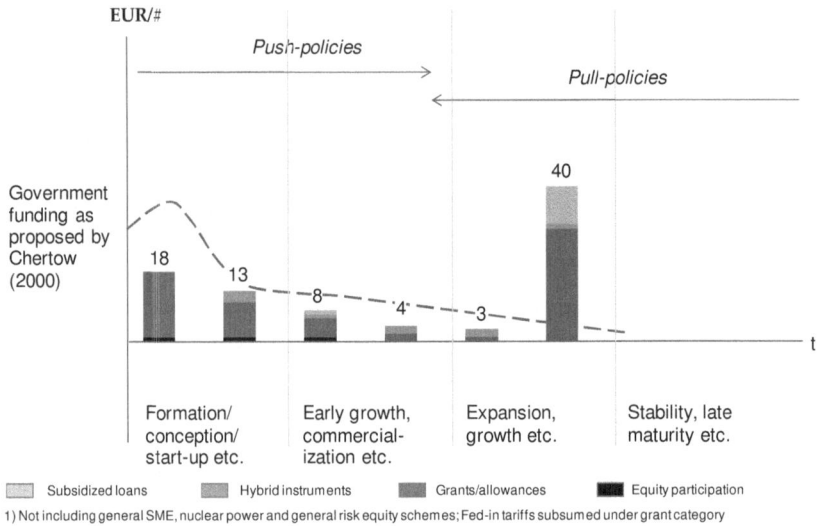

Figure 3-17: Number of greentech-related policy instruments available on the German market, by type

3.5.3 Empirical evidence on the effectiveness of firm-level support programs

As noted earlier, there is, to the author's best knowledge, no study that assesses the effectiveness of greentech support programs at the level of the firm. However, some studies do assess the impact of government support on the firm within the wider body of SME literature. On a general level, this literature will therefore contribute to an understanding of the impact of government support at firm level. It can be assumed that financing of greentech firms does not differ from that of other firms. Accordingly, policy support will probably have no other effect on greentech firms than it would have on other firms. The only difference might be that the high social-to-economic return ratio in the greentech sector leads investors to be more reluctant than usual, leading governments to increase the scale of their intervention.

Literature on government support programs is rather mixed. Essentially, some scholars argue that the best form of government support is for governments to stay out of the market. Others point to the need for SME-oriented support mechanisms in view of pronounced capital market imperfections. Based on hypotheses derived from these imperfections, Hyytinen and Toivanen (2005) use Tobit regression to assess the impact of policy support on firms' R&D expenditure and growth orientation. The authors' findings suggest that policy support can indeed enhance R&D and foster growth by significantly reducing the financial constraints faced by firms. The R&D observation is supported by Klassen, Pittman, and Reed (2004), who estimate the effect of tax credits on R&D for financially constrained firms in Canada and the US. They report that up to three private dollars are invested for every lost tax dollar. The relationship intensifies for constrained firms. Additionally, González and Pazó (2008) present evidence from Spain suggesting that public R&D support has a positive effect on private R&D spending without crowding out private investment. This is consistent with an international study conducted by Wu, Popp, and Bretschneider (2007). The authors report that public R&D spending correlates positively with private R&D spending in industrialized countries. The growth observation is supported by Becchetti and Trovato (2002) who, using a sample comprising Italian SMEs, report that government subsidies positively affect a firm's growth rate even when the sample is corrected for the survivorship bias. The authors thus conclude that government intervention can effectively alleviate financial constraints on firms. Honjo and Harada (2006) are the only authors to report a positive effect from support programs on assets. However, the various tax breaks, loans and subsidies do not seem to enhance either employment or sales. Another study by Dahlqvist, Davidsson, and Wiklund (2000) specifically focuses on start-up firms that were shown to experience the most constraints. Their evidence implies that government-supported firms are likely to deliver superior performance to their non-supported competitors. In addition, this finding underlines the argument that support systems focus on the seed and start-up phases to eliminate the liabilities inherent in smallness and information opacity. Some studies specifically analyze problems associated with the debt capital markets, claiming that public loan guarantees are best suited to resolving imperfections in these markets. For example, Cowling and Mitchell (2003) analyze death probabilities for UK firms and demonstrate that credit rationing reduces firms' survival propensity and duration, whereas government loans effectively have the opposite effect. The costs of the program seem reasonable, since up to 70% of the loans are paid back. In this context, Zecchini and Ventura (2009) argue that public guarantees in particular effectively overcome firms' information opacity. In fact, firms using public guarantees demonstrate consistently and significantly higher leverage ratios. This observation seems to hold for German firms as well. (Börner and Grichnik (2010)) It is also consistent with a study conducted by Kang, Heshmati, and Choi (2008), who report that credit guarantees issued by the government positively impacted firm sales growth, but that guarantees were granted primarily to profitable firms that might have been able to find financing themselves.

However, not all studies have reported such a positive effect. The view that government support necessarily leads to the desired outcome is therefore not accepted across the board. Based on existing literature from Germany on start-up support programs, Witt and Hack (2008) argue that support mechanisms can do more harm than good. They paint a rather negative picture, although pointing out that much more research is needed to finally assess this issue. A comparable result is reported by Svensson (2007) in a very unique analysis of innovation. The study shows that patents backed by government support are less likely to be commercialized than non-backed patents. The findings suggest that government terms are too soft and that governments are unable to pick winners due to adverse selection, moral hazard and issues relating to information asymmetry.

3.5.4   Conclusion – Policy effectiveness at the level of the firm

As we have seen , there is no study that examines the impact of greentech support programs at firm level. Moreover, research seems to have neglected the huge variety of different support programs. This dissertation can, of course, not completely fill the gap. Further research is therefore necessary. Nevertheless, it seeks to provide an initial overview of those programs that have been implemented to date. The overall structure, measured in terms of the number of programs (as volume data is not available), seems to be fairly consistent with the theoretical considerations put forward by Chertow (2000) with regard to US firms. Most programs focus on seed and start-up firms and fundamental technology development projects. The number of programs decreases up to the expansion phase of the corporate lifecycle. Older firms, though limited in size, at least have a broader set of programs to choose from. Although the effectiveness of these programs has not sufficiently been analyzed, general SME research gives some indications. First, it seems that the impact of support mechanisms can be positive, i.e. encouraging growth and innovation, but also negative. The actual effect seems to depend on the design and environmental context. This would also explain why some studies report significantly positive support coefficients while others report the opposite. It is therefore not fully clear what these impact factors may be, because research is scant and often too heavily aggregated. For example, most studies use a dummy for general government support but do not distinguish between different instruments. The questions whether and, in particular, how government support systems reduce financial constraints and foster development remain open to discussion. Nevertheless, one incontrovertible fact is that government support instruments will certainly have some effect on financial constraints, firms' innovation and firms' growth.

4   THEORETICAL FRAMEWORK

The previous chapters lay the foundation for the theoretical framework that is the subject of this chapter. The aim of this framework is to integrate the findings synthesized from literature on innovation, growth and finance into a single, comprehensive framework. The chapter is organized as follows: First, it provides an interdisciplinary summary of the most important conclusions from previous chapters. Second, it discusses problems relating to the integration of these three distinct fields of research. Third, it develops the theoretical framework and derives research hypotheses. The chapter ends with a brief conclusion.

4.1   Interdisciplinary summary of the main findings from previous chapters

If there is one thing that we can be sure about, it is that corporate innovation, growth and financing are interdependent and interrelated in some way. It is surprising that these three fields are still regarded as distinct disciplines in business research. This fact has, of course, led to a lack of integrated concepts that embrace all three topics and, in so doing, provide an appropriate theoretical framework for this dissertation. This omission has naturally not gone unnoticed by scholars. O' Sullivan (2007), for example, argues that contemporary research on innovation has largely neglected the role of finance. Although early seminal works on innovation (e.g., Rogers (2003)) and firm development (e.g., Penrose (1997)) noticed the importance of considering the availability of funds, it seems that only more recent studies (e.g., Harhoff (1998)) have also tried to integrate certain aspects of this observation into specific models. As we have seen, there appears to be no empirical agreement on the applicability of basic theories in literature on either innovation, growth or finance. Naturally, therefore, there can be no agreement on a potential interdisciplinary and integrated concept either. The lack of convergence across these three fields is reflected by the different choices of dependent variables in the 87 empirical studies reviewed in the previous chapters. In essence, some studies use dependent variables that other studies use as regressors. The underlying problem is the question of causality, i.e. what affects what. This question is addressed separately in the next section. The overall empirical results in the three literature bodies innovation, growth and finance are synthesized in Table 4-1. Besides the obvious differences, there is also common ground. First, most studies seem to use firm size and firm age as explanatory variables to control for firm characteristics. This lends fundamental support to lifecycles and financial growth cycles as structural concepts. Second, most studies control for industry and country effects, indicating the importance of the environment and specific context within which a firm operates. Third, profitability seems to be important when it comes to engaging in innovation activity, enabling growth and choosing financial instruments. Finally, the effect of government support systems seems to be positive with respect to firm performance, but has been analyzed primarily in capital structure research. With regard to differences in results, the section below focuses, from the authors' point of view, on the most important issue, namely causality.

| | Dependent variables used in respective research areas | | | | | | | | | |
| | Innovation | | Growth | | Finance | | | | | |
Determinants	PD/ Non-incr.	PC/ Incr.	R&D intens.	Sales growth	Int. Cap.	Ext. Cap.	EQT	D	STD	Fin. Constr.
R&D										
R&D expenditures	+	+	-	+	+	-	+	-	-	+
R&D intensity	+	+	/	+	+	-	+	-	-	+
R&D productivity	-	-	-	/	/	/	/	/	/	/
Innovation	/	/	/	+	+	-	+	-	-	+
Firm characteristics										
Firm size	+	+	-	+	-	+	+	+	-	-
Firm age	+	+	-	-	-	+	+	+	-	-
Profitability	+	+	+	+	+	-	-	-	-	-
Growth	+	+	+	/	+	+	-*	-*	-	
Internationality	+	+	+	+	/	/	/	/	/	/
Legal structure (limited=1;0)	+	+	+	+	+/-	+/-	+/-	+/-	+/-	+/-
Focused strategy	+	+	+	+	/	/	/	/	/	/
Tangibility	/	/	/	/	+/-	+	+/-	+	-	/
Resources										
General resources	+	+	+	+	/	/	/	/	/	/
Equity	+	-	+	+	/	/	/	/	/	+/-
Debt	-	+	+	-	/	/	/	/	/	-
Financial constraints	-	-	-	-	+	-	+	-	+	/
Governement support										
Grants/allowance	+*	+*	+*	+*	/	/	/	/	/	-*
Guarantees	-**	+**	+**	-**	/	+**	/	+	+	-*
Controls										
Industry	+/-	+/-	+/-	+/-	+/-	+/-	+/-	+/-	+/-	+/-
Country	+/-	+/-	+/-	+/-	+/-	+/-	+/-	+/-	+/-	+/-

* Evidence mixed, tendency   ** Relationship implied by capital structure research
n=87 studies with more than 200 different models

Table 4-1: Comparison of empirical study results in the various research areas

4.2   The chicken and egg problem – A causality dilemma in interdisciplinary research

The fundamental question of causality is: "What came first, the egg or the chicken?" Or, to transpose the same idea onto the given context: Was it innovation, growth or finance? Besides the problems of endogeneity and simultaneity in statistical methods, it seems that research has not addressed this question specifically. Driven by their respective disciplines, researchers have tended to adopt the prevailing wisdom. The concept of financial constraints serves as a good example. On the one hand, innovation and growth research model financial constraints as independent variables, implying that constraints negatively affect firms' innovation activity and growth. (e.g., Klassen and McLaughlin (1996); Musso and Schiavo (2008)) On the other hand, finance research uses financial constraints as dependent variables, implying that firms' innovation activity and growth affect the accessibility of capital. (e.g., Westhead and Storey (1997); Ploetscher and Rottmann (2002)) The relationship is therefore ambiguous. To overcome this ambiguity, at least in empirical models, researchers have developed powerful statistical methods such as instrumental regressions, switching models, etc. (Hamilton and Nickerson (2003)). In the case of financial constraints and innovation, Savignac (2008) shows that the effect of financial constraints on innovation is heavily biased due to the endogeneity of the constraints variable. The author overcomes this problem by using two-stage regression to predict first financial constraints and then innovation. Essentially, the regression model arrives at a statistically significant and non-biased negative relationship. However, it assumes that financial constraints impact innovation, but not vice versa. The problem of the egg and the chicken still remains. This is an extraordinarily important finding with respect to the theoretical framework of this dissertation. As can be seen in Figure 4-1, evidence about financial constraints and innovation as well as about growth is well mixed and is not consistent due to the circularity of relationships.

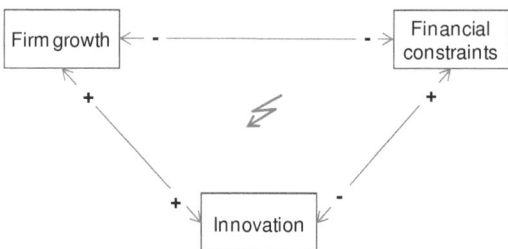

Figure 4-1: The circularity and inconsistency problem in integrating financial constraints, innovation and growth

The two contradictory arguments put forward are as follows: First, some authors, such as Savignac (2008), argue that financial constraints negatively affect innovation and growth. By implication, this must mean either that the firm sample is composed of highly comparable firms, or that studies use horizontal analysis designs at the level of the individual firm, because only then do financial constraints prove to reduce firms' performance. However, this is done neither in the study by Savignac (2008), nor in other studies employing direct measures (e.g., Canepa and Stoneman (2008); Westhead and Storey (1997)). One may thus contradict this view by arguing that innovative growth firms are more financially constrained because they need larger amounts of capital to finance comparatively larger investment amounts. The logic behind the second argument is as follows: Innovative firms possess less tangible assets and, hence, a greater degree of information opacity. Innovative firms are therefore more heavily constrained than non-innovative firms, implying a positive correlation between innovation and constraints. This argument seems particularly appealing as firms need to have an innovation concept before they can assess capital demand and apply for it. It is thus the innovation that "comes first". With respect to growth, the direction of causality likewise switches, but the correlation seems to have the same sign as in the first line of argumentation (section 3.3.). The reasoning here is that growing firms are more attractive to investors and have increasing internally generated funds. Both reduce financial constraints, implying a negative relationship between growth and constraints. A similar problem occurs with respect to capital instruments. Research into innovation and firm evolution investigates the impact of certain financing instruments on R&D, innovation and growth, etc., while finance research considers the impact of these presumably dependent variables on the capital structure. As in the constraints case, the correlation results are again inconsistent. For example, innovation research postulates a positive relationship between debt and incremental innovation and R&D, while finance research finds a generally negative association between innovation activity and debt. The problem of causality is not easy to solve and should be kept in mind. Indeed, there is no statistical method that can ultimately prove the direction of causality. (e.g., Schendera (2008)) The actual causal relationship therefore remains the subject of discussion and uncertainty.

4.3   Formulating an integrated framework

To overcome the problems cited above, the dissertation now takes a step back and focuses on capital endowment resulting from firms' capital demand and supply respectively. As both are dependent on firm characteristics, the framework utilizes an evolutionary perspective represented by the organizational and financial lifecycle. It integrates the synthesized findings from the previous chapters by identifying and combining theoretical elements with empirical validity. In cases where empirical evidence is contradictory, it relies on theoretical propositions; and in cases where empirical knowledge surpasses theoretical models, it relies on empirical findings. The fundamental idea of the framework draws on the arguments set forth in previous chapters regarding the one-firm and one-or-more-technology world. However, the propositions made refer to all greentech firms, which are segmented into cohorts in line with their respective state of development. The work that follows is structured into three main sections: First, it analyzes innovation, growth and induced capital demand. Second, it analyzes innovation, growth and capital supply. Third, it merges capital demand

and supply to assess potential constraints. The section ends with a brief summary and conclusion.

4.3.1   Innovation, growth and capital demand

To start with, consider the growth cycle model derived in section 3.3.4. Firm development can be broken down into four states of development: (1) formation and start-up, where firms begin to implement a certain concept and start operations; (2) early growth, where firms start commercializing their products on a larger scale to a broader customer group; (3) expansion, where firms expand into various markets and segments; and (4) stability, where firms have reached maturity and need to find new ways to revitalize their core business activities. In essence, the model takes a firm-level perspective without assuming a deterministic development path. Firms in the start-up state share certain demographic characteristics such as size, age, innovation and growth potential, but do not necessarily become early growers. From innovation literature, we know that innovations are inventions that are marketed successfully. Innovations can be disruptive in their effect and may thus challenge and eventually change a whole market space. These innovations are often non-incremental and therefore do not depend on the established routines and historic successes of the developing entity. Indeed, non-incremental innovations are often developed by new firms that have just entered the market space, such as the electric vehicles produced by Tesla Corporation. It seems, however, that incremental innovations are a lot more common. They depend on firms' knowledge resources and capabilities. Unlike in the case of disruptive innovations, it appears that firm size and age are positive strengths in the development of incremental innovations. Accordingly, and as was demonstrated in section 3.2.2.2, empirical studies mostly find a positive correlation between firm size and innovation. (e.g., Huergo and Jaumandreu (2004); Love and Ashcroft (1999)) One problem in empirical literature is, however, that it does not sufficiently account for the effect of an innovation. For instance, the effect of one disruptive technology can outweigh the effect of hundreds of incremental improvements to an existing technology, because disruptiveness is inherently difficult to assess and measure. Theory must therefore lead the argument, implying that firm size and age have a negative effect on disruptive innovation output due to a path dependency bias among established firms. Firms in the start-up and formation phase therefore share significant potential to generate non-incremental innovations with disruptive effects, establishing a sort of disruptive zone within the firm's innovation and growth cycle. The overall innovation potential in terms of output is greatest for expanding firms. Younger firms, although demonstrating greater relative effort in terms of higher R&D intensity, generate less innovation output than older and larger firms. It is only in the late maturity state of firm development that this potential decreases and firms suffer from substantial inertia. Although consistent with theory, the propositions put forward rather contradict the majority of empirical studies, suggesting an often linear and generally monotonic relationship between size, age and innovation. (e.g., Galende and de La Fuente (2003); Soerensen and Stuart (2000)) However, they are in line with the findings of Acs and Audretsch (1988), who suggest a non-monotonic relationship for high-tech industries.

Innovation activity is assumed to be a major driver of growth. Innovative firms thus grow faster in any state. This is in line with empirical research that studies growth as a dependent variable. Growth rates differ in different development states. Adopting the typical S-curve of technology development and picking up the concepts of the one-firm, one-technology case (Figure 3-12) implies that start-up firms grow fast, but not as fast as early growth firms, which have already passed a turning point at which the S-curve reveals a convex shape. In later states, growth rates decrease, accounting for the concave S-shape after the second turning point. The propositions made reflect not only lifecycle theory, but also most empirical work, suggesting a negative correlation between firm age and growth. (e.g., Bahadir, Bharadwaj, and Parzen (2009); Pasanen (2007)) The non-monotonic nature proposed herein is partially supported by the fact that most studies find diverging results for firm age and size with respect to growth Bahadir, Bharadwaj, and Parzen (2009), although it is undeniable that some correlation does exist between age and size. For this reason, the model assumes that older firms possess higher levels of sales then younger firms. The S-shaped relationship is represented by the Y graphs combined with the growth potential indicated in Figure 4-2. The graphs are linearized to reduce complexity in the subsequent indicative analysis. This simplification is, of course, not entirely realistic and should in particular be reflected when looking at mature firms. For example, these firms may, contrary to the model, be very small, aiming primarily to generate income for the owner and harboring no ambition to innovate or grow in the future. However, the overall target is not to derive mathematically correct and detailed formulas, but to combine findings from three distinct bodies of literature.

As can be seen in Figure 4-2, firms have varying demands for external capital in different states of development. Following on from the findings in section 3.4 and in order to understand firms' overall capital demand, a distinction must be drawn between two kinds of funds: internally generated funds and funds provided by external sources. An appropriate measure to account for internally generated funds is a firm's cash flow, denoted as CF. The cash flow is assumed to be dependent on a firm's lifecycle state, increasing over time. The modeled relationship is consistent with empirical research on capital structures, indicating a positive relationship between retained earnings and firm size Bhaird (2010) and between retained earnings and firm age Chittenden, Hall, and Hutchinson (1996). Demand for and the use of external capital can thus be seen as an inverse function of a firm's cash flow. In other words, the higher a firm's cash flow, the lower its dependence on external funding. The implications are straightforward: Younger and smaller firms are more dependent on external financing sources (including funds provided by founders and management), while older and larger firms can draw on their larger resource base and higher operating cash flow to realize optimal investment programs. The overall external capital demand function is denoted as $EXTC_D$ and represents the relative capital demand of firms in each state. A start-up firm, for example, has relatively higher external capital needs than an established firm with steady cash flows in the expansion state.

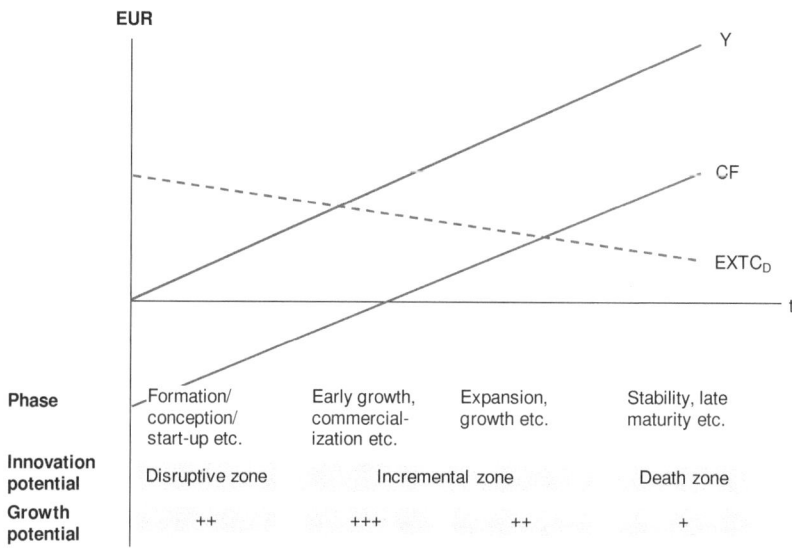

Phase	Formation/ conception/ start-up etc.	Early growth, commercial- ization etc.	Expansion, growth etc.	Stability, late maturity etc.
Innovation potential	Disruptive zone	Incremental zone		Death zone
Growth potential	++	+++	++	+

Figure 4-2: The corporate innovation and growth cycle

4.3.2   Innovation, growth and capital supply

The next step is to consider the availability or supply of funds. The proposed model draws on the financial considerations made by Berger and Udell (1998) using the modifications summarized in section 3.4.4.2. It also integrates the policy actions described in section 3.5.2. Essentially, it uses the relative availability of various financial instruments to firms (deduced from research) with respect to their current lifecycle states. The supply of funds depends on a firm's level of information opacity as well as on the type of financing instrument. There are three basic clusters, as summarized in Figure 4-3: (1) equity capital instruments, denoted as EQT_S; (2) debt capital instruments, denoted as D_S; and (3) government support programs, denoted as GOV_S. The slope of each curve is indicative and is approximated by the findings presented in the previous chapters. The segmentation of instruments is borrowed from Berger and Udell (1998), with the addition of government support programs. At the very beginning, greentech start-up firms are heavily dependent on initial funds provided as equity by the owner and on government support programs. In fact, initial funds make up the largest share of total financing and are negatively correlated to firm age and size. (e.g., Romano, Tanewski, and Smyrnios (2001); Gregory et al. (2005)) The implied limitation imposed by owner-related funds is reduced by the German government. The review of greentech-related funds conducted in section 3.5 implies that most of the funds are allocated to early states of firm development and, in particular, to R&D activities.

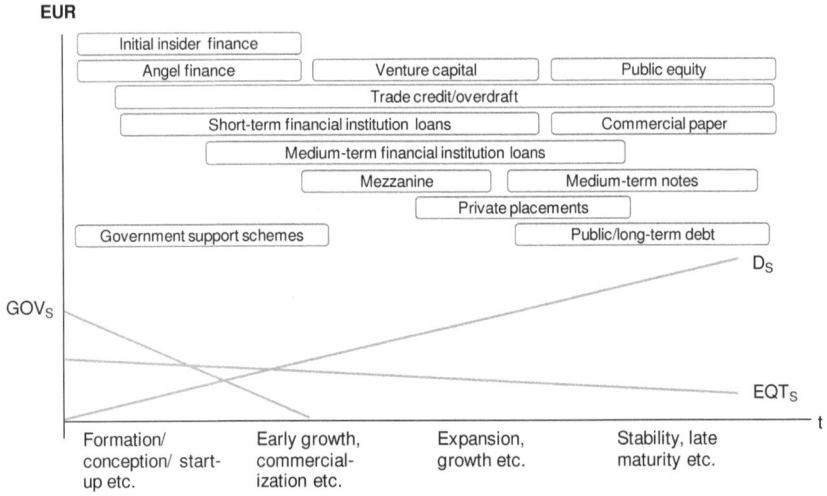

Figure 4-3: The financial growth cycle and capital supply

The curve thus reaches the x-axis when firms have already started commercializing their products. Some programs naturally focus on older and larger firms. Considering the dominance of seed programs and the vast number of firms particularly in the maturity phase, however, it appears that government support programs only show marginal effects in later development states. Additional equity is provided by business angels and, occasionally, by venture capitalists. However, availability remains low, as these forms of finance are particularly scarce in the greentech industry (Kenney (2009) and in Germany's debt capital-oriented financial markets (Achleitner and Tchouvakhina (2006). The comparably low and negative inclination of the equity curve represents the path dependency for external equity-related instruments. For instance, business angel firms often receive venture capital and venture capital firms often go public. (Bozkaya and de van Potterie (2008) The relative availability of equity capital therefore remains relatively constant, whereas funds provided by the owner and family are only sufficient in the early states of firm development. When a firm establishes a track record and grows in size, it can also obtain access to debt capital. This is generally rationed by banks due to the risks associated with information asymmetries. Firm development also correlates directly and negatively to debt maturity. Younger firms thus use short-term debt; middle-aged and medium-sized firms use short- and medium-term debt; and older and larger firms use long-term debt. Access to debt capital increases and the contractual terms offered by banks improve over time. This is consistent with empirical research, which postulates a positive relationship between debt capital and firm size or age (e.g., Börner and Grichnik (2010); Hall (2002)), as well as a negative relationship between short-term debt and firm size or age (e.g., Chittenden, Hall, and Hutchinson (1996); Hall (2002)). However, since the focus of this dissertation is on assessing the impact of financing on greentech innovation and growth, the use of certain financial instruments is only of secondary interest. In this context, and bearing in mind that capital endowment is a major driver of innovation and growth, the following analysis centers on the availability of funds to demanding firms or, to couch this concept in negative terms, the existence of financial constraints that simply result from merging capital demand and supply functions.

4.3.3 Merging capital demand and supply

A firm's development path is driven but also hampered by its financial resources (see sections 3.2 and 3.3). This argument is based on the fact that innovation and growth require funds, and that innovative growth firms will do everything possible to access these funds. Instead of considering the direct effect of financial constraints, which suffers from the causality and endogeneity bias as demonstrated earlier, the dissertation introduces a third variable, "capital endowment", which accounts for firms' capital demand on the one hand, but also for the use and supply of capital on the other hand. This directly refers to solid empirical evidence, suggesting a positive relationship between financial resources, innovation and growth. (e.g., Correa, Fernandes, and Uregian (2010); Lee (2010)) Capital endowment is the sum of internal and external funds. As demonstrated, a firm's demand for external capital depends on its ability to generate funds internally. Since most firms do not possess sufficient funds to pursue their desired investment programs, they need to acquire external capital. Obviously the existence of financial constraints is a closely related issue that occurs when the amount of capital raised falls short of the firm's requirements. Financial constraints can raise substantial barriers to firms' development. From a corporate finance perspective, they reduce the value of firms by preventing them from realizing optimal investment programs. Since internal capital is constituted by the firm's ability to generate cash, it follows that constraints are imposed in particular by the (non)-availability of external capital. The overall relative availability of capital is an accumulation of the three capital sources equity, debt and government funds. The resulting total external capital availability, denoted as TC_s, can now be matched to total relative capital requirements in each phase ($EXTC_D$). This is done in Figure 4-4, where TC_s represents the sum of GOV_s, EQT_s and D_s.

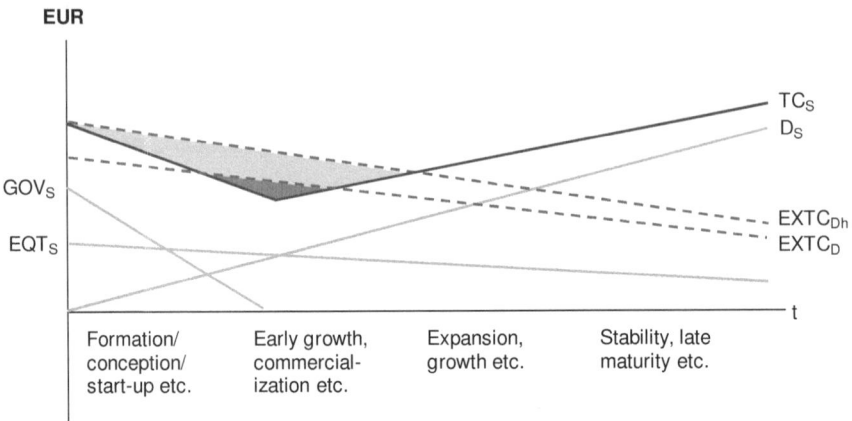

Figure 4-4: Capital demand, supply and the existence of financial constraints

Clearly, TC_S is the limiting factor for $EXTC_D$ and, hence, for firm development, whereas $EXTC_D$ depends on both CF and on the firm's investment program. The firm's actual capital endowment is then simply the minimum of either TC_S or $EXTC_D$, with financial constraints occurring whenever $EXTC_D$ exceeds TC_S. By implication, and as demonstrated in the shaded areas in Figure 4-4, firms with a higher demand for capital ($EXTC_{Dh}$) face a higher level of financial constraints, *ceteris paribus*. The relationship may be biased, however, as firms also need the desire to grow. The dissertation thus argues that (relative) capital endowment, i.e. the relative volume and number of financial instruments used by the firm, positively affects innovation and growth. In this context, $EXTC_D$ and TC_S are the components that establish a link to innovation, growth and financial economics, where capital endowment is assumed to have a positive impact not only on firms' performance but also on financial constraints. At the same time, the proposed model does not directly assume that financial constraints will affect firms' performance. The separation of constraints, innovation and growth circumvents the problems mentioned above. It also admits a financial economics perspective (see section 3.4.4.4.) and a firm-level perspective, suggesting a positive impact on innovation activity and a negative relationship between growth and financial constraints. The two effects are demonstrated in Figures 4-5a and 4-5b, respectively. Innovative firms have less tangible assets, and a large proportion of their valuation depends on rather uncertain future cash flows. The principal-agent and moral hazard problems are extraordinarily significant for the reasons identified in section 3.4. The effect of innovation activity will therefore be strongest for debt capital. Since banks would rather ration supply than raise interest rates, the relative amount of debt capital supplied ($D_{S'}$) will be lower. As a result, TC_S will be smaller for the relevant development states, as indicated by $TC_{S'}$, implying greater financial constraints for early-state firms in particular. Growth firms, on the other hand, will have better access to capital due to their proven track record and liquidity situation. (Ploetscher and Rottmann (2002)) Moreover, growth reduces dependence on external financing over time. This leads to a relatively higher level of capital supply (TCs') and a relatively lower dependence on external funding sources (EXTC$_{D'}$).

Even more important than validating the existence of financial constraints seems to be the point at which they occur and the definition of steps to overcome them. One fundamental assumption of the proposed theoretical model is that government support systems effectively alleviate financial constraints for new firms entering the market space. As was shown for the case of greentech, it appears that the majority of push programs aim at financing R&D and prototyping activities because the implicit risk is too high to attract commercial investors. Accordingly, the model postulates that every firm entering the greentech market with a positive-NPV business model qualifies to receive financing, because existing market imperfections are canceled out completely by government intervention. The argument implies that EXTC$_D$=TC$_S$, as shown in the figures. Push programs should therefore especially support innovation activities and the development of early-state firms. Government effort decreases constantly over the firms' lifecycle, assuming that commercial instruments will take their place. However, it seems that the latter do not take effect early enough. As a result, early growth firms in particular experience considerable financial constraints. Moreover, the lack of funds comes at a critical point, where potentially disruptive inventions are being commercialized and firms are growing fastest. It also comes at a point where capital requirements are high due to the need to invest in capital stock and sales channels, and where internal funds are low due to a lack of cash flow and the finite nature of owner-related funds. In this context, Kline and Rosenberg (1986) note that, especially in high-technology industries, *"not only are uncertainties over technology factors particularly great, but the financial commitments are frequently required during precisely that earliest stage when the uncertainties are greatest."* These ideas are consistent with innovation literature, which suggests that financial constraints have a negative impact on innovation (Klassen, Pittman, and Reed (2004); Savignac (2008)). They are also consistent with financial economics literature, which indicates a negative correlation between firm size, age and financial constraints (e.g., Beck and Demirguc-Kunt (2006); Canepa and Stoneman (2008)). Additionally, they underline the belief, encountered in greentech research, that enhanced early-state private capital could have a powerful impact on industry development. (e.g., Bürer (2008); Burtis (2006); Randjelovic, O'Rourke, and Orsato (2003))

a. Innovation

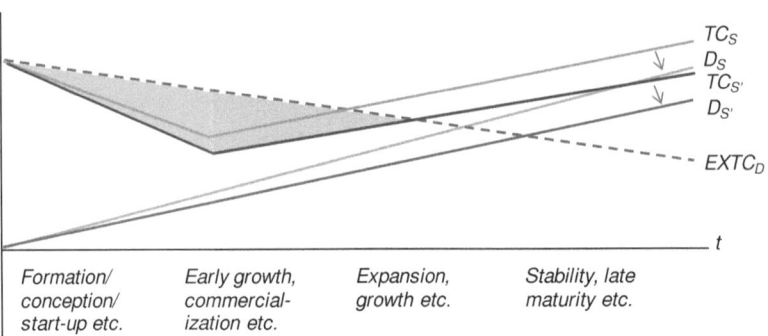

| Formation/ conception/ start-up etc. | Early growth, commercial- ization etc. | Expansion, growth etc. | Stability, late maturity etc. |

b. Growth

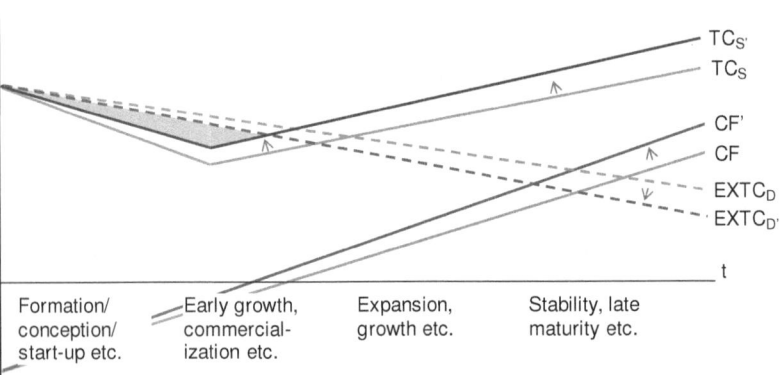

| Formation/ conception/ start-up etc. | Early growth, commercial- ization etc. | Expansion, growth etc. | Stability, late maturity etc. |

c. Policy measures

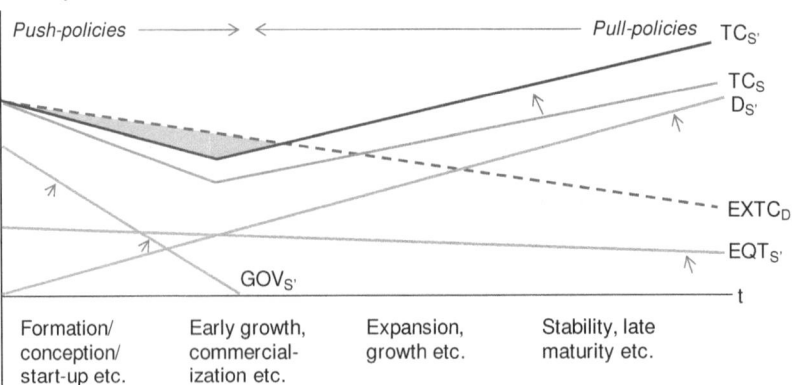

| Formation/ conception/ start-up etc. | Early growth, commercial- ization etc. | Expansion, growth etc. | Stability, late maturity etc. |

d. Collateral, banking relationship and operating efficiency

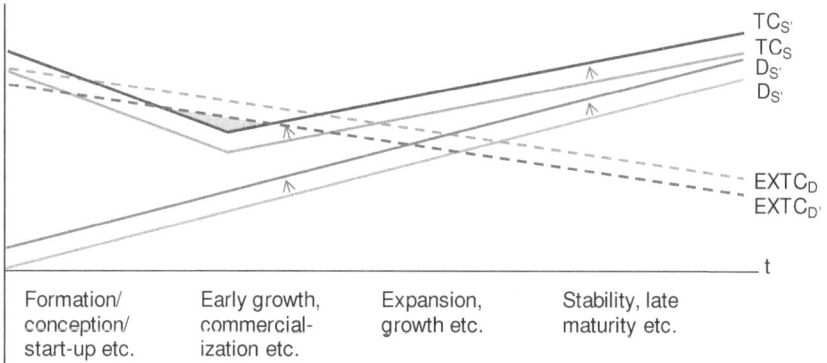

| Formation/ conception/ start-up etc. | Early growth, commercial- ization etc. | Expansion, growth etc. | Stability, late maturity etc. |

Figure 4-5: Effect of innovation activity, growth, policy and firm measures on financial constraints

The effect of government pull programs has, to the knowledge of the author, not been analyzed at the level of the firm. Several studies, summarized in the review of literature on greentech policy, have nevertheless assessed the effectiveness of these programs with respect to technology diffusion. The results suggest that pull programs in general and the German feed-in system in particular reduce investment risk, drive technology diffusion and bring not only government funds but also private funds into the sector (see section 2.2.4). It can be assumed that the same implication holds for greentech firms if support programs are effective. It follows that growth firms in markets affected by pull policies in particular will have better access to private capital than their non-affected counterparts. In other words, greentech pull programs implemented in Germany should have a positive effect on investors' mood and, hence, a negative effect on financial constraints. From a theoretical point of view, and as shown in Figure 4-5c, a well-designed government support program can significantly ease a firm's financial constraints. While extending existing push instruments would reduce the financial constraints for young firms, it appears that pull instruments foster investment in later state growth firms.

The factors discussed above are rather exogenous or depend on the characteristics of the firm. The following discussion therefore focuses on measures that firms can actively influence. In financial economics literature, there are three main concepts that reduce the financial constraints faced by firms: (1) banking relationships (e.g., Chakraborty and Hu (2006); Elyasiani and Goldberg (2004); Giudici and Paleari (2000)); (2) the provision of collateral (e.g., Coco (2000); von Nitzsch, Rouette, and Stotz (2005); Tensie Steijvers and Wim Voordeckers (2009)); and (3) the enhancement of operating efficiency (e.g., Müller and Zimmermann (2009); Watson and Wilson (2002)). The first two concepts are discussed together, as they have the same effect in that they ease financial constraints. As elaborated in section 3.4, banking relationships can overcome firms' information asymmetries and information opacity. A trust-based relationship with a bank may allow firms to access larger amounts of debt capital at lower transaction costs. The provision of collateral, directly by the owner or indirectly in the form of the firm's tangible assets, changes the ranking of creditors on the one hand, but may also signal the quality of the firm on the other. Both aspects lead to an upward shift in the debt capital supply curve as shown in Figure 4-5d, reducing the level of constraints faced by the firm. Lastly, high operating efficiency causes internally generated funds to increase. The effect is similar to a decrease in the $EXTC_D$ function, as the use of internal funds reduces dependence on external funds.

4.3.4   Implications and derivation of research hypotheses

The theoretical framework developed above leads directly to the fundamental research hypotheses postulated by this dissertation, summarizing and integrating the findings of chapter 3. As we have seen, firm development, measured in terms of innovation activity (see section 3.2) and growth (see section 3.3), is heavily dependent on the availability of capital. If capital demand exceeds the relative capital supply, firms will face financial constraints and cannot realize their optimal investment programs. Financial constraints thus appear to be a major barrier to firms' development. However, it was shown that the relationship between innovation, growth and financial constraints is complex and potentially biased due to simultaneity and endogeneity issues with the main variables. It was therefore proposed to introduce a third variable, "capital endowment". Capital endowment is observable and reflects a firm's actual financial situation. It results from the lower of either capital demand or supply, neither of which is observable. Aligned with the fact that financial resources are a major and undisputed driver of firms' development, the first two hypotheses can be formulated as follows with respect to firm development:

H1: Capital endowment positively affects firms' innovation activity and output
H2: Capital endowment positively affects firms' growth

As was demonstrated, capital endowment also constitutes the link between innovation, growth and financial constraints (see section 4.3.3.), because financial constraints depend on a firm's capital demand as well as on the relative supply of capital. Essentially, financial constraints occur where a firm's demand exceeds supply. Since supply is comparably inflexible in light of firms' characteristics, capital demand in particular drives financial constraints at the level of the firm. The third hypothesis is therefore:

H3: The existence of financial constraints depends positively on a firm's level of capital demand, which in turn constitutes it's capital endowment

Besides introducing the concept of "capital endowment", the dissertation also proposes to adopt a financial economics perspective in which the innovation and growth characteristics of a firm influence its ability to access capital and thus also affect the level of financial constraints (see section 3.4). This perspective contradicts the view, encountered in innovation and growth literature, that financial constraints negatively affect innovation activity and growth. However, since this view is incorporated through the introduction of capital endowment, the contradiction can be neglected. The fourth hypothesis is twofold:

H4a: Innovation activity increases a firm's information opacity and thus further rations the supply of debt capital in particular, leading to a higher level of financial constraints
H4b: Firm growth increases cash flow, thereby decreasing both $EXTC_D$ and investor risk. Both circumstances reduce the degree of financial constraints

Financial constraints depend on various factors. However, even more important than the question of their existence is the questions of how to diminish them. The theoretical framework identifies greentech policy instruments and firm-specific actions as potential antagonizing factors (see sections 3.4 and 3.5). This leads to the following two hypotheses:

H5: Government support programs. i.e. push and pull instruments, significantly reduce firms' financial constraints
H6: Firms may improve their financing situation and reduce financial constraints by...
   H6a: ...building banking relationships
   H6b: ...providing internal or external collateral
   H6c: ...enhancing operative efficiency and thus retained earnings

The above research hypotheses reflect the main findings of the proposed theoretical framework and thereby integrate outcomes from three distinct bodies of literature. All hypotheses are thought of as marginal effects, where all other factors are held constant. Firm demographics and other determinants of innovation, growth and financial constraints are therefore implicitly considered.

4.4   Conclusion – Proposition of an integrated framework

The theoretical framework developed above seeks to integrate findings from three different fields, namely innovation, growth and financial economics. This is a singularly important undertaking, as many authors have pointed out that financial constraints constitute major impediments to innovation and growth at small firms in particular. (e.g., Storey (1996); Gompers and Lerner (2001); Gilbert, McDougall, and Audretsch (2006)) Moreover, the approach is fairly new, considering that research into the role of finance with respect to innovation and growth is scant. (e.g., Müller and Zimmermann (2009); O' Sullivan (2007); Souitaris (2003)) The underlying argumentation was developed based on considerations drawn from assessing the situation of one firm with just one technology. The findings were then structured across the development cycle of the firm. This also provided fertile soil for the integration of results from the three bodies of literature. Furthermore, the breakdown of constraints into capital demand and supply helped to develop a consistent framework that overcomes the persistent contradictions in literature and circumvents the problems of endogeneity and simultaneity of variables in scope. To account for the main shortcomings of lifecycle models, i.e. their limited predictive value and the heterogeneity of existing models, the fundamental assumptions were relaxed.

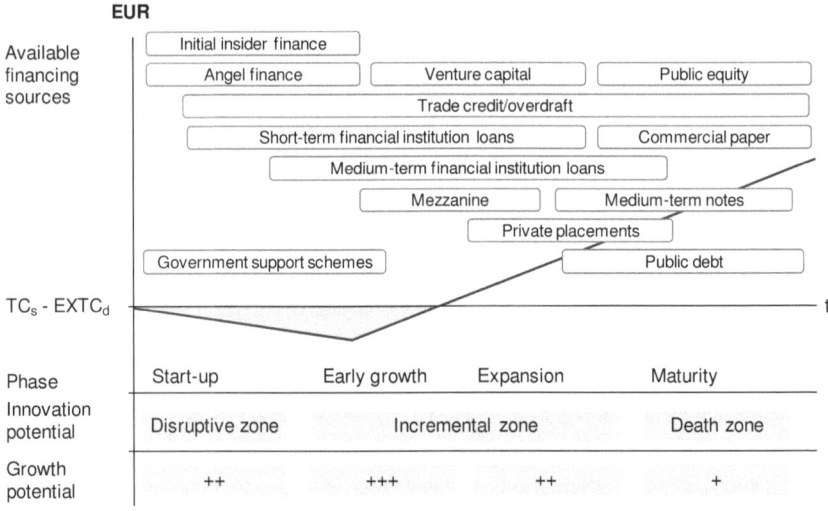

Figure 4-6: The corporate financial innovation and growth cycle – An integrated framework

Accordingly, the overall framework does not claim to have predictive value or to fit all firms. However, it serves as a tool to assess the financing situation of greentech firms as a function of their current development state and to derive implications for innovation and growth. The integrated framework is summarized in Figure 4-6. The core assumptions of the framework are derived from the specified fields of research, are structured in line with firms' state of development and are consistent with contemporary empirical findings. The framework argues that start-up firms are likely to engage in non-incremental innovation that may have a disruptive impact on the environment. Their growth potential is significant, although not as significant as that of firms in the early growth state. To implement their strategies, start-up firms are heavily dependent on the initial funds provided by "insiders", such as founders, friends and family. Due to the high degree of information opacity, these firms have limited access to external funds. External sources that are actually available are less affected by principal-agent and moral hazard problems due to the alignment of objectives (business angel capital), reduction of maturity times (short-term loans), or simply due to their dependence on firms' primary business activity (trade credit). Start-up firms are a major target for government support programs; and it appears that most greentech-related support programs focus on firms in this state. Early growth firms already possess a commercial product. The primary focus is therefore to sell that product to a broader customer base. The relative growth potential is highest in this state. However, in order to grow, firms must invest in infrastructure, resulting in very high capital needs. Increasing demand implies that these firms are likely to be financially constrained. At the same time, innovation potential becomes incremental as innovation activity primarily focuses on improving existing products. Due to the persistence of a potentially successful business model and proven track record, these firms have broader access to capital than start-up firms. Mature firms can increase their debt capital, and both VC and mezzanine capital become valid means of financing. Expanding firms are most likely to be large, old and thus transparent. These firms can access the entire range of financing instruments. At the same time, dependence on external finance is lowest as they can draw on healthy cash flows. The degree of financial constraints is lowest.[15] Expanding firms have the highest innovation output, although most of it is of an incremental nature. Mature firms still have access to a broad range of financing instruments. However, they are threatened by younger competitors that could render their products obsolete. Additionally, these firms suffer from substantial inertia, leading to low innovation potential and low relative growth. Firms that do not reinvent their ways of doing business may even exit the market.

Finally, six main research hypotheses have been derived to further empirically analyze greentech innovation and diffusion from a financial economics and firm-level perspective.

---

[15] The linearized model is misleading here as it suggests increasing returns for mature firms. The actual relationship may be represented by the S-curve, as discussed earlier.

First, capital supply and, hence, endowment are major drivers of firm innovation. Second, capital endowment is a major driver of firm growth. Third, the demand for capital and thus endowment correlates positively to financial constraints. Fourth, innovative firms experience greater financial constraints, while growth firms are less constrained than their counterparts. Fifth, policy intervention can effectively alleviate persisting financial constraints in the greentech market. Finally, firms can improve their financial situation by establishing banking relationships, providing collateral and improving operating efficiency. These hypotheses are tested using empirical analysis, which is the subject of the next chapter.

5   EMPIRICAL ANALYSIS

5.1   Research methodology

5.1.1   Overall empirical strategy and approach

The empirical strategy aims to test implications derived from the theoretical framework in order to assess the impact of financing on innovation and growth in the greentech sector. The empirical analysis is composed of two main parts. The first part analyzes the main characteristics of firm cohorts in each state of development, i.e. start-up, early growth, expansion and maturity with. This part focuses especially on innovation, growth and financing at firm level. The methodology is primarily based on non-parametric testing methods for independent samples and descriptive analysis. The second part seeks to translate the theoretical framework into a testable empirical model. Essentially, capital endowment is regressed toward innovation, growth and financial constraints while taking into account relevant control factors derived from literature. Analysis uses two logit models and one OLS model to test the research hypotheses. The overall chapter is structured as follows: First, it outlines the sampling approach and the underlying database. Second, it shows how the data was validated. Third, it defines all relevant variables and describes how they are measured. Fourth, it presents the methodology for and results of the comparative analysis. Finally, it provides an overview of the empirical models, presents the findings, assesses validity and reliability, and ends with a brief conclusion. The combined assessment, discussion of findings and implications for theory and practice are the subject of the chapter 6.

5.1.2   Sample and response

The greentech market in Germany is dominated by small and medium-sized firms. Most of these firms are either privately held or limited corporations. Only very few are listed and thus required to publish their data. For example, a selection of German greentech (alternative energy, alternative fuel, water, waste and disposal) firms in the Thomson ONE Banker database[16] yields 45 firms, of which 11 also have other activities. In light of the rather specific research questions proposed by this dissertation, collecting data from public databases such as Thomson does not appear very promising. The dissertation thus uses a questionnaire-based data collection approach. The empirical sample is drawn from three major databases: (1) the IHK-UMFIS database (IHK (2009)); (2) the BMU GreenTech made in Germany database (Büchele et al. (2007)); and (3) the BMU GreenTech made in Germany 2.0 database (Büchele

---

[16] http://banker.thomsonib.com/, accessed September, 2010

et al. (2009)). The UMFIS database (1) is the largest database for greentech and environmental firms in Germany. It is compiled by the German Chamber of Commerce and Industry. The overall database consists of more than 10,000 in the environmental sector. The database used in this dissertation reflects a status quo of November 2009 and comprised 10,467 firms with various informations for each firm (address, phone, main fields of activity etc.). In total of 8,038 firms, so these firms were included in the constructed data set. The first BMU database was created in 2007 and comprises 1,457 firms. The information for each firm is a lot more detailed, including information about the management and several firm characteristics. The second BMU database includes 1,241 firms and features comparable content to (2). Additionally, the data set was enriched by firms that were not included in the database, but were known to the author through his work in the greentech industry over the past five years. The three databases were merged using various procedures provided by Microsoft Excel and subsequently checked for redundancies and incorrect entries. Most of the information provided by (2) and (3) could be integrated into (1), providing a more detailed set of information. The combined raw data set was composed of 8,023 firms. After checking for firm characteristics and excluding firms that did not match the definition provided in section 2.1, the data set was reduced to 7,568. Most of the firms excluded operate in other areas (such as car dealers, painters, etc.) and, for some unknown reason, were listed in (1). After sending out the questionnaire, another 1,810 entries were deleted due to unknown recipients and sending errors. This relatively large number of unknown recipients reflects the dynamism of the greentech industry on the one hand, but also the fact that (1) and (2) in particular also contained older data. The final data set thus contained 5,770 firms, of which 2,327 firms included direct management contact information. In a recent study, albeit with a slightly different orientation, the German bureau of statistics identified 5,808 firms in the environmental sector in Germany (Destatis (2009)). In light of this fact, it is fair to claim that the constructed data set can be considered as a total population survey basis.

The questionnaire was sent out on November 12, 2010. A final reminder was sent out on December 3, 2010. The e-mail was programmed in HTML to ensure good readability on all e-mail programs and systems. The mail explained the background to the study and provided personal contact information for both the author and the Ph.D. supervisor. Every mail also contained a personalized but anonymous link to the online questionnaire, allowing participants to interrupt and resume completion at any time. The survey was closed on December 17, 2010. A summary of the results was later sent to all participants. Overall, there were 901 visitors to the site, of whom 536 completed the questionnaire. The final response rate was thus 9.2%, which is a satisfactory result for an online survey. (Diekmann (2010)) One of the major problems, and also an explanation for the relatively high break-off statistic, is that more than half of the firms were addressed generically without personal contacts, although the cover letter explicitly requested that surveys should be filled out by the CFO or management. Some respondents answered that they could not participate in the survey because they did not work in finance. Although they were asked to forward the message to an appropriate colleague, it seems likely that this had a negative impact on the overall response rate. The average time taken to complete the questionnaire was around 16 minutes, which is consistent with the estimate of 20 minutes referred to in the cover letter. It is interesting to note that more than 50% of all participants responded on the day they received the questionnaire. Nearly 90% of the completed questionnaires were filled out within five days of receipt of the e-mail. The data thus received is unique. It is the first set of empirical evidence relating to innovation, growth and finance in the German greentech industry. The relatively high response rate appears indicative of the level of interest in and the topical nature of the subject.

5.1.3 Questionnaire development

The analysis uses a tailored design method (Dillman (2000)) to account for the wide range of questions and the specifics of the greentech industry. The overall questionnaire is reproduced in appendix (B). It is composed of six main sections: (1) firm strategy and core activities, (2) innovation and investment, (3) financing, (4) government support, (5) firm characteristics and (6) contact, wrap-up and expression of gratitude. The questionnaire is introduced with a description of the study, it's purposes and the author's contact details. The strategy for the questionnaire is reflected by the order of the sections. In the first section, respondents begin with relatively simple questions, such as the firm's main technologies and markets. Following recommendations made by Diekmann (2010), the questions gradually become increasingly challenging and detailed. The most sensitive and detailed information is requested in the very final section (5), where respondents are asked to reveal information about firm size, growth and profit. Most questions are programmed as compulsory exercises. Only highly sensitive requests, e.g. for profit information, are optional. The majority of questions are asked in a closed manner and multiple choice design to account for the confirmative nature of the dissertation. Only one question is asked as an open question. Most items are coded as six-point Likert scales. The dissertation uses a six-point range to force respondents to make decisions and to achieve greater variability of results. The number of alternatives is also chosen to be even, as the author wanted to avoid uncertain outcomes. The proposed procedure seems justifiable, considering that literature takes no clear line on even and uneven Likert scale measures (Matell and Jacoby (1972)). The overall target of the questionnaire is to collect information about dependent variables (innovation, R&D, growth and financial constraints) and independent variables (e.g., collateral, size, bank relationship). The variables used are described in the section below.

5.1.4   Measures and variables

The dissertation uses a large number of variables. Table 5-1 provides an overview, describes each variable and defines how it is measured. There are two main segments: the main variables of interest and the control variables. The main variables of interest are subsumed under the following categories: (1) innovation, (2) growth, (3) financial constraints and (4) financing and capital structure. The control variables (5) are not further categorized. The section below describes each variable and explains how it was chosen and measured.

5.1.4.1   Main variables of interest

5.1.4.1.1   Innovation

The original set of innovation variables was composed of six distinct measures, taking into account the different types of innovation (incremental, non-incremental, product and process). This set was then reduced to three main measures due to measurement overlaps and implementation-related problems that are discussed below. The first key variable *INNOVATION* is an aggregate score for a firm's innovation output. The innovation measurement draws on the definition used in section 3.1.1.1. The variable has a dichotomous definition, with 1 denoting that a firm is innovative and 0 denoting that a firm is not innovative. An innovative firm must have brought at least one new or significantly improved product/process to the market in the past three years and is currently operating innovation-related projects. The variable was implemented by a six-item block in the questionnaire. The items were borrowed from the study by Vaona and Pianta (2008) and aggregated to a single innovation variable, as previously done in comparable manner by Savignac (2008). The second variable *PCINNOVATION* has a similar structure. However, it exclusively refers to process-related innovations. The third variable in the given context is *R&D* and denotes research and development spending expressed as percentage of sales in 2009. This measure is also called R&D intensity Bah and Dumontier (2001) and has previously been used as a standard measure in many innovation-related studies. (e.g., Del Monte and Papagni (2003); González and Pazó (2008)) Since R&D is a more input-oriented measure, it plays an exceptionally important role as an independent variable, e.g. as a determinant of financial constraints. R&D expenditures thereby takes into account both formal and informal expenditures to avoid the non-reporting bias as pointed out by Gil (2010), particularly for smaller firms.

Variable	Variable description	Measurement
INNOVATION	Company introduced a new or improved product over the last three years and has running innovation-oriented projects	1, if firm is innovative; 0, if otherwise
PCINNOVATION	Company introduced a new process over the last three years and has running process innovation-oriented projects	1, if firm is innovative; 0, if otherwise
R&D	Research and development expenditures ratio (formal and informal) in 2009	In %, measured as R&D/sales
GROWTH	Growth of sales p.a.; average between 2008-2010	In %
GROWTH'	Quantile rank of GROWTH	In %, measured as percentage of firms with lower GROWTH
GROWTHO	Growth orientation	1, if expected growth exceeds past growth; 0, if otherwise
INVESTCONSTRAINT	Aggregation of five single measures for financial constraints concerning economic risk, general capital shoartage, high cost, insufficient time and resources and insufficient support programs	Average Likert-scale with a range of 1 to 6
INVESTCONSTRAINT'	Average INVESTCONSTRAINT > 4	1, if firm is innovative; 0, if otherwise
TRADECREDIT	Usage of trade credits (corresponds to denied trade discounts, not part of financial basket)	Six point Likert scale, if firm utilized trade credits; 0, if otherwise
TC	Capital endowment as cumulated measure of all capital instruments used, including their respective intensity	Calculated as the average use of all instruments with a range of 1 to 6
FRIENDSEQT	Equity capital of friends and family	Measured as % of financing basket, calculated from six point Likert-scale (0 if instrument is not used and a maximum of 1)
SHEQT	Shareholder equity	
PUBEQT	Public equity	
RETAINED	Retained earnings	
BA	Business angel capital	
VC	Venture capital	
GOVEQT	Governement equity, grants, allowances and equity	
OVERDRAFT	Bank overdraft	
CREDITCARD	Credit card debt	
SHD	Shareholder debt	
FRIENDD	Debt capital of friends and family	
STD	Short-term debt	
MTD	Mid-term debt	
LTD	Long-term debt	
GOVD	Subsidized debt	
PUBD	Public debt	
INTEQT	FRIENDSEQT, SHEQT (excl. GOV)	
EXTEQT	BA, VC (excl. GOV)	
BANKD	STD, MTD, LTD (excl. GOV)	
INTD	FRIEND, SHD, CREDIT, OVERDRAFT	
GOVC	GOVEQT, GOVD	
PUBC	PUBEQT, PUBD	
INTCAP	INTEQT, PRIVD, RETAINED	
EXTCAP	VCBA, BANKD, GOVC, PUBC	
GOVGRANTS	Government grants and allowances for greentech firms (in GOVE)	Measured as % of financing basket, calculated from six point Likert-scale in relation to GOVEQT and GOVD
GOVSUBLOANS	Government subsidized loans for greentech firms (in GOVD)	
GOVVC	Governement venture capital for greentech firms (in GOVE)	
GOVHYBRID	Governement hybrid instruments for greentech firms (in GOVE/GOVD)	
GOVSME	General SME oriented support programs (in GOVD)	

Variable	Variable description	Measurement
PULL	Dependence of business model on government pull-measures, such as EEG, KWK	Likert-scale average with a range of 1 to 6
LTD_INTEREST	Net interest rate for long-term debt	ln % p.a.
MSTD_INTEREST	Net interest rate for mid-term debt	
SEED	Firm is in the seed state, where it prepares or begins business operation, seeks capital, sources resources, develops technology etc.	1, if firm is in respective stage 0, if otherwise
START_UP	Firm is in the start-up state, where it runs first operations, acquires first employees, sells locally, begins to generate revenues etc.	
EARLY_GROWTH	Firm is in the early growth state, where it establishes specialized functions, grows rapidly, commercializes innovations to attract a larger market segment, spreads activities regionally etc.	
EXPANSION	Firm is expansion state, where firm grows fast, formalizes organization, optimizes costs and processes, specializes on proved product characteristics etc.	
HOLD	Firm is maturity state, that can be reached at any size and age of the firm, where it maintains proven actitivities, serves mainly returning customers, grows moderately etc.	
PHASEOTHER	state if the above do not fit	
LEGAL	Company is incorporated (GmbH, KG, AG etc.)	1, if firm is incorporated; 0, if otherwise
AGE	Firm age expressed in years since foundation	Number of years
COLLATERAL	Outside collateral provided by the founder, third parties or other sources	1, if firm used collateral; 0, if otherwise
RELATION	Assessment of relationship with banks	Likert-scale with 1=rather bad and 6=very good)
RELATION2	Number of house banks	Ordinal measure with four categories
EMPLOY	Firm size	Number of employees
SALES	Sales category with ragard to net sales in 2009	Ordinal measure with eight categories
PROFIT	Profit margin category with regard to EBT/sales in 2009	Ordinal measure with eight categories
RENEWABLE	Environmentally sound energy technologies	1, if firm works in respective sector; 0, if otherwise
EFFICIENCY	Energy efficiency technologies	
WATER	Sustainable water technologies	
MOBILITY	Sustainable mobility technologies	
RESOURCES	Sustainable resource/new material technologies	
WASTE	Waste management and recycling	
OTHER	Other greentech fields, e.g. natural conservation; emission reduction; noise reduction; coaching and teaching	

Table 5-1: Overview of measures and variables

5.1.4.1.2 Growth

Growth is represented by the most common measure in literature, i.e. the rate of growth of sales. *GROWTH* is measured as average growth per year during the period 2008-2010. Like all variables, growth data too was collected directly from study participants, as in Börner and

Grichnik (2010). The use of sales as the baseline variable avoids potential problems of undercounting (Shepherd and Wiklund (2009)), e.g. where firms have outsourced activities. *GROWTH'* is a quantile rank transformation of *GROWTH* to account for outliers. The detailed definition is discussed later in section 5.3.4. Finally, *GROWTHO* accounts for the firm's willingness, ambition and orientation towards growth as well as its actual opportunities. It is defined as the difference between the expected growth and the past growth rate. If the result is positive, the firm is classified as growth-oriented, i.e. *GROWTHO=1*. The importance of growth options and orientation when researching growth rates has been argued by several studies (e.g., Heyman, Deloof, and Ooghe (2008); Thornhill, Gellatly, and Riding (2004)) and is integrated in the growth model defined below.

5.1.4.1.3 Financial constraints

As was pointed out in section 3.4.4.1, financial constraints have been measured directly or indirectly. Both approaches have comparative advantages and disadvantages. The dissertation uses a modified direct measure adopted from Savignac (2008) that is comparable to the constraints categories reported in Bozkaya and de van Potterie (2008). There are five sub-categories, each of which is presented as a six-point Likert scale. As in the study conducted by Savignac (2008), the second variable *INVESTCONTRAINT'* is transformed into a categorical variable with a cut value of four especially to identify firms subject to relatively high levels of constraints. However, the model formulated later for financial constraints also tries other cut values to assess the sensitivity of the results. Another financial constraints-related measure is trade credit, which corresponds to unused trade discounts. This variable was collected in order to validate *INVESTCONSTRAINT* as proposed by Ploetscher and Rottmann (2002). *TRADECREDIT* is zero if the firm generally accepts trade discounts, and is plotted on a Likert intensity scale if it does otherwise.

5.1.4.1.4 Financing and capital structure

The present study covers all potential internal and external financing sources. The first variable *TC* represents the overall use of all instruments and is thus a proxy for capital endowment, matching capital supply and capital demand. *TC* is measured as the average use across all instruments. The use of single instruments is thereby measured as their relative intensity on a six-point Likert scale. This approach was successfully employed by Börner and Grichnik (2010). In a second step, and for the purposes of comparative analysis, all instruments were transformed into relative shares, where the lowest value corresponded to no use at all. All instruments were considered as belonging to one financial basket. The relative share of each instrument could then be calculated as the relevant Likert value minus one divided by the total sum of Likert values greater than one, i.e. $\frac{(L_{ij}-1)}{\Sigma(L_{ij}-1)}$, where L represents the Likert value for instrument j and firm i. The transformation was, of course, not performed to calculate *TC*, as *TC* also represents the overall intensity. The accumulation of relative values would indeed yield one in all cases. The financing instruments examined are summarized in Table 5-1. With respect to equity, two different sets of instruments were identified. The first is internal equity (*INTEQT*), which is composed of equity provided by friends and family (*FRIENDSEQT*) and the owner (*SHEQT*). The second is external equity (*EXTEQT*), which comprises business angel capital (*BA*), venture capital (*VC*), government equity (*GOVEQT*), related funds and public equity (*PUBEQT*). Debt too is composed of two elements. The first is internal debt (*INTD*), provided by owners (*SHD*), friends and family *(FRIENDD)*, credit cards (*CREDIT*) and overdrafts (*OVERDRAFT*). The second is external debt *(EXTD)*, which is sourced with banks (*STD, MTD, LTD*), capital markets (*PUBD*) and government agents *(GOVD)*. In addition, government-supported funds were further divided into grants and allowances (*GOVGRANTS*), subsidized loans (*SUBLOANS*), government venture capital (*GOVVC*), hybrid instruments (*GOVHYBRID*) and general SME support programs (*GOVSME*). Up to now, only push programs have been considered. To account for demand-pull initiatives, another variable is therefore introduced. *PULL* gauges the extent to which a firm is dependent on certain demand-pull programs, such as the EEG or KWK on a six-point Likert scale. Obviously, *PULL* is not part of the financial basket but is a separate measure. Finally, and driven by practical importance to study participants, the interest rates for mid/short-term (*MSTD_INTEREST*) and long-term (*LTD_INTEREST*) debt were collected. The variable was measured for 2009 and in % p.a.

5.1.4.2   Control variables

States of development are a major component of the theoretical framework. They are the structuring element that links together innovation, growth and financial economics literature. Accordingly, they were integrated into the questionnaire on a self-categorization basis. This approach has been used by several studies (e.g., Kazanjian and Drazin (1989)) and has the advantages of being up to date, possibly being able to integrate customized firm features and, in particular, being useful for large sets of firms. Categorization was aligned with the theoretical framework, while phase descriptions were, with two exceptions, adopted using the taxonomy proposed by Hanks et al. (1993). The first exception is that the questionnaire allowed respondents to choose an additional category called "other". The second exception is that the category "maturity" not only comprised Hank's cluster D, but also the clusters E and F, which comprise older firms of a small size. The underlying idea and argument was summarized earlier in Figure 3-12, which shows that firm development may stagnate in any state leading or leapfrog other states to achieve instant maturity. The questionnaire originally also subdivided the start-up state into seed and start-up states. However, this distinction was not maintained and was ultimately merged due to theoretical considerations and its limited statistical relevance. The states are primarily used as structuring elements for comparative analysis purposes and not so much for statistical modeling. For the latter purpose, eight main variables were derived from literature: First, *LEGAL* denotes whether a firm is incorporated and therefore has limited liability. Second, *AGE* indicates the number of years that the firm has been on the market. Third, *COLLATERAL* is a dummy variable that features sub-categories borrowed from Bhaird (2010). Fourth, *RELATION* and *RELATION2* represent the firm's degree of satisfaction with its banking relationships and the number of house banks it uses. Fifth, *EMPLOY* represents the number of employees and the approximating size of the firm. Sixth, *SALES* is a sales-based measure for size. Seventh, *PROFIT* is implemented as an ordinal variable where each category stands for a certain margin interval. The margin is measured as earnings before taxes divided by sales for the year 2009. Finally, account was taken of potential heterogeneities in the greentech industry by integrating the technology lines and sectors defined earlier by using dummy variables, where one denotes that a firm is active in its specific sector.

5.1.5 Conclusion – A mixed method approach for exhaustive coverage

The proposed empirical approach combines descriptive and comparative research with rigorous statistical modeling. The structure offers several advantages. First, it allows both descriptive models (e.g., Berger and Udell (1998)) and the integrated framework presented in chapter 4 to be tested. Second, descriptive statistics can drill down deep into financial instruments, their sources and the reasons for financial constraints, etc. The results are therefore more intuitive and are thus of considerable practical value. Third, statistical modeling allows comparative observations to be validated. It enriches the results by adding substantial inferential value, validity and reliability. The data collected by means of the questionnaire accounts for the limited availability of public data. However, the firm database used probably represents one of the most comprehensive databases for greentech firms. The implemented measurements are mostly drawn from similar studies to ensure validity and reliability. The Likert scales, although implying less informational depth than cardinal measures, seem most appropriate as they are easy to understand and fill out. The relatively high response rate seems to confirm this conjecture.

5.2 Comparative analysis

5.2.1 General sample characteristics and the non-response bias

The previous section showed that the data set can be considered to be a full population sample. However, one common problem with mail surveys is the non-response bias. The underlying argument is that firms that do not respond to the questionnaire may possess structurally different characteristics and KPIs to those that do respond. The sample may then not properly represent the underlying population. A method to test for the existence of this bias is suggested by Armstrong and Overton (1977). The basic idea here is to compare the first third of respondents with the last third. The latter cluster is assumed to come closest to mimicking the characteristics of non-responding firms. The dissertation follows this approach and tests for distribution differences across the main variables of interest: *INNOVATION, GROWTH', INVESTCONSTRAINT* and *TC*. Since none of these variables is normally distributed, two independent samples and a non-parametric U-test (Bortz, Lienert, and Boehnke (2008)) are employed. None of the variables shows a significant z value. The existence of a non-response bias is therefore rejected, indicating that the sample properly represents the underlying population. The overall sample comprises 536 firms, of which seven are excluded from the comparative and regression analysis due to missing data about the state of development. Five of the excluded firms are in the process of liquidation because of their owners' retirement plans. The average greentech firm in the sample employs 287 people and generates sales of around EUR 26 m. Most firms focus on services (69%) and production (35%). Only a small number of firms regard trade (13%) as their core activity. The representation of technology lines is rather diverse and covers all defined branches. Only 8% of all firms engage in activities that were labeled "other", indicating the appropriateness of the proposed industry definitions. Activities subsumed under this category were either non-core activities in other business fields, such as construction or medical technologies, or were concerned with noise and emission control. Future research may take this into account and add another category. The largest proportion of firms is active in the field of environmentally sound energy technologies, waste management and recycling, and energy efficiency technologies. As was to be expected, more than half of the firms pursue an innovation-oriented strategy whose aim is to differentiate services and technologies that will eventually allow them to charge premium prices. On the other hand, only 11% of the firms are pursuing cost strategies to gain a competitive advantage. Besides the two generic strategies that were originally proposed and extended by Porter and van der Linde (1995), the dissertation utilized a third category to account for income-oriented firms with limited growth and profit ambition. The last type of firm is occasionally described as a sort of employment substitute to provide a living (Storey (1996)). The overall composition of firms and their strategic focus is shown in Figure 5-1.

Interestingly, it seems that greentech firms do not have superior sustainable ambitions. Of the firms studied, only 3% value sustainability objectives more highly than economic objectives. This underlines the idea that greentech *"contrasts to previous definitions in the academic literature on environmental entrepreneurship that center on "environmental entrepreneurs" being identifiable due to their motivation in creating positive environmental impacts and change"* (O'Rourke (2009). A full 10% of greentech firms explicitly state that sustainability is of no relevance to their business strategy. This figure is supported by comments from study participants. Sustainability, although an important driver for greentech market development, is seen as an unstable objective of policymakers whose commitment is not sufficient to justify long-term business decisions. One firm argued that sustainability is simply *"not appreciated enough"* by relevant stakeholders. Another firm even claimed that not all supported technologies are *"actually sustainable but rather have a negative impact on the environment"*. Most importantly, 38 firms criticized the fact that policy is *"unstable"* and not *"sufficiently long-term oriented"*. The resulting business risk seems to lead to a situation where firms do not rely on the value of sustainability as a product attribute. It is therefore speculated that this leads to sustainability assuming only minor importance in firms' strategy development process. Another interesting comment was made by two non-energy firms, which noted critically that the public discussion of sustainability is *"strongly and too much focused on the energy sector"*.

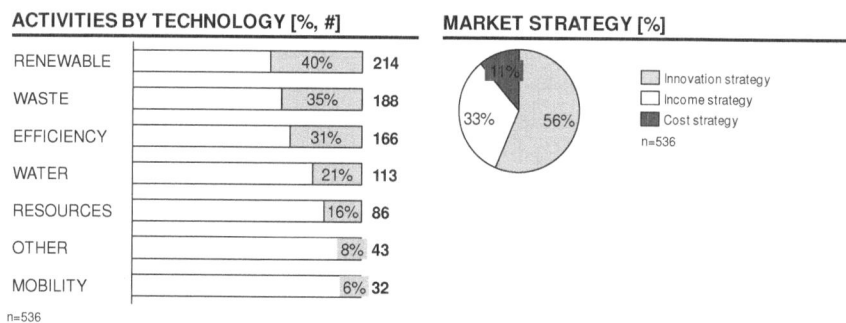

Figure 5-1: Technology lines and strategic focus

The degree of internationality is relatively low. This is consistent with practitioners' observation that the German greentech industry must become more international in order to become more competitive. (e.g., Büchele et al. (2007); Büchele et al. (2009)) However, internationalization is a difficult endeavor and was noted by two firms in particular as a major problem. One firm suggested that potential barriers could be overcome by *"government support towards international expansion"*. The other firms suggested that support is particularly needed for *"international patenting"*. However, 28% of all respondent firms plan to engage in international activities. The most common international activity is exporting (31%), followed by joint ventures and international affiliates. The sustainability orientation and internationality of German greentech firms is summarized in Figure 5-2.

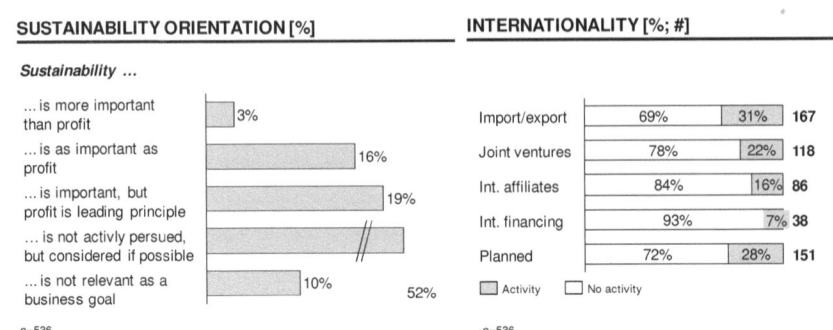

Figure 5-2: Sustainability orientation and internationality

5.2.2 General innovation activity of greentech firms

The large proportion of innovation-oriented strategies is also reflected in the innovation output. Nearly half of all firms have developed and marketed an innovation over the past three years. However, the distinction between non-incremental, incremental, product and process innovation as suggested by literature on the subject Carayannis, Gonzalez, and Wetter (2003) yielded no significant results. Practical implementation is particularly difficult for the first three types. In fact, the average for the first three types is approximately identical; nor was there any significant divergence in the distribution of data. Only process innovation appears to follow a somewhat different pattern and may prove to have different implications. For this reason, the sections that follow focus primarily on the general measure of innovation. While innovation is primarily used as a dependent variable to account for actual innovation output (see section 3.2.2.1), R&D in particular measures the innovation cost or input. Ignoring a pronounced standard deviation, the average R&D cost is nearly 11% of sales – a rather high figure that is above the average found in Büchele et al. (2009). However, the latter study also finds that greentech firms are planning to raise R&D expenditures by 8% p.a. The findings are summarized in Figure 5-3.

The innovation orientation and strategic focus of greentech firms is underpinned by their innovation output. 42% of all respondent firms claim to have brought entirely new products to the market space within the past three years. The questionnaire specifically asked for new, non-incremental products, i.e. not based on existing ones or improved versions. However, the figure of very nearly 43% for incremental innovations raises the suspicion that the distinction between types of innovation may not have been fully understood and might be "a bit too academic" in practice. The distinction between product and process innovation appears more promising, as process innovations were fostered by only 22% of firms. The share of innovative firms is slightly lower than in the study conducted by Vaona and Pianta (2008), from whom the measure was adopted. While the authors report a relative share of 51% with respect to product innovation, the relative share of process innovators is 45%. Greentech firms thus seem to be more product-oriented and less process-oriented.

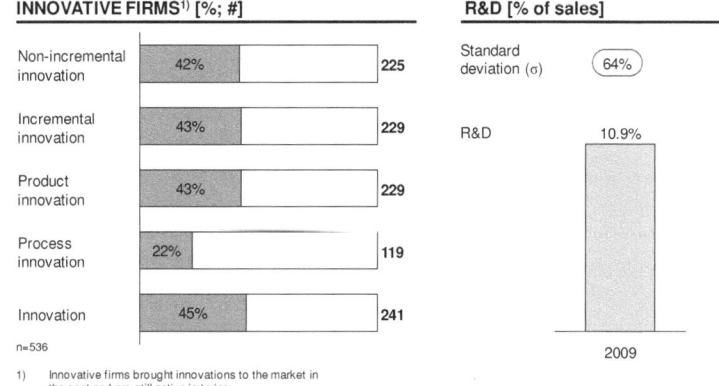

INNOVATIVE FIRMS[1] [%; #]

Non-incremental innovation	42% 225
Incremental innovation	43% 229
Product innovation	43% 229
Process innovation	22% 119
Innovation	45% 241

n=536

R&D [% of sales]

Standard deviation (σ)   64%

R&D   10.9%

2009

1)   Innovative firms brought innovations to the market in
     the past and are still active in trying

Figure 5-3: Innovation output and R&D effort

5.2.3   General growth of greentech firms

Many studies have highlighted the extraordinary growth and dynamism of the greentech industry. (e.g., Wolff (2009); Wüstenhagen and Bilharz (2006); WWF (2009)) Despite the transient impact of the financial crisis Yarow (2009), it seems that this trend has continued after all. As shown in Figure 5-4, the surveyed firms report average sales growth of nearly 15% for 2008-2010. The reported growth figures are likely to be organic rather than acquisitive, since greentech is a fairly new and innovation-driven business populated mostly by small firms Büchele et al. (2009) that are more likely to pursue organic growth paths (e.g., Penrose (1997)). Either way, this distinction was not drawn in the questionnaire. Expected growth is even higher, being estimated at nearly 20% over the next three years. The standard deviation is as high as in the innovation case. Interestingly, albeit consistent with expectations (see section 3.3.3.3), innovation is a strong driver of firm growth. As can be seen in the figure, innovative firms' average growth rate is three times higher than that of non-innovative firms. The difference is significant at the 1% significance level.

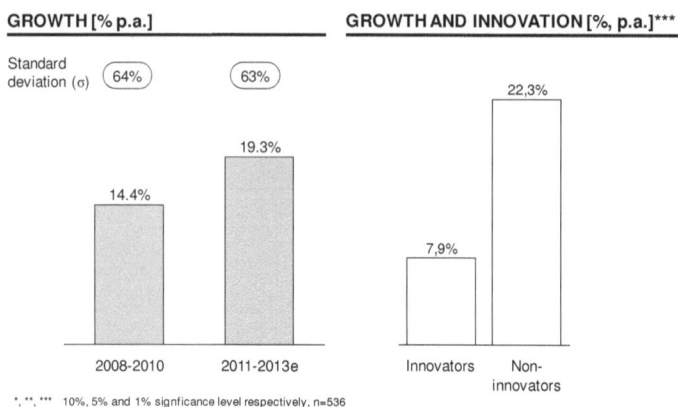

GROWTH [% p.a.]

Standard deviation (σ) 64% 63%

14.4%
19.3%

2008-2010   2011-2013e

GROWTH AND INNOVATION [%, p.a.]***

22,3%
7,9%

Innovators   Non-innovators

*, **, *** 10%, 5% and 1% signficance level respectively, n=536

Figure 5-4: Average sales growth

5.2.4 General financial constraints faced by greentech firms

As discussed above, the present study uses a direct measure of financial constraints. Financially constrained firms thereby claim that, within the past three years, either not all investments or not the full investment amount could be realized due to a lack of funding. The Likert scale measure shows the average of three sub-measures summarized in Figure 5-5.

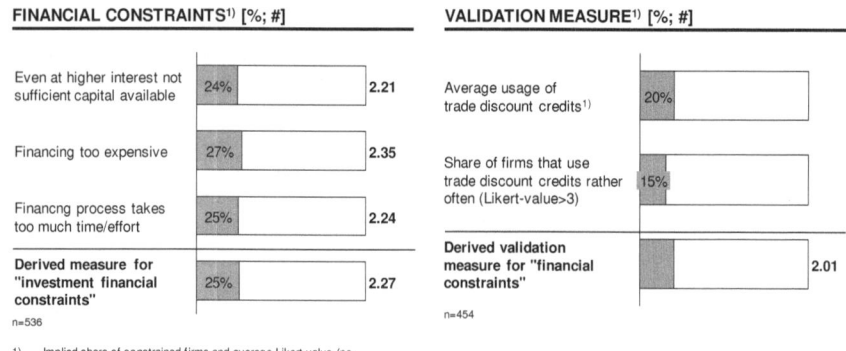

FINANCIAL CONSTRAINTS[1) [%; #]

Even at higher interest not sufficient capital available — 24% — 2.21

Financing too expensive — 27% — 2.35

Financng process takes too much time/effort — 25% — 2.24

Derived measure for "investment financial constraints" — 25% — 2.27

n=536

VALIDATION MEASURE[1) [%; #]

Average usage of trade discount credits[1) — 20%

Share of firms that use trade discount credits rather often (Likert-value>3) — 15%

Derived validation measure for "financial constraints" — 2.01

n=454

1) Implied share of constrained firms and average Likert-value (no relevance, 6=very high relevance)

Figure 5-5: Existence of financial constraints and main reasons

In the chart, the Likert scale is also transformed to the implied average number of financially constrained firms by subtracting the lowest possible scale, i.e. one, and dividing by the sum of remaining Likert scale points, i.e. five. Essentially, 24% of greentech firms claim that they could not raise sufficient capital even at higher interest rates. 27% find financing to be too expensive, while 25% complain that the financing process is too complex and takes up too much time. Overall, it appears that a quarter of German greentech firms experience financial constraints. This result is comparable to the 19% for the most severe measure reported by Savignac (2008), but is a lot lower than the number of young Belgian technology firms found by Bozkaya and de van Potterie (2008) to be experiencing financing difficulties. The reported existence of financial constraints seems realistic, considering that 20% of the analyzed firms use trade credits which, according to Ploetscher and Rottmann (2002), serve as a good additional indicator.

5.2.5   Comparative analysis of greentech firms

5.2.5.1 Methodology

To analyze specific characteristics of greentech firms and in order to match findings to the theoretical framework, the following analysis categorizes findings across the proposed states of development. Four different states give rise to four distinct cohorts of firms. It is important to note that these four states are of an ordinal nature, i.e. one firm is only part of one group. The theoretical framework uses graphical analysis to represent the main characteristics of each group, showing the implications for the individual firm. The linearized patterns imply a continuous scale that is not reflected in the following data set, for two reasons: First, the underlying assumptions of the proposed framework have been relaxed (e.g., the model has no predictive value). Second, the curves in the framework have been linearized to reduce complexity. The following analysis assesses the main variables of interest as well as firm demographics with respect to firms' development states. For each state, the data is aggregated and arithmetic means are reported. For some topics, different cases are specified and the observations for each case are counted, e.g. the share of firms that use policy-based support programs. In any event, the data is at least aggregated for each state of development, resulting in four distinct sets of state characteristics. These are then compared to each other to validate propositions drawn from the theoretical framework. The validity of the comparative analysis is tested using statistical tests of means. First of all, histograms are generated and checked for normal distribution for all cardinal variables ($R\&D$, $GROWTH$, etc.). Additionally, Kolmogov-Smirnov goodness-of-fit tests for normal distributions are run for validation purposes. (Bortz, Lienert, and Boehnke (2008)) Essentially, none of the variables in scope is normally distributed. In other words, the hypothesis that variables are normally distributed is rejected for all cardinal cases. Comparative analysis must therefore rely on non-parametric methods. These tests allow for testing group differences for non-normally distributed and non-parametric, i.e. dichotomous or polychotomous, variables. (Siegel and Castellan (2003)) An appropriate methodology to test whether a number of independent samples are rooted in the same population is proposed by Kruskal and Wallis (1952). The test is also known as the

H test, after Kruskal-Wallis H. (e.g., Horrell and Lessig (1975)) The general idea is to test for distribution differences between samples based on ranks. The observations from the different (non-empty) groups are jointly organized and ranked. The test statistic (1) is computed according to Kruskal and Wallis (1952) as:

(1) $H = \frac{12}{N(N+1)} \cdot \sum_{i=1}^{C} \frac{R_i^2}{n_i} - 3(N+1)$

where C represents the number of samples, i.e. states of development, n the number of observations in the ith sample, N the number of observations in all samples combined, i.e. $N = \sum n_i$, and R_i the sum of ranks in the ith sample. The test statistic assumes that no ties exist. As this assumption is not realistic, however, ties must be considered in particular for continuous samples. If ties occur, every observation is assigned the mean of the ranks for which it is tied. (1) is then divided by:

(2) $1 - \frac{\sum T}{N^3 - N}$

where $T = (t-1)t(t+1) = t^3 - t$ represents the ties t within each group. The generalized test statistic, i.e. the one that is valid for tied and non-tied ranks, can then be defined as (3):

(3) $H' = \frac{12}{N(N+1)} \cdot \sum_{i=1}^{C} \frac{R_i^2}{n_i} - 3(N+1) \Big/ 1 - \frac{\sum T}{N^3 - N}$

where H is chi-square distributed with C-1 degrees of freedom. The one sided p value for the test statistic is $p = P(\chi_{C-1}^2 \geq H')$. It follows that, when $p < \alpha$, the null hypothesis is rejected. As noted above, the following analysis compares states of development. Each state is thus modeled as an assigned set of firms, where sets are independent from each other. The test statistics are programmed in SPSS and reported in graphical form. The results from tests for normality are not reported, since they were rejected in all relevant cases. The analysis will first focus on firm demographics and then follow the overall structure of the dissertation by looking at innovation, growth, financial structure, financial constraints and the factors that can alleviate the latter. It then discusses findings and leads to regression analysis.

5.2.5.2   Firm characteristics

The theoretical framework argues that firms can be structured in line with their state of development, while the state of development can be characterized by the key demographics of the firm. The survey data shows a generally good match with theoretical propositions (see Figures 5-6, 5-7).

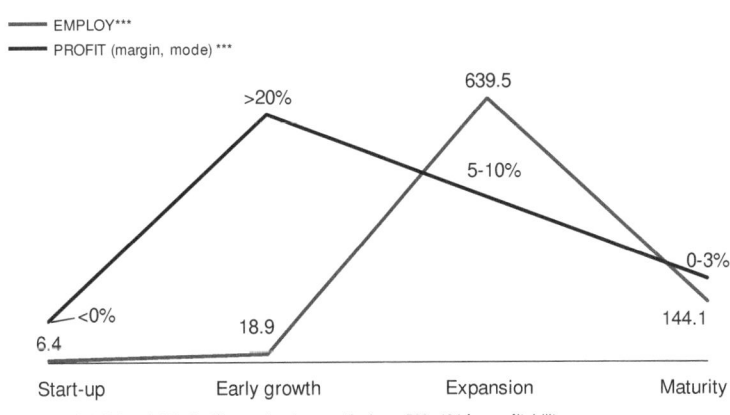

Figure 5-6: Key demographics: Employment and profit

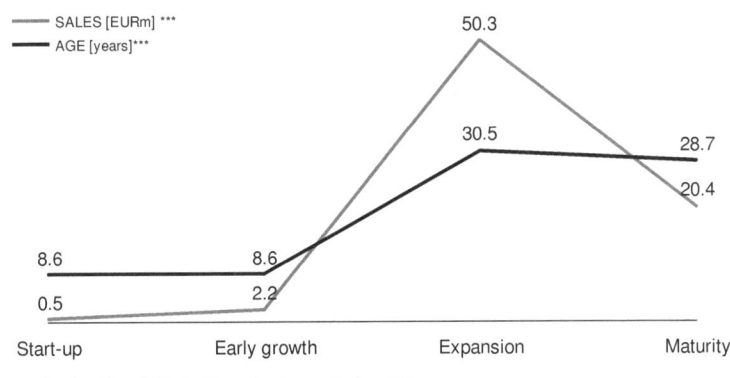

Figure 5-7: Key demographics: Sales and firm age

Greentech start-up firms are small, with an average of 6.4 employees and an average implied sales volume of EUR 0.5 m. They have been on the market for 8.6 years on average. The latter number seems high, although it remains comparable to other studies using categories from below five to below 15 years. (e.g., Bhaird (2010); Kazanjian and Drazin (1990)) However, the sample contains three out of 28 firms that refocused their business objectives after ownership changes or the closure of other activities. This may also explain why start-up and early growth firms nearly have the same average age. As proposed by the theoretical framework, greentech start-ups do not generate funds internally, which is indicated by the negative profit margin. Surprisingly, profit margins for early growth firms are significantly higher. This contradicts the idea that early growth firms do not possess sufficient internal funds, although it may indicate that profit margins and cash flow in particular diverge for younger and smaller firms. Early growth firms are four times larger in terms of sales and three times larger in terms of employees. Expanding firms are, as expected, the largest group, with more than 600 employees and over EUR 50 m in sales. Their after-tax profit margin typically lies between 5-10% and they have been on the market for around 30 years on average. Mature firms are smaller than expanding firms but have approximately the same age. This is in line with the argument that firms at any state may enter maturity. Nonetheless, the fact that mature firms represent the second largest in the sample lends some support to older lifecycle models (e.g., Greiner (1998); Levitt (1965)), which assume sequential development. The data also demonstrates the problem of linearization, which was discussed earlier. The problem is, as anticipated, most severe for mature firms that are smaller than expanding firms. The data also reflects the S-curve pattern of the one–firm, one-technology case adopted from product lifecycle frameworks (e.g., Klepper (1996); Bass (1969)) and derived in section 3.1.5. In this context, it appears that, even with more firms, the market, approximated by average firm size, is largest in the expansion state and decreases thereafter. It thus seems that well-established product and technology lifecycle frameworks also work with firms and support some of the implications drawn by organizational lifecycle models. All measures are statistically significant at the 1% confidence level.

5.2.5.3 Innovation activity and growth

The theoretical framework argues that younger firms are less innovative compared to older firms in terms of innovation output. However, young firms need to put relatively more effort into the development of innovations, measured in terms of relative R&D expenditures, and achieve higher relative growth rates. Although the distinction between innovation types does not reveal sufficient measuring strengths, the overall innovation output hypothesis (e.g., Freeman and Soete (2004); Huergo and Jaumandreu (2004); Vaona and Pianta (2008)) is supported by the data. 60% of greentech firms in the expansion state are currently running innovation projects and have introduced an innovative product or process during the past three years. In essence, expanding firms are more innovative than firms in the early growth and start-up states, while mature firms reflect the lowest innovation activity. The non-monotonic relationship is consistent with the findings of Pasanen (2007) and the literature review conducted by Bahadir, Bharadwaj, and Parzen (2009). The R&D effort is, again as expected, generally greater for smaller firms that need to invest a larger proportion of their sales. R&D expenditures, measured as a percentage of sales, are thus highest for start-up firms, at 32%. This rate decreases significantly as the firm grows in size and age, dropping to 14% for early growth firms, 9% for expanding and 5% for mature firms. The pattern implies the existence of economies of scale for R&D. Only mature firms seem to buck this trend, as they firms have both lower expenditures and lower innovation output. These findings are partially consistent with Müller and Zimmermann (2009), who find that increasing absolute R&D expenditures correlate negatively to R&D intensity. They are also consistent with the majority of studies, which identify a positive relationship between firm size and absolute R&D expenditures (e.g., Graves and Langowitz (1993); Klassen and McLaughlin (1996)), but which to some extent contradict studies that assume decreasing productivity on R&D expenditures. (e.g., Hansen (1992); Love and Ashcroft (1999))

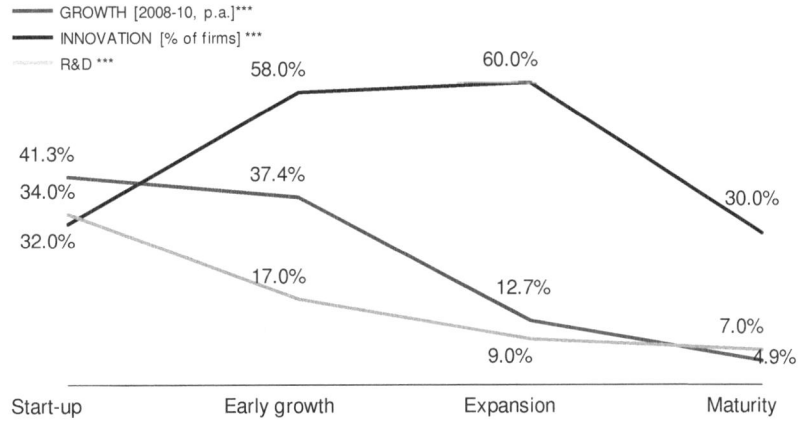

Figure 5-8: Innovation, R&D and growth

Additionally, there seems to be a strong pattern of growth rates across firms' development states. Young firms thus grow faster, while older firms grow more slowly. Moreover, two interesting observations can be made: First, the level of growth seems rather high in all states, reflecting the current dynamism of the greentech industry even among mature firms. Second, start-up and early growth firms evidence the highest rates. While the latter point is expected, these firms are growing from a smaller base than, say, expanding firms. The data presents three patterns that reflect a strong moderating impact on the part of development states (see Figure 5-8). All three patterns are also consistent with the theoretical framework proposed in this dissertation, as well as with literature suggesting a negative relationship between firm size and growth (e.g., Sutton (1997); Musso and Schiavo (2008)). The differences between states of development are significant at the 1% confidence level.

5.2.5.4   Financial structure

Alongside the demographic and innovation-related pattern, the theoretical framework assumes that the development states of greentech firms also reflect a certain usage of financing instruments. The average financing basket and relative use of financing instruments across the four development state cohorts is shown in Figure 5-9. As expected, internal equity provided by the owner, friends and family (*INTEQT*) is the most important financing source in the formation and start-up state. Contrary to the theoretical framework, however, it also remains important throughout all states, with no large differences occurring. Start-up firms and early growth firms use internal equity with approximately the same intensity. Only expanding firms seem to make noticeably less use of internal equity. The differences are not statistically significant. *INTEQT* is the most important financing source and should be considered for any firm in any state, and not only in the start-up phase, as proposed by Berger and Udell (1998). Moreover, this observation supports the pecking order concept (Myers (1984)), suggesting that firms choose to capitalize internal funds before turning to external sources. This view is also supported by the use of retained earnings, which increasingly become available over time, as will be shown later. An even more meaningful picture is drawn by the *INTD* aggregate, which represents debt-related and bootstrapping instruments, e.g. loans provided by the owner, family and friends as well as bank overdrafts and credit card debt. The intensity of use decreases over the course of the development states to maturity. Start-up firms show the highest intensity of use and expanding firms the lowest. As expected, it appears that later-state firms can rely on external and, ultimately, cheaper financing means than earlier-state firms. The comparative results are statistically significant at the 1% confidence level. This is also reflected in the debt aggregate (*D*), which comprises short-term, medium-term and long-term loans provided by banks. In line with the information opacity hypothesis (Berger and Udell (1998)), debt only represents 12% and 11% of the start-up and early growth firms financing basket, respectively. Expanding and mature firms make more intensive use of debt, at 20% and 16%, respectively. These findings are largely consistent with descriptive studies on the financial growth cycle (e.g., Hogan and Hutson (2005); Gottschalk et al. (2007); Beck and Demirguc-Kunt (2006)) and rigorous capital structure studies of internal funds (e.g., Bhaird (2010); Gregory et al. (2005)) and external/bank financing (e.g., Cassar (2004); Egeln and Licht (1997)). However, contrary to the theoretical framework, which suggests that government support programs have an offsetting impact during early-development states, it seems that expansion-oriented firms in particular benefit from push programs. Government capital (*GOVC*) thus represents 12% of the financial structure at expanding firms, compared to 11% at start-ups, 9% at early growth firms and 8% at mature firms.

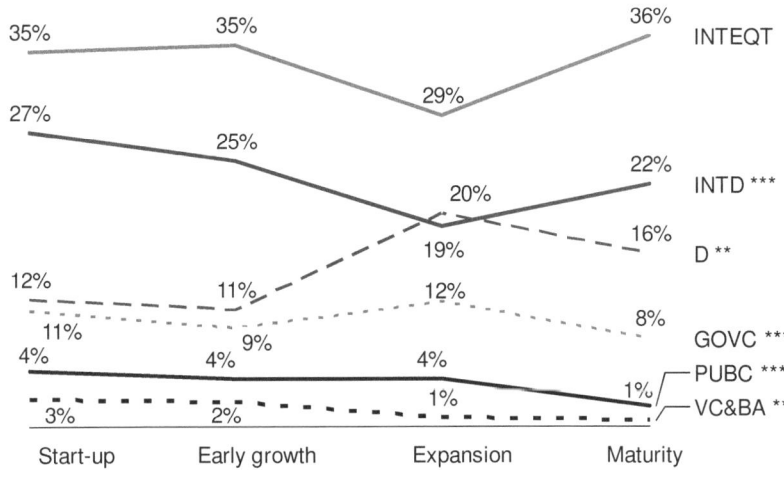

*, **, ***   10%, 5% and 1% significance level respectively, n=529

Figure 5-9: Aggregated financing baskets

However, the observation requires further assessment on a more detailed level. It may also be biased by the low number of items (n=28) in the category concerned. The share of government funds characterizes the regulation-driven industry environment and is high compared to other studies. For example, the seminal work by Berger and Udell (1998) reports a share of less than 1%. Venture capital and business angel capital is represented most strongly in start-up and early growth firms. This is fairly consistent with the proposed framework, considering that no seed category was included in the analysis. The low average intensity overall also highlights a point that has been discussed extensively in greentech-related research (Kenney (2009)), namely the potential lack of early-state private equity in the greentech industry. Public capital, on the other hand, is used by firms in all states of development and at comparable intensities. This may be rooted in the circumstance that greentech has emerged as a favorable investment category, and that the German greentech market has a specific investment vehicle – i.e. closed-end funds – dedicated to specific greentech projects (Enzensberger, Fichtner, and Rentz (2003)). Except for *INTEQT* and *BA*, all comparisons are statistically significant at the 1% or 5% confidence levels, as indicated on the graph. However, in order to provide a deeper understanding of greentech firms' capital structure choices, the following analysis takes a closer look at equity and debt and at government support instruments.

5.2.5.4.1 Equity financing

SHEQT is the major driver for *INTEQT* and represents the main source of equity for firms in all states, as demonstrated in Figure 5-10[1].

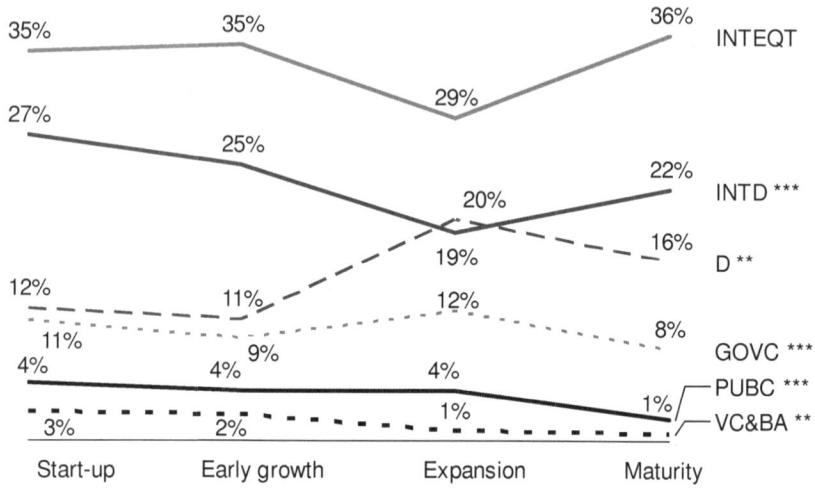

*, **, ***   10%, 5% and 1% significance level respectively, n=529

Figure 5-10: Equity financing

Mature firms reveal the highest ratio, representing around 30% of the overall financial basket. The pattern is comparable to that of *INTEQT*, although start-up firms use *SHEQT* rather less intensively than early growth firms. Although the results are not significant, this is surprising considering that early growth firms already have a functioning product and a better track record. However, it underlines the assumption that early growth firms face the highest gap between capital demand and supply. This is also supported by the availability of internally generated funds, i.e. *RETAINED*. Consistent with the theoretical framework, retained earnings are lowest when a firm enters the market space. They then increase constantly over time until firms reach maturity and margins decrease due to fiercer competition and organizational inertia. Retained earnings and cash flow are thus the most important sources from which to financing expansion-oriented activities. The implications are the same as those noted above for the aggregated instruments. Contrary to the profitability distribution, they lend substantial support to the EXTC$_D$ curve in the theoretical framework. It is also interesting to note the importance of equity provided by friends and family, which is used more intensively on average than government grants and allowances. Only expanding firms do not seem to be dependent on funds provided by the founders' social network. In line with the studies conducted by Bozkaya and de van Potterie (2008), Gottschalk et al. (2007) and Hogan and Hutson (2005), it appears that capital provided by friends and family should explicitly find consideration in relevant frameworks in general and in the framework of this dissertation in particular.

5.2.5.4.2 Debt financing

It has been shown that greentech firms depend heavily on equity financing provided by the owner, family and friends. This does not seem to be different for debt financing, where shareholders and their social networks provide up to 15% of the overall financing basket. However, utilization decreases more sharply than for equity and accounts for only 8% in the expansion state. As expected, though, the use of external debt increases over time, with the highest ratio being found among expansion-oriented firms. The congruence of firm development and debt maturity implied by the theory (Myers (1977)) is supported to some extent, whereas the *MTD* curve inclines more gently than the *LTD* curve. In addition, *MTD* is used more intensively by earlier-state firms, while expanding firms use *LTD* more intensively. This observation is supported by the fact that later-state firms also possess higher proportions of tangible assets, which are used in studies to assess internal collateral and the degree of information opacity (e.g., Chittenden, Hall, and Hutchinson (1996); Voulgaris, Asteriou, and Agiomirgianakis (2004)). The findings contradict the pecking order principle (Myers (1984)), which would imply relatively greater use of debt by firms exposed to severe financial constraints due to insufficient internal funds. However, they do support both the information opacity hypothesis put forward by Berger and Udell (1998) and empirical studies suggesting a positive correlation between firm size, firm age and the use of debt (e.g., Börner and Grichnik (2010); Colombo and Grilli (2007); de Haan and Hinloopen (2003)). All comparative results are significant at the 1% confidence level. The only exception is overdraft finance, which is not significant. In the theoretical framework, overdrafts were shown to be a major financing instrument for small and potentially constrained firms. It is available to a large population of different firms and is thus represented through the overall development cycle. The data shows that early-state firms in particular make extensive use of overdraft facilities. For them, it is indeed the second most important source of debt. This aligns with the findings of Freel (1999), for example, who report that around 40% of small UK ventures make use of overdraft debt. However, the statistical evidence is indicative at best. The overall findings are summarized in Figure 5-11.

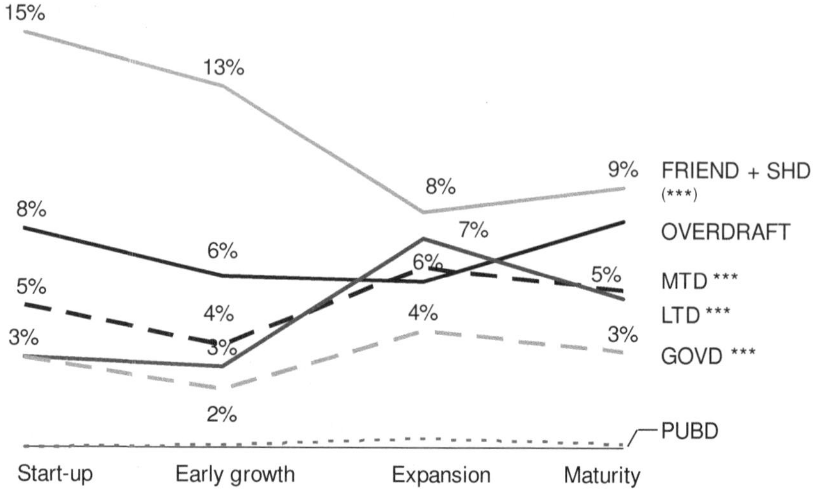

*, **, *** 10%, 5% and 1% significance level respectively, n=529

Figure 5-11: Debt financing

5.2.5.4.3 Government-backed financing

The apparently disorderly picture of government funds and their allocation is not resolved by a more detailed examination of the instruments deployed. In line with the theoretical considerations presented in section 3.5, start-up and growth firms use government funds – in this case primarily grants and allowances (*GOVEQT*) – more intensively than they use retained earnings, thereby partially offsetting the limited ability of young and small firms to generate cash flow. The use of *GOVEQT* is thus highest for start-up firms, thereafter decreasing constantly up to maturity, a pattern that gives partial support to the gap theory formulated by Chertow (2000). However, the differences between the various states are only marginal and do not reflect either the structures or the objectives of German support programs (see section 3.5). In particular, the difference in use between start-up and expansion firms is anything but striking. Additionally, the heterogeneous nature of the late-development state makes it even more difficult to identify a clear pattern. Nevertheless, both results are significant at the 5% and 1% confidence levels, respectively. Government-backed loans are – again contrary to expectations – extensively used by expanding firms. However, the effect is biased as the questionnaire and, hence, the data also included general SME-oriented subsidized grants. Once again, further analysis is necessary to flesh out valid implications. This is done in the section 5.2.2.6.

5.2.5.5   Financial constraints and applications for capital

Besides assessing the mere existence of financial constraints, it is important to understand their importance to each state of development. The theoretical framework postulates that constraints may be particularly apparent for early growth firms that have a substantial demand for capital but face limited supply. This hypothesis is evidently reflected by the data. With respect to the financial constraints measure *INVESTCONSTRAINT*, the Likert-implied percentage of firms constrained to some degree is 26% in the early growth state. In other words, 26% of early growth greentech firms claim that investment projects have been canceled or delayed due to a lack of capital. Start-up firms too face significant constraints, with an implied share of 25%. The level of constraints decreases to around 10% for expanding firms and around 13% for mature firms. Both the relatively high proportion of start-up and early growth firms that experience constraints and the sharp drop among expanding firms clearly support the hypotheses formulated in the theoretical framework. At least in this respect, the evidence seems to support Chertow (2000), given the assumption that her theory can be adapted to firms. Figure 5-12 provides an overview and also presents results for the validating measure "use of trade credit". The use of trade credit is greatest for early growth firms, confirming the results of the first measure. However, usage is rather low for start-up firms. This is probably due to the fact that start-up firms have only limited operating activities and, hence, accounts payable. These firms are simply not able to draw on trade credit. This argument underscores the flaws inherent in the use of such a measure in studies of financial constraints, as was done by Ploetscher and Rottmann (2002), for example. Moreover, comparisons across developmental states are not statistically significant for this measure.

A third proxy for assessing financial constraints is the rate of rejection for certain capital forms. The measure has been used in several studies and contexts. (e.g., Buttner and Rosen (1992); Freel (1999)) Obviously, there are some differences compared to the financial constraints variables. First, the main argument differs from financial constraints as defined above. While financial constraints measure the general lack of funds, rejection rates only assess the ability to successfully access specific types of financing. A firm could, for example, be rejected several times but still receive the full amount of capital it requires. Second, the rejection of applications for capital is also an indicator of how hard a certain firm tries and could therefore even correlate negatively to financial constraints. Finally, rejection rates are connected to a certain financial instrument. A firm's financing situation can only be assessed overall based on a cumulated variable, however. Taking into account theses differences, analyzing capital rejection rates does not ultimately serve as an indicator of financial constraints, but rather provides an additional insight into the sources of and reasons for financial constraints.

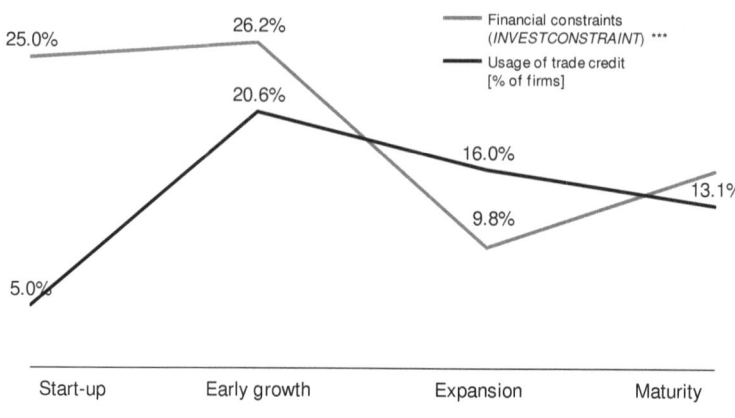

*, **, *** 10%, 5% and 1% significance level respectively, n=529 (A) and 449 (B)

Figure 5-12: Financial constraints

The analysis distinguishes between equity and debt capital sources. Figure 5-13 shows the rejection rates for the major equity capital sources discussed above across states of development. The latter are denoted by the capital letters A (start-up), B (early growth), C (expansion) and D (maturity). The first two rows stand for the total number of firms that applied for a particular capital source and the number that applied in each development state.

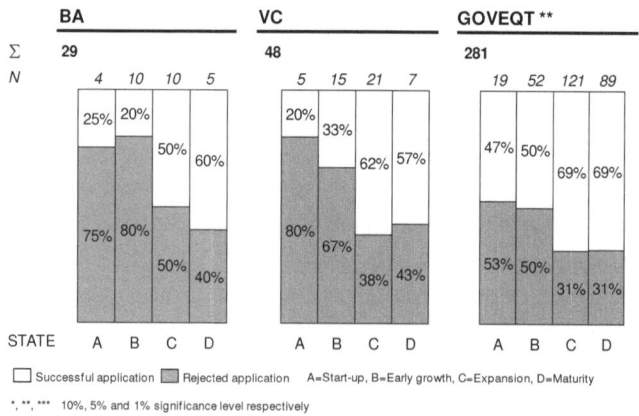

*, **, *** 10%, 5% and 1% significance level respectively

Figure 5-13: Applications for equity capital – Rejection and success rates

Apart from the fact that the availability of venture and business angel capital is apparently low in the greentech market, it also seems that only very few firms actually apply for it. Of the 529 firms in the data set, only 48 applied for VC and 29 for BA capital. Although the rejection rate is high at up to 80%, it is still lower than average use, which is of less than 5%. The implication is straightforward: Firms should step up their efforts to apply for such funds. Moreover, the results may indicate that the debate about a lack of VC (e.g., Burtis (2006); Diefendorf (2000); Randjelovic, O'Rourke, and Orsato (2003)) may actually be a debate about whether firms actually ask for it. This view would also support the claim that VC is not the appropriate instrument to finance greentech-specific, capital-intensive investments, as argued by Kenney (2009). However, the evidence says nothing about firms' motives and must be considered indicative. Although the questionnaire asked whether the firm had applied for one of the capital forms at any point in the past, the results still contradict the propositions of the financial growth cycle model Berger and Udell (1998) and, hence, the theoretical framework of this dissertation. It thus seems that VC, BA and even – to a lesser extent – GOVEQT sources prefer to target later-state firms. Greentech start-ups and early growers evidence the highest rejection rates of all. The VC rejection rate for early growth firms, for instance, is nearly twice as high as that for expanding firms. It seems that external equity sources cannot substitute for the lack of debt available to informational opaque firms. The comparative analysis is statistically significant for GOVEQT but not for VC and BA, probably due to the low number of applicants. Regarding debt capital, the overall results seem to better match the findings of the capital structure analysis and the theory, with one exception (see Figure 5-14). The lowest rejection rates for STD, MTD and LTD are reported by mature firms. It appears that track record, represented by firm age, is more important than growth and profitability, which were shown to be highest for early growth firms. The finding lends fundamental support to the idea that credit rationing can be reduced by building banking relationships over time (Elyasiani and Goldberg (2004)). This idea is picked up again in the next section. Start-up firms and early growth firms show the highest rejection rates for debt capital. Of the 39, 37 and 34 early growth firms that applied for STD, MTD and LTD respectively, 54%, 49% and 53% did not receive funding. By comparison, only 26%, 25% and 26% of expanding firms and 24%, 18% and 15% of mature firms experienced similar problems. Contrary to expectations, it seems that GOVD is even harder to access. The proportion of rejected applications on this score is 64% for start-ups, rising as high as 89% for early growth firms. In addition, the rejection rate for later-state firms is higher than, say, the rejection rate by commercial banks. The findings suggest that government support systems do not seem to improve access to financial funds for young, developing firms. According to one study participant, one major problem in this context seems to be the fact that GOVD is distributed across commercial banks that ration credit in the first place. The observation is reinforced by the fact that, of all capital instruments, most firms (281) applied for government grants and allowances, although nearly as many applied for STD and MTD. The rejection rates for commercial bank debt are generally lower than those incurred when applying for commercial equity. With respect to available support programs, it seems that grants and allowances, being the less costly instruments, are easier to access than subsidized loans.

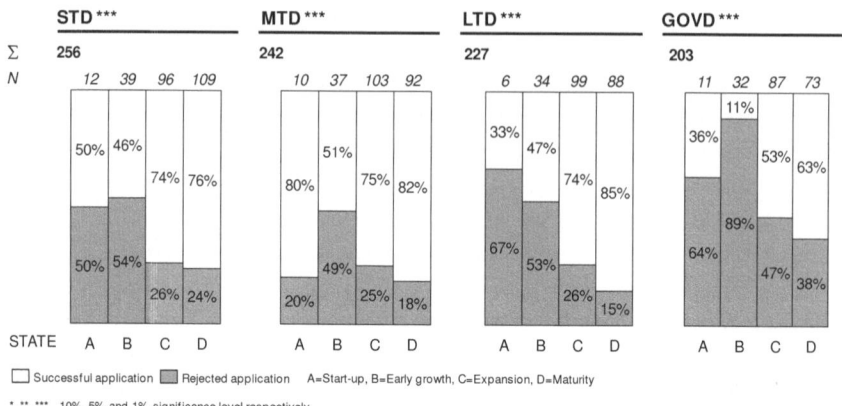

Figure 5-14: Applications for debt capital – Rejection and success rates

The results are surprising when one considers that, overall, government action is an important pillar of the greentech market space, which depends heavily on regulation and government support programs to offset the problem of internalizing externalities. Moreover, study participants pointed to certain critical issues with respect to implementation. Government action is "difficult to observe" and particularly to follow for small and young firms with limited resources. At the same time, this is precisely the type of firm that needs support most urgently. One firm also argued that policymakers must take a more active role in educating society, for example "providing information about sustainability, technology choices, their objective utility, the potential for job creation, etc." which might in turn increase investor confidence. The following section follows up on this by focusing on what government instruments and firms themselves can actually do to alleviate potential constraints.

5.2.5.6 Alleviation of financial constraints

As was demonstrated in the previous sections, an aggregated analysis of government support programs did not yield sufficient insights to assess the effectiveness of their implementation. Accordingly, the sections that follow focus on the lowest level of granularity, examining the individual use of specific government instruments. This analysis also looks at firms' own levers, i.e. banking relationships and collateral. It is be complemented by a regression analysis (section 4.3.) that directly links alleviating factors to financial constraints. It is assumed that both steps, taken together, will yield sufficient insights into the overall problem.

5.2.5.6.1 Policy support programs

As we have already seen, there are two main types of government support instruments. One involves supply-push programs that aim to foster progress at the very start of the innovation value chain and, by implication, to foster early-state firm development. The other involves demand-pull programs that aim to diffuse existing technologies and, by implication, to foster the growth of later-state firms. The theoretical propositions made herein were shown to be consistent with existing greentech support instrument objectives in Germany. It is therefore reasonable to expect that start-up and early growth cohorts in particular will reveal a higher relative share of supported firms. The aggregated analysis described above is assumed to have experienced interference from general SME loan programs. It is therefore necessary to distinguish between existing instruments. In line with the programs identified in section 3.5, the empirical analysis discusses five distinct push instruments, one of which is intended for SMEs in general, while the other four are greentech-specific: (1) general SME support programs (*GOVSME*), (2) grants and allowances (*GOVGRANTS*), (3) subsidized loans (*GOVSUBLOANS*), (4) hybrid instruments (*HYBRID*) and (5) government venture capital (*GOVVC*). Contrary to expectations, firms that have received government funding over the past three years are not primarily those that are in early states of development. Instead, it seems that the highest share – up to 50% of supported firms – is found in the expansion state. This may reflect the fact that government agencies concentrate on the development stage of technologies rather than actual firms. Of course, the technology perspective is in line with most contemporary research on greentech. However, as was shown in the theoretical framework, it is the firm that faces financial constraints; and it is the firm that has to overcome barriers to innovation. This view is supported by the fact that young firms in particular have the potential to generate innovations that produce disruptive effects and may ultimately lead to sustainable development. In essence, expanding greentech firms experience the lowest average degree of financial constraints, as was shown above, but include the highest share of supported firms. Clearly, there is no certainty about causality. It may thus be that expanding firms experience lower levels of constraints because they actually receive government support. Regression analysis may provide additional insights. At this point, however, it can be argued that early-state firms – the ones that experience the most constraints – are underrepresented in government support programs. Figure 5-15 provides an overview of supported firms with respect to development states and support instruments.

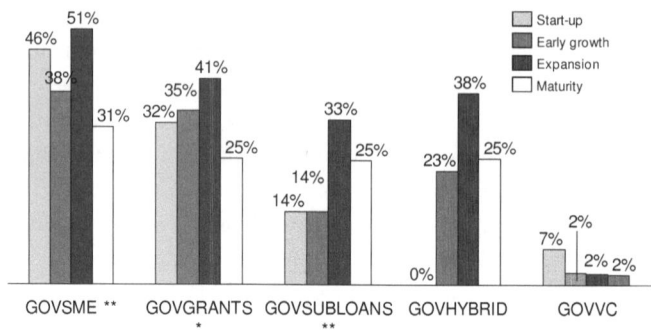

1)   Measured as share of firms that use the instrument independent from intensity of usage
*, **, ***   10%, 5% and 1% significance level respectively, n=529

Figure 5-15: Share of government-supported firms by instrument type[1)]

Five further interesting observations can be made. First, the overall share of supported firms is rather high, reflecting the strong dependence of the greentech sector on government support. Second, *GOVSME* represents the most important measure across all states of development due to its broad focus on firm size. Third, mature firms are supported despite their demographic characteristics and lower average performance. Fourth, government grants are the second most important instrument and have a particular impact on early-state firms. Lastly, although credit rationing is a well-known problem in the credit markets for small and young firms, it appears that early-state firms are underrepresented in subsidized loan programs for the greentech industry. On the other hand, the share of supported firms does not account for the relative prominence of instruments in firms' financial baskets. It is therefore necessary to assess the relative share of government support instruments in the overall capital structure. This share is calculated as a percentage of all financing instruments. Figure 5-16 summarizes average shares for development-state cohorts (also including non-supported firms). The result does not change the fact that expanding firms use government funds more intensively than their younger competitors. The results presented above, which contradict the postulated theory, are thus reinforced. However, the relative importance of support programs is comparatively high for start-up firms and nearly as high for early growth firms, representing 10.7% and 9.2% of the overall financing mix, respectively. These findings underpin the theoretical proposition that early growth firms receive less support than start-up firms, even though the availability of external instruments is not sufficient. Exceptions to this observation are grants and allowances (*GOVGRANTS*), which are used most intensively by early growth firms. General SME programs are also most important for start-up firms that, on average, derive around 5% of their financing from this source. The findings for *GOVSME, GOVHYBRID, GOVSUBLOANS* and *GOVGRANTS* are significant at the 1% and 5% confidence levels, respectively.

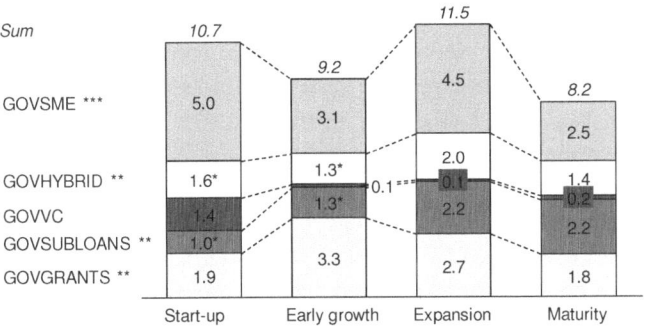

1)   Relative share different from sum of single shares – Calculation based on the relative proportion of
     government capital to other financing instruments
*, **, ***   10%, 5% and 1% significance level respectively, n=529

Figure 5-16: Share of government instruments within capital structure

The last programs to be considered are demand-pull programs, one example being the German feed-in law for renewable energy. Demand-pull programs are examined separately as they affect a firm's financing situation only indirectly. As elaborated above, demand-pull programs aim to promote large-scale diffusion and are expected primarily to affect later-state firms. Contrary to expectations, however, it appears that pull programs are more important for early-state firms than for later-state firms, as can be seen in Figure 5-17.

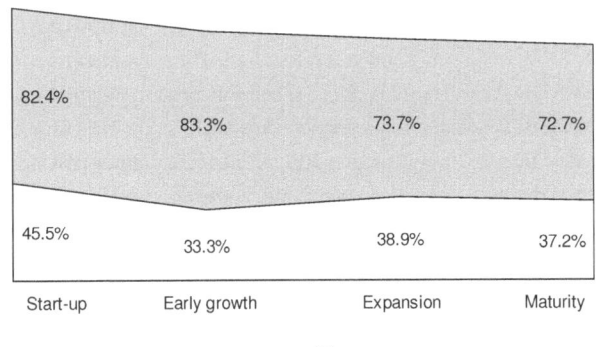

☐ Renewable energy technologies (RENEWABLE)   ☐ All technologies
*, **, ***   10%, 5% and 1% significance level respectively, n=529

Figure 5-17: Share of firms with pull policy-dependent business models

As anticipated, the effect is stronger for renewable energy firms. Around 40% of firms in all sectors claim that their business model is dependent on government pull policies. The share is highest for start-up firms (46%), followed by expanding firms (39%), mature firms (37%) and early growth firms (33%). The picture changes when only renewable energy firms are considered – and the contradiction of the theoretical propositions becomes even sharper. Start-ups (82%) and early growth firms in particular (83%) are heavily dependent on these government initiatives in this segment. The share decreases to 73% for mature firms. Although the comparative results are not statistically significant, there is one major implication: Pull programs have an effect on all firms in all development states. However, it is difficult to draw more detailed implications. For example, the impact of pull instruments on a large number of firms could be explained by the technology lock-in hypothesis (e.g., Cowan and Hultén (1996); Nelson, Peterhansl, and Sampat (2004)) or, on the other hand, by market creation effects that allow for many different technology alternatives (e.g., Ayres (1991); O'Rourke (2009)). To gain a further insight into the matter, it seems sensible to evaluate the effect of pull programs on financial constraints. This is done in section 5.3.5.

5.2.5.6.2 Firm-based measures

As discussed above, the availability of debt is affected most by firms' transparency. One way to achieve transparency was shown to be relationship banking. A good banking relationship grows over time and may reduce not only the extent of credit rationing by banks, but also the interest rates granted to a firm. The sample implies that these arguments also hold for greentech firms. Firms with reasonably satisfactory banking relationship are significantly older and larger than their less satisfied counterparts. On average, only 61% of start-up firms are contented with their relationship to banks. The proportion is 14% higher for expanding firms, and even the heterogeneous group of mature firms shows a 12% higher ratio. This is also reflected in the average interest rate on debt. Start-up and early growth firms pay with 7% for *LTD* and 8% *MTD*, the highest average interest rates of all. The average interest rates for expanding firms are only 5% and 7%, respectively. *LTD* comparisons are significant at the 10% confidence interval, while *MTD* comparisons are not statistically significant. However, this may be due to the smaller sample size, as around 50% of all firms did not provide this information. All in all, the findings strongly support the assertion that good banking relationships significantly reduce financial constraints over time. (see Elyasiani and Goldberg (2004) for a comprehensive literature review) Figure 5-18 summarizes the observations and comparative test results.

Another means to overcome information asymmetries is collateral (see sections 3.4.4.3 and 3.4.4.4). Collateral may be provided by the firm on the basis of tangible assets or by third parties such as the owner. According to existing literature, the provision of collateral is expected to improve a firm's chances of receiving external financing.

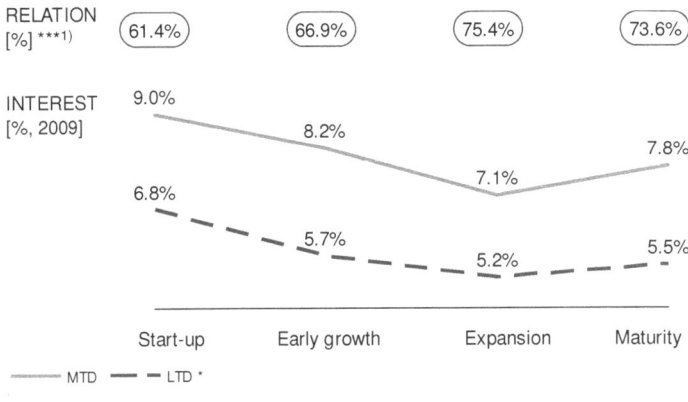

*, **, *** 10%, 5% and 1% significance level respectively, n=529, 279 (MTD) und 234 (LTD)

Figure 5-18: Bank relationship satisfaction and average interest rate paid

This issue primarily arises in debt capital markets, where banks face uncertain firm performance but receive only returns with a defined upper ceiling. It is further expected that early-state firms in particular will need to provide external collateral, because they only have limited tangible assets, face greater uncertainty, have a relatively high degree of information opacity and may generally present higher risks to investors (see section 3.4.). These expectations are supported by the data, as shown in Figure 5-19.

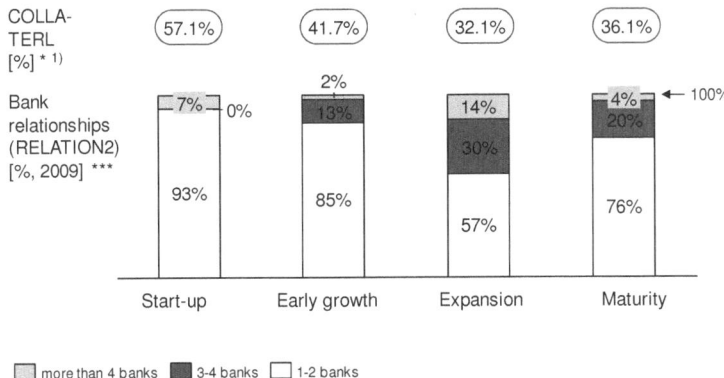

1)   Share of firms that had to pledge collateral in order to receive debt capital
*, **, *** 10%, 5% and 1% signficance level respectively, n=529

Figure 5-19: Collateral and the number of bank relationships

Around 57% of all start-up firms were required to provide collateral in order to access debt capital. This requirement is relaxed over time: 42% of early growth firms, 32% of expanding firms and 36% of mature firms also had to provide external collateral. The results are significant at the 10% confidence interval. The presented findings are consistent with the study by Coco (2000) and imply that early-state firms in particular use collateral to signal firm quality and thereby overcome information asymmetries. As suggested by Tensie Steijvers and Wim Voordeckers (2009), the analysis only considered external collateral, i.e. collateral provided by shareholders and third parties. Interestingly, collateral also seems to reduce the number of banking relationships. As was noted earlier, collateral changes the payout structure for liquidation proceeds in the event of bankruptcy. Collateral is assigned to only one creditor at a time and is generally limited, implying that younger firms in particular will be more dependent on a smaller number of banks. Accordingly, 93% of start-up firms collaborate with either one or two banks. The ratio again decreases over time, the figure being 85% for early growth firms, 57% for expanding firms and 76% for mature firms. Expanding firms establish the largest number of banking relationships, whereas 30% maintain three to four relationships at a time. Another firm-based measure to overcome financing constraints is the accumulation of earnings, i.e. retained earnings. This issue has already been discussed, however, and will not be repeated here.

5.2.6   Conclusion – Theoretical framework supported

To sum up, the theoretical framework seems to be of considerable explanatory value. It can be used as roadmap for managers and policymakers. Managers may, for example, apply for capital sources that are currently underrepresented, but that involve comparatively low rejection rates, such as venture capital for expanding firms. Policymakers may, for example, identify firms with high innovation potential and accordingly increase the availability of support funds. The framework remains indicative, however, and is not fully congruent with the data. One strong assumption is the linearization that was, as expected, shown to have drawbacks when explaining the characteristics of mature firms. Four modifications can be derived from the analysis. First, (disruptive or incremental) innovation potential could not be modeled entirely, although general innovation output did increase through the various states up to expansion. Second, actual growth did not fully reflect the S-curve pattern across firm states. Third, government funds do not seem to specifically support early-state firms. Finally, funds provided by the founder's social network seem to play a more important role than anticipated and thus merit separate investigation.

5.3   Regression analysis and testing of the research hypotheses

This section uses regression models to test the hypotheses formulated in chapter 4. The analysis is the second component of the empirical analysis and is structured as follows: First, the theoretical framework is converted into a rigorous and testable empirical model. Second, relevant variables are introduced and descriptive statistics are reported. Third, statistical models are specified and evaluated and results are presented. This is done separately for innovation, growth and financial constraints, because the overall model is composed of three individual models, corresponding to the structure of chapter 3. The first model assesses the impact of capital endowment on innovation (hypothesis 1), the second the impact of capital endowment on growth (hypothesis 2), and the third the impact of capital endowment (hypothesis 3), innovation and growth (hypothesis 4), policy support (hypothesis 5) and firm-based measures (hypothesis 6) on financial constraints.

5.3.1   Overall empirical model – Financing greentech innovation and growth

According to the bodies of literature referenced in this dissertation, there are three main components to be considered in the empirical model: (1) the impact of capital endowment on firms' innovation activity; (2) the impact of capital endowment on firms' growth; and (3) the impact of capital demand/endowment on financial constraints. As discussed earlier, the overall empirical model uses capital endowment as a proxy for capital demand and supply. In line with sections 3.2, 3.3 and 3.4, it is the supply of capital in particular that drives firms' performance. Demand for capital, on the other hand, increases the likelihood that a firm will experience financial constraints. In practice, however, demand and supply are hard to measure or even to keep apart, because the observable use of capital (capital endowment) by a given firm results from the minimum use of either. It follows that capital endowment is the component that links the three literature streams together and is thus the most fundamental component of the empirical model, as summarized in Figure 5-20.

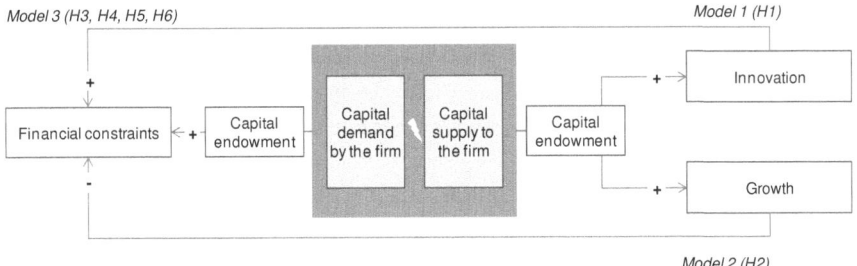

Figure 5-20: Empirical approach

As already indicated, the overall approach adopts a financial economics perspective by assuming that innovation and growth influence financial constraints and not vice versa (see chapter 4). The model thus assumes that innovation activity correlates positively and growth correlates negatively to financial constraints. By using capital endowment as a determinant, the approach also avoids common problems of endogeneity and simultaneity between financial constraints and measures of firms' performance. The first model tests hypothesis 1, assuming that capital endowment positively affects innovation. It controls for the key innovation determinants derived from literature (see section 3.2.2.2). These comprise relative R&D spending as an aggregate measure of expenditures, intensity and productivity, firm age, firm size and technology line dummies to account for firms' environment. In line with the findings presented in Figure 3-8, all determinants are expected to correlate positively to innovation, while technology lines correct for industry-specific effects. To avoid simultaneity, growth is not included in the model. Essentially, the author believes that growth has a lesser effect on innovation than innovation has on growth. This argument is implicitly supported by a large number of articles treating growth as dependent variable. (For a literature review, see Bahadir, Bharadwaj, and Parzen (2009)) There are two further differences compared to Figure 3-8. First, the model does not correct for country because the analysis is concerned only with Germany. This is also the case for the other two models. Second, the model does not use the broad range of resource and financing variables mentioned in literature. Capital endowment and profitability are used instead. As argued in chapter 4, there appears to be a clear relationship between capital endowment and profitability. Be that as it may, profitability is not directly subsumed under capital endowment as would be the case with retained earnings. However, the putative correlation is considered in the actual analysis when testing the robustness of the model. The second model tests hypothesis 2, assuming that capital endowment drives firms' growth and, hence, their development. Like the first model, it controls for other relevant factors by drawing on the determinants of growth identified in section 3.3.3.3. These factors are relative R&D spending (in accordance with the casual relationship assumption), growth orientation, firm age, firm size and technology lines. R&D spending is assumed to characterize investment in growth and a positive relationship is expected. Growth orientation approximates a firm's willingness to grow as was shown in Figure 3-11 and is also expected to have a positive impact. Age and size represent major demographic characteristics of the firm. Unlike the innovation case, however, a positive relationship is only expected for size, whereas age is assumed to be negatively correlated. The differences in the model compared to the determinants identified in Figure 3-11 are similar to those mentioned for the innovation case with respect to Figure 3-8. The third and final model tests hypotheses 3 to 6 and is thus the most capacious but also complex one. With respect to hypothesis 3, and as derived in chapter 4, this model assumes that capital demand, approximated as capital endowment, adds to the financial constraints faced by greentech firms. A positive relationship is thus expected. The model also integrates innovation activity and growth. Innovation-focused firms are assumed to have greater informational opacity, implying a positive relationship with financial constraints (hypothesis 3). Growth, on the other hand, reduces dependence on external capital and investor's risk exposure, leading to a potentially negative relationship (hypothesis 4). Age and size are expected to correlate negatively to financial constraints. The impact of firms' legal structure is ambiguously reported in literature, but is included here as a control factor. There are two alleviating factors. First, it is assumed that government push and pull instruments reduce the financial constraints faced by the firm. The extensive government programs summarized in section 3.5 break down into equity-related push, debt-related push and demand-related pull programs (hypothesis 5). Second, it is assumed that the firm can influence the relative capital supply and thereby reduce potential constraints by establishing a long-term banking relationship (hypothesis 6a), providing collateral (hypothesis 6b) and increasing operating efficiency (hypothesis 6c). The model also controls for technology lines. The three models are summarized in Figure 5-21.

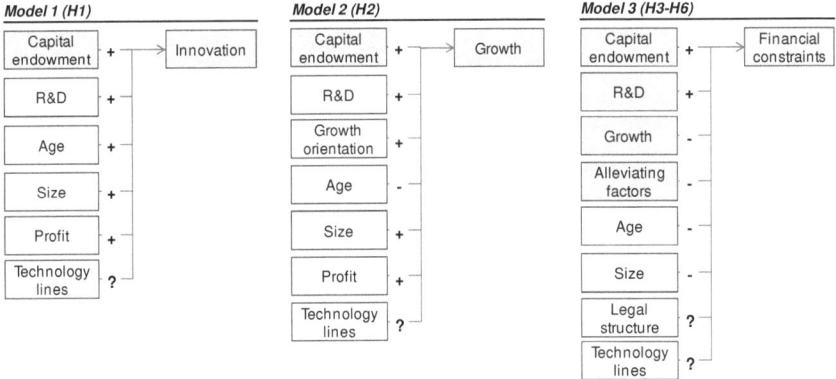

Figure 5-21: Overview of three models

The three models presented above integrate a large variety of findings from three major bodies of literature, as well as from policy support programs for the German greentech sector. Descriptive statistics for relevant variables are the subject of the next section.

5.3.2 Descriptive statistics for relevant measures and variables

The descriptive statistics for the variables used in the models specified below are reported in Table 5-2. As in the overall analysis, items with missing values were not considered. For example, 27 firms did not report on their profit situation. The descriptive statistics are comprised of the number of items (N), the arithmetic mean, median, standard deviation, and the minimum and maximum values. The model in which each variable is used is reported. For example, *INNOVATION* is used solely in model 1. The variables *INNOVATION*, *PCINNOVATION*, *GROWTH* and *INVESTCONSTRAINT* were discussed earlier; the discussion is not reiterated in the section that follows. R&D is used as an input factor for the growth and financial constraints model. Average R&D expenditures as a percentage of sales score 10.66, with a lower median indicating outliers above 5%. This is also expressed by the high standard deviation of 21.02 and the maximum value of 300. It seems that firms with a focus on research and development rely on external capital, as sales remain below R&D expenditures. This lends support to the idea that some technologies are still in a very early state of development. (e.g., Foxon et al. (2005)) Capital endowment scores an average of 2.11 out of six and has a median of 2.00. The minimum and maximum values correspond to the Likert-scale boundaries. The figures reported for *GOVEQT* and *GOVD* represent the relative share of the total financial basket. While the median is zero, the averages are 6% and 3% respectively. It appears that at least one firm is fully government-financed. The *PULL* measure scores an average of 2.36 out of six. The median is considerably lower at 1.5. This may be because renewable energy firms are a lot more dependent on demand-pull systems than the rest of the population.

Variable	N	Mean	Median	Std. Dev. (σ)	Min.	Max.	Model
INNOVATION	529	0.45	0.00	0.50	0.00	1.00	1
PCINNOVATION	529	0.22	0.00	0.42	0.00	1.00	1
R&D	529	10.66	5.00	21.02	0.00	300.00	1,2,3
GROWTH	529	0.15	0.05	0.65	-0.95	10.00	2
GROWTH'	529	0.47	0.42	0.29	0.00	1.00	2,3
GROWTHO	529	0.40	0.00	0.49	0.00	1.00	2
INVESTCONSTRAINT	529	0.16	0.00	0.37	0.00	1.00	3
INVESTCONSTRAINT'	529	0.18	0.00	0.39	0.00	1.00	3
TC	529	2.11	2.00	0.62	1.00	6.00	1,2,3
GOVEQT	529	0.06	0.00	0.09	0.00	1.00	3
GOVD	529	0.03	0.00	0.06	0.00	0.40	3
PULL	529	2.36	1.50	1.70	1.00	6.00	4
RETAINED	529	0.14	0.13	0.16	0.00	1.00	5
COLLATERAL	529	0.44	0.00	0.50	0.00	1.00	6
RELATION	529	4.68	5.00	1.26	1.00	6.00	7
AGE	529	25.07	17.00	28.65	1.00	253.00	8
EMPLOY	521	290.06	12.00	2085.80	1.00	39000.00	9
PROFIT	494	3.61	3.00	1.80	1.00	7.00	1,2
LEGAL	529	0.83	1.00	0.38	0.00	1.00	3
RENEWABLE	529	0.40	0.00	0.49	0.00	1.00	1,2,3
EFFICIENCY	529	0.30	0.00	0.46	0.00	1.00	1,2,3
WATER	529	0.21	0.00	0.41	0.00	1.00	1,2,3
MOBILITY	529	0.06	0.00	0.24	0.00	1.00	1,2,3
RESOURCES	529	0.16	0.00	0.37	0.00	1.00	1,2,3
WASTE	529	0.35	0.00	0.48	0.00	1.00	1,2,3

Table 5-2: Descriptive statistics for measures and variables

RETAINED earnings seem generally to be very important to greentech firms. The average value of 14% is comparatively high; and even the median is 13%, indicating a potentially fat-tailed distribution. *COLLATERAL* was pledged by 44% of all firms on average. Judging from a median of 0 and standard deviation of 0.5, it appears that this variable is uniformly distributed. The score of 4.68 for the *RELATIONSHIP* (with banks) reveals that a relatively high proportion of firms are reasonably happy, whereas a median of 5.0 and standard deviation of 1.26 imply a similar distribution as *RETAINED*. The government support and firm-based programs discussed above apply solely to the third model of financial constraints. The demographic control variable AGE is characterized by upper outliers. The mean is 25.07 years, while the median is only 17 years. However, this distribution makes sense, as firm age has a natural lower boundary but no real upper limit. The observation is underpinned by a high standard deviation of 28.65 and a maximum value of 253. A comparable picture is painted by the firm size variable *EMPLOY*. The smallest firm has only one employee, while the largest has nearly 40,000 employees. The average is 290.06, whereas the median is only 12. As with *AGE*, it appears that outliers for *EMPLOY* are found in particular in the upper range. With respect to the dichotomous *LEGAL* variable, more than 80% of the sampled firms are incorporated firms (corporations). This finding is in line with the median of 1, reflecting a right-skewed distribution. *LEGAL* is only integrated in model 3. The industry variables are also dichotomous and reflect the activity focus of the firms discussed in the introductory part of chapter 5. They are not discussed in detail. The following section introduces the first model for innovation and proceeds as discussed above.

5.3.3   Financing innovation of greentech firms (model 1)

Previous studies have used a broad range of methodologies with various proxies for the variable of interest. Studies using cardinal dependent variables, such as R&D, usually employ OLS regression models. (e.g., Himmelberg and Petersen (1994); Harhoff (1998); Bhaird (2010); Bougheas, Görg, and Strobl (2003); Hansen (1992)) Some authors have also used models with censored variables such as Tobit regressions (e.g., Müller and Zimmermann (2009); Love and Ashcroft (1999)) or non-linear models (e.g., Audretsch and Acs (1991); Scherer (1965)). Other studies have used a dichotomous variable to assess innovation, e.g. specifying that a firm is innovative or not innovative. These studies usually employ probability models that use maximum likelihood functions to estimate parameters. (e.g., Correa, Fernandes, and Uregian (2010); Bah and Dumontier (2001); Graves and Langowitz (1993)) The present study follows the latter approach and employs a logit model which is specified below.

5.3.3.1   Specification of the regression model

Logit models assess the relationship between categorical and independent variables. The methodology is directly related to multinomial models. However, it is of superior quality if predictors are mostly metric, whereas the multinomial model may be preferable for categorical regressors. Logit models estimate probabilities for items belonging to a certain group. Let us, for example, assume the existence of two groups, group A and group B. Group A is composed of 80 items while group B is composed of 20 items. A random pick would have an 80% chance that the item belongs to group A. Another measure to describe this probability is the relationship between A and B, i.e. the probability that a certain item belongs to group A is 80/20=4 times higher than the probability that the item belongs to group B. In other words, the odds of the item belonging to group A are 4. This is exactly the question that logit models aim to answer. (Spicer (2005)) With respect to the first model, let us again assume the existence of two groups. Group A is composed of innovative firms, whereas group B is composed of non-innovative firms. The logit model will now be defined in a way that it predicts, with respect to the main determinants of innovation, whether a firm is potentially innovative or not. The *INNOVATION* variable is thus defined as a dichotomous categorical variable, where standard linear regression would yield implausible estimates. (Backhaus (2008)) The response event can take two forms, where *INNOVATION=1* indicates that a firm is innovative and *INNOVATION=0* indicates that a firm is not innovative. Now, let us assume that z_{ij} is an unobserved variable that represents the innovation benefits for firm i in sector j given its capital endowment and capabilities:

(1) $$INNOVATION_{ij} = \begin{cases} 1, & if\ z_{ij} > 0 \\ 0, & if\ otherwise \end{cases}$$

The latent variable z_{ij} is allowed to depend linearly on firm ij's capital endowment (*TC*) and two sets of control variables. The first set reflects the main determinants of innovation derived from literature (see section 3.2.2.2), while the second set corrects for technology line and sectoral effects. z_{ij} can thus be defined as:

(2) $z_{ij} = \beta_0 + \beta_1 \cdot TC_{ij} + \sum_{l=1}^{n} \beta_l \cdot x_{ijl} + \sum_{k=1}^{m} \beta_k \cdot y_{jk} + \epsilon_{ij},$

where ϵ_{ij} is an error term that accounts for unobservable firm characteristics that influence the decision and ability of firm i to innovate. The error term is assumed to be normally distributed. The first set of control variables comprises the natural logarithms of *AGE*, *EMPLOY* and *R&D*, as is standard in literature, and *PROFIT*, i.e. $\sum_{l=1}^{n} \beta_l \cdot x_{ijl} = \beta_2 \cdot lnR\&D_{ij} + \beta_3 \cdot lnAGE_{ij} + \beta_4 \cdot lnEMPLOY_{ij} + \beta_5 \cdot PROFIT_{ij}$. The second set is composed of sector dummies, i.e. *RENEWABLE, EFFICIENCY, WATER, MOBILITY, RESOURCES* and *WASTE*, i.e. $\sum_{k=1}^{m} \beta_k \cdot y_{jk} = \beta_6 \cdot RENEWABLE_j + \beta_7 \cdot EFFICIENCY_j + \beta_8 \cdot WATER_j + \beta_9 \cdot MOBILITY_j + \beta_{10} \cdot RESOURCES_j + \beta_{11} \cdot WASTE_j$. The latent variable linearly connects the binominal variable *INNOVATION* to the independent variables. However, the relationship to z_{ij} and the probability of innovation is assumed to be non-linear.

To assess the actual propensity of firm i to innovate, probability $P(INNOVATION_{ij} = 1)$ must be calculated. The complementary outcome is $1 - P(INNOVATION_{ij} = 1)$, where P follows a logistical function. The probability function is specified in a non-transformed manner, taking into account conclusions drawn by Ge and Whitmore (2010) and avoiding potential misspecifications:

(3) $P(INNOVATION = 1) = \frac{e^z}{1+e^z} = \frac{1}{1+e^z}$

Equation (2) is estimated using the maximum likelihood method based on the Newton-Raphson algorithm as is standard in SPSS Wüstenhagen (2008). The likelihood function L for n observations $y_1, y_2, ..., y_n$ with probabilities $z_1, z_2, ..., z_n$ and case weights $w_1, w_2, ..., w_n$ is defined as:

(4) $L = \prod_{i=1}^n z_i^{y_i w_i} (1 - z_i)^{w_i (1-y_i)}$

The estimate for z_{ij} is conducted hierarchically, i.e. demographic control variables are added first, then sector dummies and finally the variable of interest. As in a stepwise regression, it is expected that $+\epsilon_{ij}$ decreases at each step, hence further observed and relevant information is included and a higher fraction of variance could be explained. Contrary to the rather technically driven standard stepwise algorithms, however, the calculation proposed here is fully based on theoretical implications. It is therefore assumed that the final model will not entirely hold significant predictors. The approach has two advantages. First, the overall model validity can be assessed for each set of variables in a consecutive manner. Second, the explanatory value of each set can be derived from the corresponding model properties.

5.3.3.2   Model performance

Evaluation of the model follows roughly the structure proposed by Peng, Lee, and Ingersoll (2002). First, the overall model is evaluated. Second, the model's goodness of fit is assessed. Lastly, the predicted probabilities are validated. The effectiveness of the overall model is evaluated by comparing estimates of the base model without predictors (often also known as the intercept-only model) to estimates of the model derived from equations (2) and (3). The underlying idea is that model predictions are improved by integrating relevant regressors. The predictions yield the categories defined in equation (1) by using a cut value of 0.5 for probabilities. In other words, firms with probabilities of greater than 0.5 are assumed to be innovative, while firms with estimated probabilities smaller than 0.5 are assumed to be non-innovative. The evaluation approach is very practical as it demonstrates the real predictive performance of the model, assuming that the sample is representative. Table 5-3 summarizes the performance of the specified model compared to the intercept-only prediction. In the table, the observed values are compared to the predicted values. The overall performance of the model is fairly high, with 75.5% correct estimates, i.e. the model assesses seven out of ten firms correctly. The high predictive value of the specified model is also reflected by the performance increase compared to the baseline. While the intercept-only model assumes that all firms are part of the non-innovative group and assesses only 53.9% correctly, the predictive power of the actual model is approximately 40% better. Apparently, innovative firms are predicted less accurately than their non-innovative competitors, partially due to their lower representation in the sample. On average, 71.9% of the innovators are classified correctly, compared to 78.6% of the non-innovators. This model assessment is of considerable practical importance. Policymakers, for example, could use the logit formula predicted below to assess whether a firm is potentially innovative or not. This may also help with the decision process of allocating support funds.

| | Predicted | | | | | |
| | Baseline model | | | Specified model | | |
Observed	Non-innovative	Innovative	% correct	Non-innovative	Innovative	% correct
Non-innovative	262	0	100.0	206	56	78.6
Innovative	224	0	0.0	63	161	71.9
Overall % correct			53.9			75.5

Table 5-3: Predictive performance of the specified model

Hosmer and Lemeshow (1989) nevertheless suggest still more rigorous methods to evaluate the overall model. Three main concepts are used in the section that follows: (1) the omnibus test of coefficients, (2) the Hosmer-Lemeshow test, and the (3) 2-log-likelihood and R^2. Rather like the F test in OLS regression, the overall model can be assessed by testing the hypothesis that all coefficients are equal to zero. This test is called the omnibus test of coefficients (1) and considers changes to the 2-log-likelihood, which follows a Chi² distribution. Peng, Lee, and Ingersoll (2002) In essence, the difference between the 2-log-likelihood values of the base and the actual model is tested for significance. As reported in the model evaluation in figure 5-4 the Chi² value of the overall model is 183, with 11 degrees of freedom. The corresponding significance level is thus below 1%. The hypothesis that $\beta_0 = \hat{a}_1 = \cdots = \hat{a}_{11} = 0$ can therefore be rejected, meaning that the predictors in the model have a significant impact on a firm's probability of innovating. The second test (2), the Hosmer-Lemeshow (H-L) test, is an inferential test to assess the goodness of fit of the model Hosmer and Lemeshow (1989). The test is based on a modified Pearson Chi² test. The basic idea is to replicate the overall population in several subpopulations and assess model fit in each group by comparing observed and expected frequencies. The null hypothesis is that the model does not have an adequately fit: A non-significant test result would indicate a good model fit to the data. However, the test may not always be reliable, for example if categories are empty or understaffed. (Schendera (2008)) For the present model, ten categories are constructed, where each is composed of a sufficient number of firms, i.e. 45-49. The model seems to fit the data well in all categories and the test is highly insignificant with a p value of 0.778. The null hypothesis, that the model does not fit the data, can thus be rejected. In OLS regressions, R^2 is a strong measure to describe how much of the overall variation in the dependent variable can be described by a specified set of predictors. Unfortunately, no such measure is available for logit models. Several scholars have therefore attempted to develop comparable measures, which are often called pseudo-R^2 and vary across different concepts. However, none of the measures developed to date corresponds to the percentage of variance explained by this model. (Hoetker (2007))

Model evaluation

Test	Chi²	df	p
Omnibus-test of coefficients	183.084	11	0.000
H-L goodness of fit	4.807	8	0.778

Modell summary

2-log-likelihood	487.680
Cox&Snell R²	0.314
Nagelkerke R²	0.419

Table 5-4: Model summary and evaluation

Table 5-4 shows the 2-log-likelihood value and pseudo-R^2 according to Cox and Snell and to Nagelkerke. Both measures are based on a comparison of the baseline with the specified model likelihood, where the smaller the ratio, the better the model and the higher the pseudo-R^2.[17] The difference between Cox and Snell and Nagelkerke is that the latter measure is adjusted and cannot be less than 0 or more than 1. In fact, Cox and Snell's R^2 never reaches 1 either, even if the model perfectly predicts the observed outcome. The specified model has a Nagelkerke R^2 of 42%, which can be considered fairly high and supports the findings from (1) and (2). Performance valuation of the model was validated by running another subsample comprising 240 randomly selected items. The subsample maintained significance (p<0.000), rejected the H-L null hypothesis and maintained good predictive value with 78.3% correct predictions in total. The model is considered robust.

---

[17] For a definition of each measure and an overview, see Hoetker (2007).

5.3.3.3   Underlying model assumptions and validity

Up to this point, the empirical results reported here have to be considered preliminary, even though the model fits the data well. So far, basic model requirements have been assumed to hold true, but have not been analyzed specifically. This is done in the section below. Logit models require four basic properties. First, predictors need to be independent of each other. Second, residuals should possess the same variance. Homoskedasticity, for instance, increases model stability but is not a necessity. Third, residuals need to be independent of predictors. Lastly, a sufficient number of cases must be available. In this context, Schendera (2008) presents a collection of "rules of thumb" for each requirement derived from literature: (1) Predictors can be considered independent if standard errors of coefficients are not too high (>2) and the correlation between predictors does not exceed 70%. (2) Homoskedasticity is confirmed when plotted residuals randomly spread around the zero. (3) Residuals are independent of predictors if they are not correlated. (4) The data set must at least comprise 50 items. More items are needed if the distribution of 1 and 0 is not symmetrical and the number of predictors is high. Spicer (2005) even recommends a minimum of 100 cases and 50 cases per independent variable. To test the first assumption, the correlation coefficients between variables are calculated. Since not all variables are cardinal, a non-parametric methodology must be adopted. Bortz, Lienert, and Boehnke (2008) propose to use Spearman's ρ for non-cardinal data sets. The methodology is based on rank correlation, where ranks are assumed to be normally distributed. Statistical significance can be tested accordingly. Table 5-5 presents the results of the correlation analysis. Industry dummies are not reported, as their correlation coefficients remain below 30%.

Spearman's rho		a	b	c	d	e	f
INNOVATION	a	1.000					
ln(RD)	b	0.334***	1.000				
ln(AGE)	c	0.042	-0.187***	1.000			
ln(EMPLOY)	d	0.303***	-0.103**	0.493***	1.000		
PROFIT	e	0.011	0.048	-0.059	-0.119***	1.000	
TC	f	0.261***	0.116***	0.134***	0.268***	-0.127***	1.000
ZRESID		0.863***	0.044	-0.02	0.028	-0.017	0.007

1%, 5% and 10% confidence level indicated by *, ** and *** respectively

Table 5-5: Correlation of predictors and residuals

None of the reported coefficients exceed the critical value, with correlations generally mirroring the results of regression analysis. Firm age and size share the highest significant correlation, followed by R&D and innovation. Interestingly, firm age and size correlate positively, while the correlation between age and innovation is not significant. Together with low standard errors for predictors (all < 1), it can be concluded that multicollinearity does not pose a problem to the model. The second requirement was confirmed by plotting standardized residuals *(ZRESID)* with the dependent variable. As expected, it appears that residuals spread randomly around zero. In order to test the third property, *ZRESID* are calculated for each item and then correlated to predictors. The result is also reported in Table 5-5. As can be seen, the correlation of residuals to predictors is random and does not exceed 5%. It can thus be concluded that autocorrelation does not exist. Finally, and with respect to the last property, it is fair to argue that the data set is sufficiently large, particularly considering that distribution among the categories is fairly symmetrical. The basic requirements for the model are given and the results reported appear to be valid. Overall, the model seems to perform well, showing a good fit with the data. In light of satisfactory model performance and the binary nature of the dependent variable, no alternative specifications were made. The next step therefore focuses on the results of the model, i.e. coefficient statistics.

5.3.3.4   Presentation of results

5.3.3.4.1 Presentation of coefficient results and regression equation

The results of the specified model lend fundamental support to the postulated relationships. Table 5-6 recapitulates the expected and actual direction of relationships and reports the beta value (β) and its corresponding standard errors *(S.E.)*, the Wald statistic and the odds $(Exp(\beta) = e^{\beta_i})$ for each independent variable.

Predictor	Expected relationship	Actual relationship	β	S.E.	Wald	p	Exp(β)
TC	+	+***	0.860	0.210	16.810	0.000	2.364
ln(R&D)	+	+***	0.775	0.107	52.066	0.000	2.171
ln(AGE)	+	-*	-0.318	0.170	3.506	0.061	0.728
ln(EMPLOY)	+	+***	1.485	0.203	53.741	0.000	4.416
PROFIT	+	+*	0.115	0.066	3.032	0.082	1.121
RENEWABLE	+/-	+	0.100	0.244	0.170	0.680	1.106
EFFICIENCY	+/-	+	0.200	0.252	0.628	0.428	1.221
WATER	+/-	+	0.124	0.275	0.203	0.652	1.132
MOBILITY	+/-	+	0.038	0.510	0.005	0.941	1.038
RESOURCES	+/-	+***	1.111	0.327	11.525	0.001	3.039
WASTE	+/-	+	0.158	0.252	0.391	0.532	1.171
β_0	+/-	-***	-4.857	0.772	39.553	0.000	0.008

with n=492, 1%, 5% and 10% confidence level indicated by *, ** and *** respectively

Table 5-6: Empirical results of the first model on innovation

The beta values represent the maximum likelihood estimate coefficients for the model. In other words, the coefficients are generated in a way that reproduces the actual group allocation in a best possible manner. The coefficient is not intuitive and is a common source of misinterpretation. It basically says that an increase of one unit in the capital endowment increases the log odds of belonging to the innovation group by the coefficient value. A negative sign would imply that the log odds of belonging to the group decrease. There have been several propositions in literature to use more meaningful values, e.g. setting other variables at their means, calculating probabilities or using odds (see Hoetker (2007) for an overview). The present study focuses particularly on odds, because these seem to be most common in literature and are easy to interpret. (Spicer (2005)) The predicted logit equation of *INNOVATION* for firm i with technology j is derived from coefficient values and can now be formulated as follows:

$$\text{Predicted logit of } INNOVATION = -4.875 + 0.860 \cdot TC + 0.775 \cdot \ln(R\&D) + (-0.318) \cdot \ln(AGE) + 1.485 \cdot \ln(EMPL) + 0.115 \cdot PROFIT + 0.100 \cdot RENWABLE + 0.200 \cdot EFFICIENCY + 0.124 \cdot WATER + 0.038 \cdot MOBILITY + 1.111 \cdot RESOURCES + 0.158 \cdot WASTE$$

The robustness of coefficient estimates was tested by running the same model with random subsets of the overall sample. The results appeared to be relatively robust with respect to model performance and fit. However, there were some variations in the significance and magnitude of the coefficients. For example, a subset involving 50% (240) randomly selected items yielded the same coefficient signs but generated insignificant p values for *ln(AGE)* and *PROFIT*.

5.3.3.4.2 Results for hypothesis 1: Capital endowment and innovation

As expected, the coefficient for TC is positive at 0.86. The standard error is 0.21. The odds of capital endowment *(TC)* are 2.36. As in the simple example provided above, the odds can be interpreted as marginal effects. For example, subtracting 1 and multiplying by 100 yields a marginal effect of 136%. An increase of one unit in the capital endowment thus increases the odds of belonging to the innovative group by 136%. The effect has a relatively high magnitude and is highly significant with a p value < 0.000 derived from the Wald statistics. However, it should be noted that even the odds are not standardized and can thus only be compared to other variables and their magnitude if they share common measurement units. The analysis controlled for the main determinants of innovation, which are discussed in the next section. The results provide fundamental support to the first hypothesis, namely that capital endowment, particularly measuring the non-observable capital supply, is a major driver of firms' innovation output. This is in line with basic innovation theory Schumpeter (1947) and empirical evidence suggesting that capital availability is an important determinant of innovation (e.g., Correa, Fernandes, and Uregian (2010); Galende and de La Fuente (2003)) (see section 3.2.). Moreover, the finding implicitly supports the idea that the provision of government capital may have a positive effect on firms' innovation activity. (Klassen, Pittman, and Reed (2004)) The capital endowment construct also avoids problems with differing impacts for different capital forms, as reported by Bah and Dumontier (2001) and Bhaird (2010). Clearly, the solution does not come free of charge, hence the model lacks any distinction between different capital forms.

5.3.3.4.3 Results of control variables

With one exception, the estimated coefficients are not surprising and reveal the relationships postulated in literature. R&D correlates positively to innovation, with an odds ratio of 2.17 at the 1% confidence level. This confirms the assumption that a firm must invest in research and development to actually generate innovation output. This is also consistent with literature (see section 3.2.2.2.1). It is further controlled for firm size, i.e. *EMPLOY* and *PROFIT,* both of which yield the expected positive relationship (see sections 3.2.2.2.6 and 3.2.2.2.4). A marginal increase of *ln(EMPLOY)* leads to an increase of 342% in the odds of belonging to the innovative group. Firms' resources and their accumulation seem, in line with most literature, to be critical to the generation of innovation output. The result is highly significant (p<0.000). The argument is supported by the *PROFIT* variable, as profit represents an important corporate resource. The coefficient for *PROFIT* is also positive, with an odds ratio of 1.121, and significant at the 10% confidence level. This means that a marginal increase in the ordinal profit measure leads to a significant increase of 12% in the respective odds. Following a similar line of thought, firms' age was also expected to correlate positively to innovation (see section 3.2.2.2.5). However, this hypothesis is not supported by the given data set. The coefficient of *ln(AGE)* is negative with a standard error of 0.17, leading to an odds ratio of less than one. In essence, a marginal increase in the age measure leads to a 27% decrease in the odds of a firm being innovative. The result is statistically significant at the 10% confidence level. The negative relationship is surprising and, in particular, contradicts the positive relationship between innovation and firm size, given that innovation research suggests that size and age are both proxies for firm knowledge and resources (e.g., Galende and de La Fuente (2003)). However, research also argues that "too much" experience and size can lead to the opposite outcome. (e.g., Müller and Zimmermann (2009)) The data seems to indicate that age is a proxy for inertia while size is an unrelated proxy for firms' resources. Most innovative firms are thus large and still rather young. Taken together, the results mimic findings by Huergo and Jaumandreu (2004), who argue that very young firms have unique capabilities and may do new things in new ways, while older firms become less flexible and may only offset the comparative advantage of young firms by accumulating a larger resource base. Another explanation for the anomaly is the fact that insufficient distinctions are drawn between innovation types, as mentioned earlier. Proper categorization may have led to the conclusion that disruptive innovations correlate negatively to size and age, while incremental innovations show the opposite sign. The mixed measures used in the analysis may thus have led to mixed results with respect to firm age and size (see Figure 5-3). To further investigate this apparent anomaly, a second regression is run on process innovation. The model is similar to the one specified above, except that $INNOVATION_{ij}$ is replaced by the process-related measure $PCINNOVATION_{ij}$. The coefficient results are reported in Table 5-7.

Predictor	Expected relationship	Actual relationship	β	S.E.	Wald	p	Exp(β)
TC	+	+**	0.465	0.210	4.914	0.027	1.592
ln(R&D)	+	+***	0.756	0.125	36.489	0.000	2.129
ln(AGE)	+	-	-0.007	0.176	0.001	0.969	0.993
ln(EMPLOY)	+	+***	1.100	0.186	34.783	0.000	3.003
PROFIT	+	+	0.084	0.071	1.406	0.236	1.088
RENEWABLE	+/-	+	0.118	0.266	0.196	0.658	1.125
EFFICIENCY	+/-	+	0.057	0.269	0.045	0.832	1.059
WATER	+/-	+	0.121	0.309	0.154	0.695	1.129
MOBILITY	+/-	+	0.453	0.494	0.841	0.359	1.574
RESOURCES	+/-	+***	0.886	0.321	7.639	0.006	2.425
WASTE	+/-	+	0.004	0.277	0.000	0.987	1.004
β_0	+/-	-***	-5.631	0.870	41.918	0.000	0.004

with n=492, 1%, 5% and 10% confidence level indicated by *, ** and *** respectively

Table 5-7: Empirical results of the first model on process innovation

As can be seen, coefficient signs and odds are more or less comparable to the first regression, indicating that the measure of innovation types is weak with respect to the degree of differentiation. This supports the approach adopted, which uses a more general single measure. However, there is also indicative evidence to support the argument with respect to the different effects of age and size. A firm's age loses significance with a p value of 0.969, while the coefficient gets closer to zero and odds diverge toward one. This may be interpreted as a tendency toward a positive relationship between process innovation and age, which would in turn support the idea that the negative age and innovation correlation could indicate a greater degree of non-incremental and disruptive innovations in the greentech industry. Nonetheless, the overall line of argument is rather speculative and is, at best, only implicitly supported by the data. Another interesting observation is that the level of magnitude for TC decreased, implying that process innovation output is less dependent on capital endowment than other innovation types.

The analysis also controls for technology lines. It appears that firms committed to sustainable resources and new materials are, in both cases, more innovative than firms from other fields of technology. Given that the category also includes white biotechnology, these findings seem plausible. The other industry dummy variables are not significant. In light of the small proportion of *RESOURCES* firms, it is fair to argue that greentech can be considered a rather homogenous industry in the given context.

5.3.4   Financing growth of greentech firms (model 2)

5.3.4.1   Specification of the regression model

The set-up for the second hypothesis is comparable to the one proposed in the innovation case. However, there is one important difference, namely that firm growth is measured as a cardinal variable. It is for this reason that most studies in the field use OLS regression to assess the determinants of growth. (e.g., Wijewardena and Cooray (1995); Becchetti and Trovato (2002); Kang, Heshmati, and Choi (2008); Honjo and Harada (2006); Lee (2010)) Clearly, models are specified differently and may deal with both vertical and horizontal data. At the same time, firm growth as a dependent variable and a variable of interest has been the subject of substantial discussion in literature due to the variety of different measures. (Bahadir, Bharadwaj, and Parzen (2009)) The model proposed here uses OLS regression and a relative measure as defined above, i.e. the rate of sales growth, following the approach already adopted by several researchers. (e.g., Sutton (1997); Carpenter and Petersen (2002); Hyytinen and Toivanen (2005); Lee, Lee, and Pennings (2001)) However, one fundamental problem arising from the nature of relative growth is its variability and, hence, its high level of variance. This is also reflected in two ways in the analyzed data set. First, the standard deviation for mean growth is relatively high at 64%, whereas less than half of all firms (175 out of 536) are in the range of $E(GROWTH) - \delta$ and $E(GROWTH) + \delta$. Second, a relatively large number of outliers evidence particularly high growth rates. For example, 43 firms have grown by more than 50% p.a. over the past three years. This finding is also reflected by the relatively high mean of 15% and the significantly lower median of 5% (see Table 5-2). The problem is important, because OLS regression is highly sensitive to outliers. Even a small number of outliers can bias the results, potentially leading to a misinterpretation of the regressors. (Fahrmeir, Kneib, and Lang (2009)) Another related problem is the potential violation of the normality assumption for error terms, reflecting the gap between the predicted and actual values. In essence, OLS models assume that residuals are normally distributed and possess variance homogeneity (homoskedasticity). (Schendera (2008)) Some studies (e.g., Del Monte and Papagni (2003); Lee (2010)) use logarithmic regression functions to cope with these problems. However, since the logarithm can only be applied for values greater than one, this approach is of limited use for the given data set. The dissertation therefore proposes to transform growth rates into quantile ranks. To do so, it follows the idea of rank transformation in order to "bridge" parametric and non-parametric statistics. (Conover and Iman (1981); Iman and Conover (1979)) Ranks are, of course, widely used in statistics and serve as the basis for non-parametric tests. However, their integration in regression models is seldom applied, with a few exceptions. Rank transformations are, for example, successfully employed in real estate valuations. (e.g., Cronan, Epley, and Perry (1986)) The proposed transformation into quantile ranks is supported by theoretical and non-technical considerations. First, the dissertation is interested in factors that lead to higher growth and accelerate corporate development, whereas the horizontal research set-up yields comparative results between firms. The quantile rank matches the idea well, as it expresses the proportion of firms that underperform the firm in question. For example, a quantile rank of 85% for a certain firm means that it has grown faster than 85% of its peer over the past three years. Second, the analysis did not account for the distinction between organic and acquisitive growth, as discussed earlier. The resultant bias, though not fully resolved, is substantially reduced by this approach. Third, the purpose of the analysis is not to forecast exact growth rates, but rather to assess whether a certain firm has the potential to grow and to identify factors that may drive this growth. Both issues are addressed by a quantile rank regression. Indeed, the analysis below shows that the quantile rank model fits the sample data well and fulfills all underlying OLS assumptions. It also shows that the explained variance (R^2) is nearly four times as high as in alternative specifications. The quantile rank (qR) of a certain growth value (v) can be defined as the relative position of the observed value in an ordered data set (1)[18]:

(1) $(v) = \frac{R(v)-1}{n-1}$,

With $0 \leq qR(v) \leq 1 - \frac{\sum_{i=1}^{n} 1\{x_i=x_n\}-1}{n-1}$,

and $R(v) = \sum_{i=1}^{n} 1\{x_i < v\} + 1$

$R(v)$ denotes the absolute rank of v and assumes an ascending order in the data set. The rank of v is therefore equal to the number of smaller values plus one. Since v represents the observed $GROWTH_i$, the quantile rank of the variable of interest is denoted as $qR(GROWTH_i) = GROWTH'_i$. It is calculated for all expressions of the dependent variable. For tied values, the same quantile rank as defined in (1) is allocated. All quantile ranks are rounded to the second decimal place, the lowest rank being 0 and the highest 1. Clearly, the OLS regression does not limit its predictions to these ranks. However, a censored model is not required because predicted values greater than 1 (or smaller than 0) can simply be interpreted

---

[18] This is only possible if the value of interest is definitely given in the sample, which holds true for the specified model. Otherwise the quantile rank would have to be interpolated.

as belonging to the highest (or lowest) rank. Contrary to the methodology developed by Iman and Conover (1979), the present study transforms only the dependent variable, but not the independent variables, for two main reasons. First, quantile ranks of growth rates can be interpreted without backward transformation. In other words, quantile growth ranks represent a new distinct variable. Regression results can thus be interpreted easily and on a stand-alone basis. For example, a certain firm with certain characteristics can be forecast to perform better than $x\%$ of its competitors. Second, independent variables also contain categorical measures that cannot be ranked. It is thus assumed that the quantile rank of firm i's growth rate in industry j is allowed to depend linearly on capital endowment *(TC)* and on a set of independent variables derived from literature and presented in the previous chapters. Overall, the second model can be specified as (2):

(2) $GROWTH'_{ij} = \beta_0 + b_1 \cdot TC_{ij} + \sum_{l=1}^{n} b_l \cdot x_{ijl} + \sum_{k=1}^{m} b_k \cdot y_{jk} + \epsilon_{ij}$

where ϵ_{ij} is an error term accounting for unobservable effects. The error term is assumed to be normally distributed with an expected value of zero. The model further assumes homoskedasticity, i.e. variance homogeneity, and the stochastic independence of error terms. The dependent variable $GROWTH'_{ij}$ is further assumed to depend on two sets of control variables. The first set (x_l) comprises the natural logarithm for *R&D*, *AGE* and *EMPLOY*, a measure of growth orientation *(GROWTHO)* and the ordinal variable *PROFIT*, i.e. $\sum_{l=1}^{n} b_l \cdot x_{ijl} = b_2 \cdot lnR\&D_{ij} + b_3 \cdot GROWTHO_{ij} + b_4 \cdot lnAGE_{ij} + b_5 \cdot lnEMPLOY_{ij} + b_6 \cdot PROFIT_{ij}$. The second set (y_k) is composed of sector dummies, i.e. $\sum_{k=1}^{m} b_k \cdot y_{jk} = b_7 \cdot RENEWABLE_j + b_8 \cdot EFFICIENCY_j + b_9 \cdot WATER_j + b_{10} \cdot MOBILITY_j + b_{11} \cdot RESOURCES_j + b_{12} \cdot WASTE_j$. The model also assumes that the independent variables are correlated to the dependent variable but not, or only to a limited extent, to each other, i.e. there is no multicollinearity. Equation (2) is estimated via SPSS using the ordinary least squares methodology. The model primarily tests the second hypothesis, arguing that capital endowment, having been controlled for relevant factors, is a major enabler for firm growth and development.

5.3.4.2 Model performance

The evaluation of the overall model is aligned with the structure proposed for the first model. First, the overall model fit and explanatory value are assessed. Second, underlying model assumptions are checked. Finally, coefficient results are reported and tested for reliability. This section deals with the first issue. According to Hair (2006), there are three main measures for assessing model fit: (1) R^2, (2) the standard error of the estimate, and (3) the analysis of variance (ANOVA). R^2 (1) represents the coefficient of determination and is easy to calculate, unlike logistical models. It measures the proportion of variance in the data set that is explained by the specified model. In other words, it explains how well the model predicts the dependent variable by establishing a linear relationship with predictors. Technically, the measure supplies information about the model's goodness of fit by quantifying the distances between the observed values and the estimated linear regression line or (in the multivariate case) area. R^2 is therefore defined as one minus the sum of squared error terms divided by the total sum of squares. In the univariate case, the sum of squared error terms measures the cumulative distance of observed values from the regression line and the total sum of squares measures the cumulated distance from the sample mean. R^2 can thus assume values between 0 for no fit and 1 for congruency between observed data and the model's prediction. In the univariate case, R simply equals the Pearson correlation coefficient, reflecting the degree of correlation between the variables. It is important to note that the coefficient of determination depends on the number of predictions, as the latter define the degrees of freedom Hair (2006). The addition of a new significant or non-significant variable will thus always cause R^2 to increase, which raises concerns about the ability to generalize the model as a whole. The problem is particularly acute when the number of predictors converges toward the number of items in the tested sample. Sample size is discussed later, however. Right now, the major concern is to adjust R^2 for the effects described. This can be done by integrating the correlation of predictors to total sample size, leading to the adjusted coefficient of determination. A second measure to assess the predictive quality of a linear model is the standard error (2) of the estimate. This can be seen as a sort of standard deviation for the regression model, because it measures the variation of observations with respect to the regression line. According to Schendera (2008), a model adjusts well to the data if the standard error is below the standard deviation for the dependent variable. Table 5-8 summarizes the performance and overall fit of the specified model.

Model evaluation

R	R^2	adj. R^2	S.E.
0.613	0.376	0.360	0.233

ANOVA

	Sum of squares	df	Mean square	F	p
Regression	15.460	12	1.288	23.805	0.000
Residual	25.653	474.000	.054		
Total	41.112	486			

Table 5-8: Model summary and performance

The multiple R of the specified model is 61%, leading to an R^2 of 38%. The adjusted R^2 differs little, as the number of observations is fairly high and utilized predictors are scaled down to relevant factors. The proportion of explained variance is thus 36%, meaning that the identified predictors explain 36% of the variations in $GROWTH'$. The model performs rather well compared to other studies in the field. Becchetti and Trovato (2002), for example, report an adjusted R^2 of 5.4% with a sample of 1,832, supporting the transformation of the dependent variable proposed above. The standard error for the estimate is 23%. Considering that the standard deviation of $GROWTH$ is 64%, this appears sufficiently low. Moreover, the standard deviation of 29% for the dependent variable $GROWTH'$ is considerably higher, indicating that the model is well aligned with the data. The third indicator ANOVA provides a statistical test of overall model fit. This corresponds to the omnibus test of coefficients conducted in the first model, with the null hypothesis that all coefficients are equal to zero, i.e. $\beta_0 = \beta_1 = \cdots = \beta_{11} = 0$. The test is related to R^2 because it assesses the significance of the change in explained variance. The hypothesis thus corresponds to testing if R^2 is equal to zero. In this context, the given model reduces the prediction error by 38% (15.460/(15.460+25.653)). The F value reveals statistical significance with respect to the degrees of freedom. The p value is <0.000 and thus highly significant. The hypothesis that the identified independent variables do not have a linear correlation to GROWTH' can thus be rejected at the 1% confidence interval. In conclusion, it seems that the specified model is appropriate and shows good predictive power. All three tests consistently provide supporting evidence. However, assessing model performance is only one step toward a rigorous statistical analysis. The second step is to evaluate whether the underlying OLS assumptions are fulfilled, because only then it is possible to derive inferential implications.

5.3.4.3 Underlying model assumptions and validity

The underlying assumptions for OLS are very strict and exceed those of binary logistic regressions. Hair (2006) mention five main areas. First, the phenomenon of interest must be linear. Second, error terms must possess a constant variance. Third, error terms must be independent of each other. Fourth error terms must also be independent of predictors. Fifth, error terms must fulfill normality assumptions. Two additional assumptions are pointed out by Schendera (2008). In line with logit model assumptions, predictors must be independent and the sample must be of sufficient size. The linearity (1) and homoskedasticity (2) assumptions can be tested together using graphical means. Plotting standardized residuals with standardized observations of the dependent variable yields a rhombic distribution around zero. There is no observable pattern, implying a linear relationship. Moreover, the graph is symmetrical, indicating that the variance of residuals is homogeneous. The rhombic shape (instead of a rather round, cloudy shape) is due to the censored nature of the dependent variable. The analysis uses the Durbin-Watson autocorrelation test to determine whether residuals are independent of each other (3). (Durbin and Watson (1951)) This procedure sequentially compares residuals, testing the null hypothesis that autocorrelation is zero. The null hypothesis must be rejected if the Durbin-Watson statistic deviates from 2. Values <2 indicate positive autocorrelation and >2 negative autocorrelation. The proposed model yields a value of 1.83, implying that autocorrelation does not present a problem. The next requirement, i.e. that residuals must be independent of predictors (4), is assessed together with the sixth requirement, namely that predictors must be mutually independent. The approach adopted is similar to the one in the previous section, i.e. the dependence of residuals and the existence of multicollinearity are rejected if correlation coefficients remain below the critical threshold of 0.7. The results of correlation analysis are reported in Table 5-9.

Spearman's rho		a	b	c	d	e	f	g
GROWTH'	a	1.000						
ln(RD)	b	0.168***	1.000					
GROWTHO	c	-0.498***	0.073*	1.000				
ln(AGE)	d	-0.288***	-0.187***	0.071	1.000			
ln(EMPLOY)	e	-0.03	-0.103**	-0.036	0.493***	1.000		
PROFIT	f	0.303***	0.048	-0.188***	-0.059	-0.119***	1.000	
TC	g	0.062	0.116***	0.085**	0.134***	0.268***	-0.127***	1.000
ZRESID		0.794***	0.007	-0.036	-0.073	0.026	0.048	0.02

1%, 5% and 10% confidence level indicated by *, ** and *** respectively

Table 5-9: Correlation of predictors and residuals

GROWTH' correlates significantly to R&D, growth orientation, age and profit. The correlation to size and capital endowment is surprisingly low and not significant. As in the previous model, the highest correlation among predictors is between firm age and size. However, all correlations remain below the critical threshold of 0.7. Accordingly, multicolinearity does not seem to detract from the model's results. The same holds true for the standardized residuals, which are clearly independent of regressors. The null hypotheses that residuals are not independent and predictors are correlated can therefore be rejected. To test the fifth assumption, namely whether residuals are normally distributed (5), residuals are plotted in a histogram and a one-sample Kolmogorov-Smirnov test is run using the normality parameters proposed by Bortz, Lienert, and Boehnke (2008). The histogram shows a good match with the normal curve and it seems that the assumption is satisfied. To confirm this finding, a Kolmogorov-Smirnov test is conducted. The basic principle of this test is to compare the maximum differences between observations and postulated values for a certain distribution. In the case in point, the result of the Kolmogorov-Smirnov test is highly insignificant ($p=0.267$). The null hypothesis that the residuals are not normally distributed is therefore rejected. The indicative result from the graphical investigation could be confirmed.

Finally, it was suggested that a necessary requirement for OLS regression analysis is sufficient sample size to avoid overfitting the predictors. (Hair (2006)) According to Spicer (2005), a rule of thumb is that the overall sample must be greater than 50 plus eight times the number of independent variables. Following this line of thought and including industry controls, the specified model would require a sample of at least 146 firms. This number is easily exceeded by the 487 firms in the given data set. The reduced number of firms compared to the original sample of 536 is due to missing values with regard to profit. The model also proved to be the best when compared to alternative specifications. For example, one attempt sought to assess the non-transformed growth variable using standard OLS regression. Although coefficients were significant with comparable signs but a lower magnitude, the model failed to fulfill the assumptions of residual distribution and heteroskedasticity. Even when corrected for outliers with large COOK distance values, this still did not resolve these problems. The analysis was also conducted using log-linear ordinal regression models, transforming the growth variable into decentiles and quintiles. Although the model did fit the data quite well, not all the thresholds yielded significance. Overall, it seems that the specified model fits the sample well and that all OLS requirements are met without restriction. The next section therefore presents the results of the analysis.

5.3.4.4   Presentation of results

5.3.4.4.1   Presentation of coefficient results and regression equation
The model results lend fundamental support to most of the postulated relationships. Table 5-10 summarizes the empirical results and recapitulates the postulated effects for each variable. The summary reports coefficients (b), standard errors $(S.E.)$, t statistics (t), p values (p) and standardized coefficients (β). The coefficients were generated in a way that minimizes the distances between observed and estimated values. A positive sign indicates that an increase of one unit in the given variable will increase the dependent variable by b. A negative sign denotes an inverse relationship of the same kind. The standard error $(S.E.)$ reflects the potential variation of coefficients across different data sets and samples from the same population. The standard error in the estimate can therefore be regarded as an indicator of the model's reliability, with a smaller standard error leading to smaller confidence intervals and vice versa. Although it is a lot easier to interpret the coefficients than in probability models, the differences between variable units may cause confusion. The standardization of coefficients resolves this issue and makes the variables comparable. The standardized coefficient (β) thus expresses the relative impact of a certain variable on $GROWTH'$ compared to the other factors. In other words, it indicates by how many standard deviations the dependent variable changes when the independent variable is increased by one standard deviation. A standardized coefficient of zero indicates that the given variable has no effect on the dependent variable.

Predictor	Expected relationship	Actual relationship	b	S.E.	t	p	β
TC	+	+***	0.057	0.018	3.154	0.002	0.122
ln(R&D)	+	+***	0.035	0.009	3.946	0.000	0.152
GROWTHO	+	-***	-0.267	0.022	-12.06	0.000	-0.451
ln(AGE)	-	-***	-0.074	0.015	-4.822	0.000	-0.218
ln(EMPLOY)	+	+**	0.036	0.015	2.306	0.022	0.105
PROFIT	+	+***	0.033	0.006	5.460	0.000	0.205
RENEWABLE	+/-	+	0.006	0.023	0.239	0.811	0.009
EFFICIENCY	+/-	-	-0.007	0.024	-0.298	0.766	-0.011
WATER	+/-	-*	-0.045	0.026	-1.702	0.089	-0.064
MOBILITY	+/-	-	-0.028	0.049	-0.575	0.566	-0.024
RESOURCES	+/-	+	0.013	0.031	0.412	0.681	0.016
WASTE	+/-	-	-0.028	0.024	-1.173	0.242	-0.045
β_0	+/-	+***	0.472	0.062	7.618	0.000	0.000

with n=492, 1%, 5% and 10% confidence level indicated by *, ** and *** respectively

Table 5-10: Empirical results on growth

A positive value indicates a positive impact and a negative value a negative impact, where the upper and lower boundaries are +1 and -1 respectively. It is important to note that β is not equal to a marginal factor. For instance, a 1% change in x leads to a 5% change in y. (Greenland, Schlesselmann, and Criqui (1986)) This is often misunderstood and even claimed to be wrong in standard textbooks (e.g., Schendera (2008)). Clearly, the magnitude of the effect does not necessarily say anything about factor relevance. Some factors may thus have a high coefficient but may not necessarily be important to the model. The relevance of each factor can be assessed by the t statistic and the derived p value. In the specified model and the given data set, both the variable of interest and the control variables (with the exception of industry dummies) are entirely significant at the 1% or 5% confidence levels. All factors can thus be interpreted and seem to be relevant factors. The regression equation predicting the quantile rank of growth for firm i in industry j can be derived from coefficients as follows:

Predicted value of $GROWTH' = 0.472 + 0.057 \cdot TC + 0.035 \cdot \ln(R\&D) + (-0.267) \cdot GROWTHO + (-0.074) \cdot \ln(AGE) + 0.036 \cdot \ln(EMPLOY) + 0.033 \cdot PROFIT + 0.006 \cdot RENEWABLE + (-0.007) \cdot EFFICIENCY + (-0.045) \cdot WATER + (-0.028) \cdot MOBILITY + 0.013 \cdot RESOURCES + (-0.028) \cdot WASTE$

The robustness of coefficient estimates was tested by running the same model with a randomly generated subset of the overall sample. The results appeared to be robust with respect to model performance and fit. Using a sample of 240 selected items yielded the same coefficient signs and maintained significance for all relevant variables (excluding industry dummies). Only the degree of significance for capital endowment was reduced to the 10% confidence interval.

5.3.4.4.2 Results for hypothesis 2: Capital endowment and growth

In essence, capital endowment, in particular when it approximates capital supply, enhances firms' growth. The effect remains significant when controlled for other factors. The coefficient value is 0.057. For an increase of one unit in capital endowment, the quantile rank of growth will thus increase by 0.057 units. This effect is stronger than that of firm size and nearly as strong as the effect of ln(R&D). The standardized coefficient has a value of 0.122 and is significant at the 1% confidence level. The second research hypothesis is therefore strongly supported by the model and thus largely in line with both theory (Penrose (1997)) and existing empirical research (see section 3.3.3.3.4). As in the innovation model, capital endowment is abstracted from different capital forms and can therefore not take into account potential divergences in the effects of equity (e.g., Himmelberg and Petersen (1994); Audretsch and Elston (2002)) and debt (e.g., Scellato (2007)). On the other hand, it does not yield ambiguous results, unlike the study by Honjo and Harada (2006). Moreover, the capital endowment measure represents an aggregate of financing intensity and the number of financial instruments used. In this context, the confirmed hypothesis indicates that growth not only needs capital but will use any type of capital to satisfy demand. Ultimately, therefore, the difference between capital instruments does not appear to be expedient. Similar to the innovation model, the results indirectly support studies indicating that government support may lead to greater firm growth and thus industry development. (e.g., Becchetti and Trovato (2002); Kang, Heshmati, and Choi (2008))

5.3.4.4.3 Results for control variables

As assumed by the theoretical framework, R&D seems to be an enabler of growth. This supports the argument that innovation-oriented greentech firms grow faster than the industry average (see section 3.3.3.3). The result not only supports the majority of reviewed empirical evidence, but also seems to reflect the advantages of using both formal and informal R&D expenditures, as small firms in particular tend not to report their R&D spending Gil (2010). The corresponding bias could therefore not be observed in the present study. Firms' age hampers growth rates, supporting the argument that younger firms grow from a smaller base and are more likely to enter growing markets with early-stage technologies (see section 3.1). The effect for age – a β factor of -0.218 – shows a relatively high magnitude and is also significant at a higher confidence level than size. In line with the innovation model, the effects of size and age seem to point in different directions. Larger firms thus grow faster than smaller firms. However, the effect is neither as strong (β=0.105) nor as significant (p=0.022). The positive and significant coefficient is in line with implications drawn from the literature review conducted by Bahadir, Bharadwaj, and Parzen (2009), but contradicts Gibrat's law, which claims that size and growth are independent (Gibrat (1931). It also contradicts basic lifecycle models (e.g., Hanks et al. (1993); Greiner (1998)) that assume a strong correlation between firm size and age as a firm's main demographic characteristics. Nonetheless, it supports the modified lifecycle framework elaborated in chapter 4. Profit enhances growth as it increases a firm's ability to invest. Needless to say, the motivation to grow with profitable projects is greater than the motivation to grow with unprofitable projects. The β is 0.205 at the highest confidence interval. This result is congruent with the postulated relationship and, hence, with existing empirical research. (e.g., Kang, Heshmati, and Choi (2008)) Only one determinant – growth orientation – did not demonstrate the anticipated relationship. Section 3.3.3.3 postulated that the willingness to grow and thus a dedicated growth orientation is a major prerequisite if firms are indeed to grow. The relationship was therefore expected to be positive, i.e. for growth-oriented firms to grow faster than their average peers. Contrary to this hypothesis and to contemporary literature (e.g., Pasanen (2007)), empirical results suggest that growth-oriented firms actually grow more slowly than their less growth-oriented counterparts. The relationship is characterized by a strong β of -0.451. To gain a better understanding of why the effect changes direction, it is important to recall the actual measurement process. In this context, growth orientation was defined as a categorical variable, reflecting the difference between expected and past growth rates. A firm expecting higher growth rates in the next three years than in the past three years would be categorized as a growth-oriented firm and vice versa. Referring to this issue, one study participant argued that the past three years were "not representative for [her] firm due to the consequences of the financial crisis". The measure may thus be biased because it primarily represents firms that suffered more than they would normally have done from the recent financial crisis. The negative effect of the crisis might have a particularly strong impact on fast-growing firms and may thus overcompensate for the actual measurement of growth orientation. Essentially, GROWTHO seems to measure more than one phenomenon and can thus not be interpreted consistently. Finally, as in the innovation model, it seems that most technology dummies are not significant, giving further support to the homogenous industry hypothesis. There is one exception – sustainable water technologies – in which firms seem consistently to grow more slowly at the 10% confidence interval.

5.3.5   Determinants of financial constraints for greentech firms (model 3)

5.3.5.1   Specification of regression model

Researchers have used a variety of models to test for determinants and the existence of financial constraints, depending on the specification of the dependent variable. Some studies (e.g., Freel (1999); Egeln and Licht (1997); Beck and Demirguc-Kunt (2006)) also refer primarily to descriptive statistics. On the one hand, scholars focusing on investment sensitivities as constraint proxies usually employ cardinal dependent variables. These variables are often predicted using OLS regression models. (e.g., Wolf (2006); Bougheas, Görg, and Strobl (2003)) On the other hand, studies employing binominal or ordinal measures for financial constraints often utilize probability models. The actual specification varies broadly and there seems to be no "best" approach. Consequently, the specification depends on the individual research situation. For example, Hyytinen and Pajarinen (2008) analyze disagreements between rating agencies to assess information opacity using fixed-effect models. Other surveys have used probit or logit models to assess the probability that a directly measured event, such as self-assessed constraints, will occur. (e.g., Guiso (1998); Piga and Atzeni (2007); Vos et al. (2007)) This dissertation follows the latter approach, using a binominal proxy for *INVESTCONSTRAINT'*. Following this line of thought, and to test hypotheses 3 through 6, a logit model is specified. In essence, *INVESTCONSTRAINT'* can, like *INNOVATION*, take one of two forms, where *INESTCONSTRAINT'=1* indicates that a firm is subject to financial constraints and *INVESTCONSTRAINT'=0* indicates that a firm is not subject to financial constraints. The model presumes that z_{ij} is an unobserved variable that represents the information opacity and operating risk for firm i in sector j, given that:

$$(1) \quad INVESTCONSTRAINT'_{ij} = \begin{cases} 1, & if\ z_{ij} > 0 \\ 0, & if\ otherwise \end{cases}$$

As in the innovation model, the latent variable z_{ij} is allowed to depend linearly on TC_{ij}, but also on $lnR\&D_{ij}$, $GROWTH'_{ij}$, $GOVEQT_{ij}$, $GOVD_{ij}$, $RELATION_{ij}$, $COLLATERAL_{ij}$, $RETAINED_{ij}$ and $PULL_{ij}$. On the other hand, TC_{ij} is the average of all financial instruments used by a firm. Double counting is not a problem, however, as capital endowment is measured using a different unit to government support programs and retained earnings (absolute versus relative). Nor is any multicollinearity expected due to the large number of capital instruments used by each firm. In keeping with the second model on growth, $GROWTH'_{ij}$ is calculated as the quantile rank of the sample in order to eliminate outliers. The model also corrects for other main determinants derived from literature (see section 3.4.4.4) as well as sectoral effects. z_{ij} can therefore be defined as:

$$(2) \quad z_{ij} = \beta_0 + \beta_1 \cdot TC_{ij} + \beta_2 \cdot \ln(R\&D_{ij}) + \beta_3 \cdot GROWTH'_{ij} + \beta_4 \cdot GOVEQT_{ij} + \beta_5 \cdot GOVD_{ij} + \beta_6 \cdot PULL_{ij} + \beta_7 \cdot RELATION_{ij} + \beta_8 \cdot COLLATERAL_{ij} + \beta_9 \cdot RETAINED_{ij} + \sum_{l=1}^{n} \beta_l \cdot x_{ijl} + \sum_{k=1}^{m} \beta_k \cdot y_{jk} + \epsilon_{ij}$$

where ϵ_{ij} is an error term accounting for unobservable effects. Again, the error term is assumed to be normally distributed. The first set of control variables (x_l) comprises the natural logarithm of *AGE* and *EMPL* and the dummy variable *LEGAL*, i.e. $\sum_{l=1}^{n} \beta_l \cdot x_{ijl} =$

$\beta_{10} \cdot \ln(AGE_{ij}) + \beta_{11} \cdot \ln(EMPLOY_{ij}) + \beta_{12} \cdot LEGAL_{ij}$. The second set (y_k) is composed of the sector dummies mentioned earlier, i.e. $\sum_{k=1}^{m} \beta_k \cdot y_{jk} = \hat{a}_{13} \cdot RENEWABLE_j + \hat{a}_{14} \cdot EFFICIENCY_j + \hat{a}_{15} \cdot WATER_j + \hat{a}_{16} \cdot MOBILITY_j + \hat{a}_{17} \cdot RESOURCES_j + \hat{a}_{18} \cdot WASTE_j$. The probability that firm i might experience financial constraints $P(INVESTCONSTRAINT'_{ij} = 1)$ and it's complement $1 - P(INVESTCONSTRAINT'_{ij} = 1)$ is calculated via the logistical function:

(3) $\qquad P(INVESTCONSTRAINT' = 1) = \dfrac{e^z}{1+e^z} = \dfrac{1}{1+e^z}$

Similar to the *INNOVATION*-model equation, (2) is estimated using the maximum likelihood method based on the Newton-Raphson algorithm. A total of four submodels are run for z_{ij} in a stepwise approach. The submodels are additive and are summarized below. All models include the error term $+\epsilon_{ij}$, which is expected to decrease in each step as additional relevant information is observed and included. The last submodel (d) is equal to equation (2) and is crucial, because it includes all relevant factors.

(a) $\qquad z_{ij} = \beta_0 + \beta_1 \cdot TC_{ij} + \sum_{l=1}^{n} \beta_l \cdot x_{ijl} + \sum_{k=1}^{m} \beta_k \cdot y_{jk} + \epsilon_{aij}$

(b) $\qquad z_{ij} = (a) + \beta_2 \cdot \ln(R\&D_{ij}) + \beta_3 \cdot GROWTH'_{ij} + \epsilon_{bij}$

(c) $\qquad z_{ij} = (b) + \beta_4 \cdot GOVEQT_{ij} + \beta_5 \cdot GOVD_{ij} + \beta_6 \cdot PULL_{ij} + \epsilon_{cij}$

(d) $\qquad z_{ij} = (c) + \beta_7 \cdot RELATION_{ij} + \beta_8 \cdot COLLATERAL_{ij} + \beta_9 \cdot RETAINED_{ij} + \epsilon_{ij}$

The first model (a) tests the third research hypothesis and only includes capital endowment to represent capital supply and demand to build a bridge between findings from literature on financial economics, innovation and growth. The second model (b) tests the fourth and fifth research hypotheses. It is based on financial economics literature, suggesting a causal relationship between innovation/R&D and growth toward financial constraints. The third model (c) also includes government equity- and debt-related funds as well as the firm's dependence on government pull programs. It ultimately tests exogenous factors that alleviate financial constraints. The fourth model (d) also includes firm-related measures.

5.3.5.2 Model performance

The third model uses the same quantitative methodology as the first model. The evaluation of performance, the assessment of underlying assumptions and the reporting of results are therefore aligned. With respect to performance, this section first assesses the predictive power of the overall model, then tests the model's goodness of fit and finally validates the predicted probabilities. These three steps are performed for all four submodels in order to isolate effects and provide a better understanding of the impact of each set of variables. To assess the predictive power of the model, predicted outcomes are compared with observed outcomes. The overall rate of correct predictions serves as a measure of the quality of the model. As in the innovation model, a cut value of 0.5 is implemented to categorize items. The results for each submodel are reported in Table 5-11.

	Observed	Predicted		
		Non-constrained	Constrained	% correct
Intercept only model	Non-constrained	445	0	100.0
	Constrained	81	0	0.0
	Overall % correct			84.6
Model (a)	Non-constrained	439	6	98.7
	Constrained	75	6	7.4
	Overall % correct			84.6
Model (b)	Non-constrained	437	8	98.2
	Constrained	73	8	9.9
	Overall % correct			84.6
Model (c)	Non-constrained	436	9	98.0
	Constrained	67	14	17.3
	Overall % correct			85.6
Model (d)	Non-constrained	430	15	96.6
	Constrained	61	20	24.7
	Overall % correct			85.6

Table 5-11: Predictive performance of the specified model

The classification table can be read like a cross tab of the observed and estimated values. If none of the predictors are considered, the model yields 84.6% correct estimates. This relatively high value is, however, attributable to the relatively small share of firms that actually experience constraints. It thus directly corresponds to the relationship between the two considered groups. In other words, the share of constrained firms is 15.4%, which is equal to 1-84.6%. As can be seen, the model's performance in respect of constrained firms increases constantly with each step (submodel), indicating that the specified variables are relevant predictors of financial constraints for greentech firms. However, the overall rate of correct predictions only improves if alleviating factors are added (c). This finding is rooted in particular in the asymmetric distribution of the dependent variable. If the same model is run with a lower strain factor for financial constraints[19] and thus a higher percentage of constrained firms, the baseline prediction deteriorates (to 82.1% overall) but improves with each step, rising to 85.2% for the full model. The share of correctly classified constrained firms also increases with each step, from 12.6% for submodel (a) to 32.6% for submodel (d).

---

[19] As mentioned earlier, financial constraints are originally measured as ordinal variables on a six-point Likert scale and then recoded into two categories. The strain factor can be decreased by lowering the cut value. The proportion of constrained firms will rise accordingly.

The overall model performance with respect to correctly classified constrained firms is thus better at 24.7% for the first measure and 34.4% for the lower strain measure. The higher degree of strain seems more appropriate as the analysis primarily focuses on potential financing gaps. However, the results of the second regression are reflected for validation. Besides the general tendency of the stepwise addition of variables, it is interesting to note that only key demographic variables, i.e. firm size, age, legal structure and capital endowment, do not improve overall model performance. At 84.6% for the overall classification and 0% for the constraint classification, the share of correctly predicted values remains at the baseline, even though firm age and size showed a strong effect on both innovation and growth. The integration of capital endowment (a) does not improve overall predictive power, but yields the first improvement in predicting potentially constrained firms (from 0% to 7.4%). The effect of adding R&D and GROWTH' to the model (b) is smaller than expected, improving the rate of constrained firms by 2.5 percentage points, but not improving the overall rate of correct predictions. Integrating government support programs (c) nearly doubles the correct prediction ratio for constrained firms and raises the overall rate to 85.6%. The relative impact of government support instruments with respect to overall model performance is the highest of all submodels. Moreover, it is the only set of variables that actually increases the correct classification rate for non-constrained firms. The final model (d) also includes firm-internal alleviation factors, i.e. relationship banking, collateral and retained earnings, improving the prediction rate for financially constrained firms to 24.7%. However, the additional variables cannot improve overall model performance as the share of correctly predicted unconstrained firms decreases. Although the specified model is helpful for assessing financially constrained firms, it appears that accurate prediction is harder here than in the innovation case and may show less of a clear pattern. For example, the innovation model classified 71.9% of the innovative firms correctly, while the financial constraints model classified only 24.7% of constrained firms correctly. As in the innovation case, the predictive power assessment is complemented by a goodness-of-fit analysis based on the (1) omnibus test of coefficients, (2) the H-L test and the (3) 2-log-likelihood and pseudo-R^2. The first approach (1) tests the hypothesis that all regression coefficients do not impact financial constraints, i.e. $\beta_0 = \beta_1 = \cdots = \beta_{11} = 0$. A significant result causes the hypothesis to be rejected and indicates validity of the specified model. The second approach (2) assesses the goodness of fit of the model by replicating ten subpopulations and comparing the expected and observed frequencies. Contrary to the first approach, a significant result indicates that the model is not properly aligned, i.e. the null hypothesis that the model does not fit the data cannot be rejected. In essence, a well-fitted model shows a significant omnibus and non-significant H-L test result. Both requirements are met for all specified submodels (see Table 5-12).

Model evaluation

Model	Test	Chi²	df	p
(a)	Omnibus-test of coefficients	52.574	10	0.000
	H-L goodness of fit	8.857	8	0.354
(b)	Omnibus-test of coefficients	62.778	12	0.000
	H-L goodness of fit	2.737	8	0.950
(c)	Omnibus-test of coefficients	82.150	15	0.000
	H-L goodness of fit	11.310	8	0.185
(d)	Omnibus-test of coefficients	105.366	18	0.000
	H-L goodness of fit	1.156	8	0.997

	Model			
Model summary	(a)	(b)	(c)	(d)
2-log-likelihood	399.336	389.132	369.760	346.544
Cox&Snell R²	0.095	0.113	0.145	0.182
Nagelkerke R²	0.165	0.195	0.251	0.315

Table 5-12: Model summary and performance

The 2-log-likelihood decreases with each step, corresponding to an improving model fit for each step and submodel. The Nagelkerke R^2 accordingly increases from 0.165 (a) to 0.315 (d). It seems that the inclusion of firm-internal alleviating factors improves the explanatory value of the model above all, with an increase of more than 10% in R^2. This seems to confirm the observation that alleviating factors are very important when assessing financially constrained firms. However, pseudo-R^2 must be assessed with caution, as discussed in section 4.3.3.2. Moreover, the skewed distribution of *INVESTCONSTRAINT* may lead to overestimation and, hence, to less reliable measures. (Backhaus (2008)) Running the model over the less ambitious constraint measure generally yields higher pseudo-R^2 values, supporting the argument that a less strained measure increases model fit and predictive power. Overall, model (d) shows the best fit, as expected, and should thus serve as the basis for final testing and discussion of the formulated hypotheses. Although it seems harder to predict the existence of financial constraints than to predict innovation, the specified model fits the data set well and may reveal the most important drivers. In anticipation of promising results, the next section discusses the validity of the model and its fulfillment of the underlying assumptions.

5.3.5.3   Underlying model assumptions and validity

As mentioned earlier, logit models require four main conditions if they are to be valid: (1) independence of predictors, (2) homoskedasticity of residuals, (3) independence of residuals from predictors and (4) a sufficient sample size. Each condition is assessed along the same lines as section 4.3.3.3. Table 5-13 summarizes correlation coefficients, i.e. Spearman's ρ, to account for the non-parametric nature of variables (Bortz, Lienert, and Boehnke (2008) and test for the independence of predictors.

Spearman's rho		a	b	c	d	e	f	g
INVESTCONSTRAINT	a	1.000						
ln(AGE)	b	-0.146***	1.000					
ln(EMPLOY)	c	-0.147***	0.489***	1.000				
LEGAL	d	-0.033	0.173***	0.387***	1.000			
TC'	e	0.223***	0.092**	0.182***	0.147***	1.000		
ln(R&D)	f	0.136***	-0.192***	-0.101**	-0.038	0.096**	1.000	
GROWTH'	g	-0.05	-0.277***	-0.014	0.05	0.045	0.167***	1.000
GOVEQT	h	-0.117***	0.027	0.219***	0.114***	0.091**	0.198***	0.036
GOVD	u	-0.025	0.109**	0.267***	0.076*	0.2***	-0.006	-0.065
PULL	j	0.092**	-0.129***	0.017	-0.088**	0.164***	0.117***	0.094**
RELATION	k	-0.252***	0.179***	0.311***	0.149***	-0.064	-0.087**	0.104**
COLLATERAL	l	0.124***	-0.015	-0.016	0.059	0.297***	0.09**	0.016
RETAINED	m	-0.133***	0.114***	0.182***	0.143***	-0.189***	-0.069	0.072*
ZRESID		0.628***	0.148***	0.224***	0.068	-0.129***	-0.049	0.064

Spearman's rho		h	i	j	k	l	m
INVESTCONSTRAINT	a						
ln(AGE)	b						
ln(EMPLOY)	c						
LEGAL	d						
TC'	e						
ln(R&D)	f						
GROWTH'	g						
GOVEQT	h	1.000					
GOVD	u	0.457***	1.000				
PULL	j	0.097**	0.129***	1.000			
RELATION	k	0.019	0.061	-0.072*	1.000		
COLLATERAL	l	0.085**	0.178***	0.12***	-0.075*	1.000	
RETAINED	m	-0.064	-0.048	-0.01	0.159***	-0.118***	1.000
ZRESID		0.197***	0.045	-0.029	0.227***	-0.13***	0.067

1%, 5% and 10% confidence level indicated by *, ** and *** respectively

Table 5-13: Correlation of predictors and residuals

Essentially, none of the coefficients exceeds the critical absolute threshold of 0.7. The existence of colinearity can be screened out (1). The correlation between firm age and size remains highest, followed in terms of magnitude by correlations with the dependent variable. Interestingly, *RELATION* is the independent variable with the highest magnitude. A well-established and carefully cultivated relationship with the house bank may be a more effective way to overcome potential constraints than government support programs and even the accumulation of retained earnings. Even so, *GOVEQT* and *GOVD* seem to reduce constraints, although only *GOVEQT* is significant at the 1% confidence level. The effect of *GOVD* is rather marginal. This is in line with its strong positive correlation with firm size. In other words, larger firms are less likely to face financial constraints, while they in particular are more likely to receive *GOVD*. Also, capital endowment correlates significantly to *INVESTCONSTRAINT*. The effect is only marginally weaker than for the *RELATION* measure. This observation essentially supports the empirical approach and the third research hypothesis of this study. As we have seen, homoskedasticity (2) makes the model more stable but is not an indispensable requirement for valid logit-models. (Hosmer and Lemeshow (1989)) In the given model, it seems that residual variances are higher for the constrained category. The homoskedasticity assumption cannot be fully supported. The analysis was adapted from the innovation model and conducted graphically. With respect to the third condition (3), there are some significant correlations between independent variables and the standardized residuals (*ZRESID*), i.e. *AGE*, *SIZE*, *TC*, *GOVEQT*, *RELATION* and *COLLATERAL*, as reported above. However, these correlations are marginal and probably attributable to the limited sample size (Schendera (2008)), as well as to the close correlation between relationship banking and firm size. Endogeneity therefore does not seem to be apparent in the data set, a finding that validates the proposed model specification. Lastly (4), the number of cases included (N=529) meets the recommended minimum formulated by Schendera (2008) but is below the one formulated by Spicer (2005). In essence, the model performs sufficiently well with acceptable stability. The model is therefore appropriate to test research hypotheses 3 to 6, because overall reliability and validity seem sufficiently high. However, a larger sample would probably resolve heteroskedasticity issues and further reduce the minor correlation between residuals and predictors.

5.3.5.4 Presentation of results

5.3.5.4.1 Presentation of coefficient results and regression equation

The third model is composed of four submodels that are aligned with the structure proposed in section 4.3.3.3 and specified above. As indicated, each submodel tests one hypothesis and integrates further variables in the overall regression equation. The control variables are reported accordingly and are shown to be rather consistent for all models. The control variables are discussed with a particular focus on the full model, i.e. submodel (d). In line with the first logit model on innovation, the beta value (β) and its corresponding standard errors *(S.E.)*, the Wald statistic and the odds ($Exp(\beta) = e^{\beta_i}$) for each independent variable are reported in this section. Again, the analysis focuses on odds, as these were shown to be the most convenient for interpretation. Unlike for the previous two models, the results are reported for each of the following hypotheses due to the stepwise approach adopted. Although only the last complete submodel is decisive as the basis for conclusions, the stepwise approach was chosen to identify potential violations of model assumptions and assess the explanatory value of each variable set. The predicted logit equation for financial constraints, i.e. INVESTCONSTRAINT for firm i with technology j, is therefore derived from coefficient values of the last submodel (d) and equation (2):

Predicted logit of $INVESTCONSTRAINT = -0.463 + 0.968 \cdot TC + 0.227 \cdot ln(R\&D) + (-1.025) \cdot GROWTH'_{ij} + (-9.429) \cdot GOVEQT_{ij} + (-0.170) \cdot GOVD_{ij} + 0.166 \cdot PULL_{ij} + (-0.473) \cdot RELATION_{ij} + 0.298 \cdot COLLATERAL_{ij} + (-1.419) \cdot RETAINED_{ij} + (-0.413) \cdot ln(AGE) + (-0.435) \cdot ln(EMPL) + 0.375 \cdot LEGAL + (-0.290) \cdot RENWABLE + (-0.492) \cdot EFFICIENCY + 0.098 \cdot WATER + (-0.403) \cdot MOBILITY + (-0.119) \cdot RESOURCES + 0.653 \cdot WASTE$

The robustness of coefficient estimates was tested by running the same model with random subsets of the overall sample. The results appeared to be relatively robust with respect to model performance and fit. Similar to the first model, there were some variations in the significance and magnitude of the coefficients. For example, a subset involving 50% (240) randomly selected items yielded the same coefficient signs but generated an insignificant p value for $ln(AGE)$.

5.3.5.4.2 Results for hypothesis 3: Capital endowment and financial constraints

The specified models give fundamental support to the third hypothesis, namely that capital endowment, as a proxy for capital demand and supply (especially demand in the given model), not only enables innovation and growth but is also a fundamental driver of financial constraints. The more capital a firm demands, the more intensively it will use various capital sources and the more it will be likely to face financial constraints. Controlled for firm age, size, legal structure and technology lines, capital endowment shows a strong positive effect on financial constraints. The odds are 2.929 at the 1% confidence level. A change of one unit in TC will thus increase a firm's probability of belonging to the constrained cohort by 193%. Capital endowment shows a positive and significant regression coefficient. This not only supports the third hypothesis, but also supports the overall empirical approach that introduced a third variable into the overall context. Table 5-14 summarizes the empirical coefficient results. To the knowledge of the author and in line with a recent literature review conducted by Carreira and Silva (2010), capital endowment has not been used as a predictor of financial constraints in literature up to now. The direct implication, namely that capital demand and supply are not observable but may be approximated by capital endowment, is, however, not as important as the indirect implication, namely that capital endowment links financial constraints to innovation and growth, both of which positively depend on the introduced variable. The significant result and good model performance show that the approach adopted here is a convenient and powerful tool to overcome the simultaneity and causality issues that were summarized in section 4.2. The associated problems have been overlooked by many researchers in the field. (e.g., Piga and Atzeni (2007); Westhead and Storey (1997); Wolf

(2006); Freel (1999)) However, the advantage also has its drawbacks. For instance, introducing a third connecting variable cannot model the direct interrelation of financial constraints with innovation and growth. Nevertheless, the dissertation adopts a financial economics perspective and argues that both innovation and growth affect financial constraints, but not vice versa. This assertion is tested with hypothesis 4 and is discussed in the following section.

Predictor	Expected relationship	Actual relationship	Model (a)				
			β	S.E.	Wald	p	Exp(β)
TC'	+	+***	1.075	0.207	26.834	0.000	2.929
ln(AGE)	-	-	-0.232	0.183	1.600	0.206	0.793
ln(EMPLOY)	-	-^^^	-0.845	0.236	12.831	0.000	0.429
LEGAL	?	+	0.133	0.356	0.138	0.710	1.142
RENEWABLE	+/-	-	-0.061	0.291	0.045	0.833	0.940
EFFICIENCY	+/-	-*	-0.592	0.321	3.405	0.065	0.553
WATER	+/-	+	0.074	0.325	0.051	0.821	1.076
MOBILITY	+/-	-	-0.395	0.804	0.241	0.623	0.674
RESOURCES	+/-	-	-0.169	0.396	0.183	0.669	0.844
WASTE	+/-	+	0.303	0.290	1.091	0.296	1.354
β0	+/-	-***	-2.620	0.696	14.162	0.000	0.073

with n=526, 1%, 5% and 10% confidence level indicated by *, ** and *** respectively, cofindence level indicated for lowest level per variable

Table 5-14: Empirical results of submodel (a) on financial constraints

5.3.5.4.3 Results for hypothesis 4: innovation, growth and financial constraints

By implication, it appears that the innovative and/or growth-oriented firm is also likely to be financially constrained, because all three attributes are positively associated with the observable measure of capital supply and demand. However, the direct effect of innovativeness and firm growth is assessed by adopting a financial economics perspective. The assumed causality therefore points toward financial constraints. The second submodel (b) analyzes the impact and finds the relationships to be significant, as implied by literature (see section 3.4.3.), supporting hypotheses 4a and 4b. Table 5-15 summarizes the empirical coefficient results. Innovation activity and investment, measured as the natural logarithm of R&D spending, increases the level of financial constraints with an odds ratio of 1.324 at the 1% confidence level. Firm growth, measured as the quantile rank of growth, reduces the degree of financial constraints faced by firms due to the availability of larger internal funds and, presumably, a proven track record. The odds ratio of 0.403 is thus between 0 and 1 and the regression coefficient is negative. The p value is 0.046, corresponding to a comparably higher standard error (S.E.) of 0.456. The estimate for TC did not change noticeably in the second submodel, indicating the reliability of the model and suggesting that intrinsic model assumptions are not violated.

Predictor	Expected relationship	Actual relationship	Model (b)				
			β	S.E.	Wald	p	Exp(β)
TC'	+	+***	1.045	0.213	24.144	0.000	2.844
ln(R&D)	+	+***	0.280	0.106	6.976	0.008	1.324
GROWTH'	-	-**	-0.909	0.456	3.975	0.046	0.403
ln(AGE)	-	-	-0.259	0.194	1.784	0.182	0.772
ln(EMPLOY)	-	-***	-0.784	0.239	10.771	0.001	0.457
LEGAL	?	+	0.125	0.361	0.120	0.729	1.133
RENEWABLE	+/-	-	-0.036	0.294	0.015	0.903	0.965
EFFICIENCY	+/-	-*	-0.698	0.327	4.552	0.033	0.497
WATER	+/-	+	0.060	0.331	0.032	0.857	1.061
MOBILITY	+/-	-	-0.397	0.807	0.242	0.623	0.672
RESOURCES	+/-	-	-0.259	0.401	0.418	0.518	0.772
WASTE	+/-	+	0.382	0.297	1.656	0.198	1.466
β0	+/-	-***	-2.601	0.779	11.153	0.001	0.074

with n=526, 1%, 5% and 10% confidence level indicated by *, ** and *** respectively, cofindence level indicated for lowest level per variable

Table 5-15: Empirical results of submodel (b) on financial constraints

5.3.5.4.4 Results for hypothesis 5: government measures and financial constraints

The third submodel adds government programs to the equation. Again, previously estimated coefficients changed neither in magnitude nor in statistical significance. The estimate results reveal two important and rather new insights into the greentech market and policy initiatives. First, push programs seem to be effective to some extent, though not entirely, in alleviating financial constraints. Second, pull programs and Germany's EEG law in particular seem to be making financial constraints worse rather than easing them. The fifth hypothesis is therefore rejected, at least in part. The coefficient results for the corresponding submodel (c) are reported in Table 5-16. In essence, *GOVEQT* shows a high negative coefficient at the 1% confidence interval. The magnitude is extraordinary, with an odds ratio close to zero. The marginal effect of an increase in one unit of government grants and allowances is close to -100% with respect to the likelihood that a firm will experience financial constraints. By implication, the 15% of firms (80 out of 536) that receive equity-based support funds face significantly less financial constraints then their non-supported competitors. Taken in isolation, this finding would strongly support hypothesis 5 and thus back the findings of Cowling and Mitchell (2003), Zecchini and Ventura (2009) and Börner and Grichnik (2010). Following this line of thought, government support would, by easing firms' financial constraints, also positively affect firm development as claimed by Becchetti and Trovato

(2002). On the contrary, however, government debt does not seem to have the desired effect. The *GOVD* variable is not significant and also carries a positive sign. Government debt thus does not seem to reduce financial constraints as hoped for by policymakers. This is particularly important considering that a large number of instruments are debt-oriented and that general SME programs represent the largest share of the overall number of instruments (see section 3.5.).

Predictor	Expected relationship	Actual relationship	β	S.E.	Wald	p	Exp(β)
				Model (c)			
TC'	+	+***	1.134	0.222	25.985	0.000	3.109
ln(R&D)	+	+***	0.347	0.114	9.318	0.002	1.415
GROWTH'	-	-**	-1.102	0.470	5.326	0.021	0.332
GOVEQT	-	-***	-8.256	2.534	10.619	0.001	0.000
GOVD	-	-	-0.342	2.767	0.015	0.902	0.710
PULL	-	+*	0.167	0.086	3.782	0.052	1.182
ln(AGE)	-	-*	-0.334	0.201	2.769	0.096	0.716
ln(EMPLOY)	-	-***	-0.699	0.250	7.805	0.005	0.497
LEGAL	?	+	0.281	0.374	0.562	0.453	1.324
RENEWABLE	+/-	-	-0.299	0.329	0.827	0.363	0.741
EFFICIENCY	+/-	-	-0.622	0.342	3.303	0.069	0.537
WATER	+/-	+	0.143	0.338	0.179	0.672	1.154
MOBILITY	+/-	-	-0.338	0.830	0.166	0.683	0.713
RESOURCES	+/-	-	-0.160	0.418	0.147	0.701	0.852
WASTE	+/-	+	0.432	0.302	2.049	0.152	1.540
β0	+/-	-	-2.813	0.814	11.942	0.001	0.060

with n=526, 1%, 5% and 10% confidence level indicated by *, ** and *** respectively, cofindence level indicated for lowest level per variable

Table 5-16: Empirical results of submodel (c) on financial constraints

The finding is consistent with those reported by Honjo and Harada (2006), Witt and Hack (2008) and Svensson (2007), but is naturally inconsistent with the studies mentioned above in relation to *GOVEQT*.[20] Two implications can be drawn from this circumstance. First, it appears that equity-related instruments are more effective than debt-related instruments at reducing financial constraints. Second, it seems rather important to distinguish between the various types of government instruments when assessing their performance. Interestingly, and in support of both implications, it seems that firms in markets exposed to pull programs (such as renewable energy) are more heavily constrained than their non-exposed counterparts. *PULL* has a regression coefficient of 0.166 with an odds ratio of 1.181 at a *p* value 0.063. This implies that markets in which governments boost demand discourage private investment

---

[20] Both are compared here, because literature has so far barely considered the differences between the individual instruments, e.g. grants, allowances and subsidized loans.

in firms, which may in turn lead to technology lock-in by locking in existing players of larger size and with larger resource bases. The firm-level result supports the views of Neuhoff (2005), Unruh (2000) and Unruh (2002) that demand-based programs lock out innovative and new technologies. Moreover, it provides partial evidence to support the assumption that investors tend to avoid heavily regulated and subsidized market spaces (Diefendorf (2000)). This leaves firms exposed to pull programs more heavily constrained due to limited access to financial markets. Ultimately, hypothesis 5 must be partially rejected. Not all government support programs seem to reduce financial constraints; and pull programs actually seem to have the opposite effect.

5.3.5.4.5 Results for hypothesis 6: firm-based activities and financial constraints

The last submodel (d) finally integrates firm-level activities to ward off potential retrenchments. Again, the variables regressed earlier did not change significantly and appear to be stable. The results of the final model are reported in Figure 5-17:

Predictor	Expected relationship	Actual relationship	β	S.E.	Wald	p	Exp(β)
TC'	+	+***	0.968	0.237	16.657	0.000	2.632
ln(R&D)	+	+**	0.277	0.118	5.527	0.019	1.320
GROWTH'	-	-**	-1.025	0.496	4.263	0.039	0.359
GOVEQT	-	-***	-9.429	2.686	12.322	0.000	0.000
GOVD	-	-	-0.170	2.813	0.004	0.952	0.843
PULL	-	+*	0.166	0.089	3.463	0.063	1.181
RELATION	-	-***	-0.473	0.110	18.536	0.000	0.623
COLLATERAL	-	+	0.298	0.298	1.001	0.317	1.348
RETAINED	-	-	-1.419	1.134	1.566	0.211	0.242
ln(AGE)	-	-**	-0.413	0.208	3.935	0.047	0.662
ln(EMPLOY)	-	-*	-0.435	0.256	2.887	0.089	0.647
LEGAL	?	+	0.375	0.388	0.935	0.334	1.456
RENEWABLE	+/-	-	-0.290	0.342	0.716	0.397	0.748
EFFICIENCY	+/-	-	-0.492	0.349	1.992	0.158	0.611
WATER	+/-	+	0.098	0.354	0.076	0.782	1.103
MOBILITY	+/-	-	-0.403	0.863	0.218	0.641	0.669
RESOURCES	+/-	-	-0.119	0.443	0.072	0.789	0.888
WASTE	+/-	+**	0.653	0.319	4.197	0.040	1.921
β0	+/-	-	-0.463	0.958	0.234	0.629	0.629

with n=526, 1%, 5% and 10% confidence level indicated by *, ** and *** respectively, cofindence level indicated for lowest level per variable

Table 5-17: Empirical results of submodel (d) on financial constraints

It seems that building a close banking relationship is the best thing a firm can do. The effect is negative, with an odds ratio of 0.623. This effect is both expected and highly significant, with a p value < 0.00. The result is not surprising, considering that banking relationships have been pointed out by several authors (see Carreira and Silva (2010) for a review) as a powerful way to diminish financial constraints, and that relationship lending has emerged as a field of research in its own right (e.g., Elyasiani and Goldberg (2004)). Hypothesis 5a is therefore successfully corroborated, and positive banking relationships can effectively reduce financial constraints of the firm. Hypotheses 5b and 5c are rejected, however, since collateral and retained earnings seem to have less of an effect. Only *RETAINED* carries the expected sign. Neither variable yielded significant coefficients. By implication, both therefore contradicted existing research into capital structures and collateral (e.g., Chakraborty and Hu (2006); Chittenden, Hall, and Hutchinson (1996); Colombo and Grilli (2007)) as well as into capital structures and retained earnings (e.g., La Rocca and La Rocca (2007); Heyman, Deloof, and Ooghe (2008), Magri (2007); Bah and Dumontier (2001)). The theoretical consideration that capital demand is reduced by a firm's ability to generate cash internally could thus not be found in the data set, weakening the explanatory power of the theoretical framework and, in particular, the capital demand curve (EXTC$_D$) postulated in chapter 4.

5.3.5.4.6 Results for control variables

Lastly, the discussion of control variables is based on the full model as reported in Table 5-17. Overall, coefficient magnitudes remained fairly stable throughout all submodels. Only the significance levels changed slightly. For example, the significance of the *AGE* variable increased from the 10% confidence level to the 5% level from submodel (c) to (d). The negative sign of ln(*AGE*) and ln(*EMPLOY*) is expected and supports the idea that track record and transparency positively affect capital access and financing (see sections 3.4.4.4.1 and 3.4.4.4.2). The odds are nearly identical for both variables, at 0.662 and 0.647, respectively. Both coefficients are statistically significant at the 5% and 10% confidence levels. The mixed results on legal structure in literature correspond to the non-significant *LEGAL* coefficient in the analysis. The result does not resolve the issue and is similar to that reported by Bhaird (2010). It seems that legal structure is not necessarily a determinant of financial constraints, but rather a strategic decision variable for each firm. It was thus not possible to achieve the original objective of measuring information opacity, represented by the legal structure of the firm. A better proxy might have been ownership fragmentation (used by Romano, Tanewski, and Smyrnios (2001) and La Rocca, La Rocca, and Cariola (2009), for example), which directly measures the number of agents (shareholders) that banks would need to monitor. Finally, the coefficient results for technology lines are comparable to those in the previous models. Only *WASTE* is significant at the 5% confidence interval.

5.3.6   Conclusion – Main research hypotheses confirmed

The first model on innovation seems to fit the greentech sample very well. Its predictive value is approximately 40% better than that of the baseline model. Moreover, the underlying model assumptions are met and neither collinearity nor autocorrelation posed problems. The results presented are thus reliable and provide a valid insight into the German greentech industry. Essentially, the first research hypothesis was confirmed. Capital endowment is a strong driver of innovation in the greentech industry. This assertion is underlined by the high predictive power of the TC variable. For example, a model including all control variables predicts 73.9% of all items, putting them in the correct category and possessing a Nagelkerke R^2 of 38.5%. Adding the TC variable raises the correct prediction ratio to 75.5% with a pseudo-R^2 of 41.9%. The relationships between control variables and innovation were as expected, with one exception: firm age. Contrary to the positive impact of firm size due to an increasing resource base, it appears that firm age leads to organizational inertia, which has a negative impact on innovation. Large but young firms are thus most likely to generate innovation output.

The second model on growth likewise adjusts rather well to the data after correcting for outliers by transforming the variable of interest. Around 36% of the variation of the dependent variable could be explained, improving the standard error by nearly one third compared to the intercept-only model. The transformation also creates a situation where all model assumptions are met, e.g. the relationship between $GROWTH'$ and predictors is linear, residuals are normally distributed and independent, multicollinearity does not exist, sample size is sufficient, and so on. Coefficient estimates are therefore not only valid but also reliable. All postulated relationships are essentially important in that they are statistically significant. Above all, it appears that capital endowment is a strong driver of firms' development and growth, lending fundamental support to the theoretical framework and to the second research hypothesis. The other relationships are consistent with both the theoretical framework and contemporary literature.

The third and final specified model exhibits a better data fit comparable to the first model. The model classifies constrained firms significantly better than the intercept-only model. The full model (d) classified 85.2% of all firms and 32.6% of constrained firms correctly. Model performance improved with each step and submodel, justifying the approach adopted and implying that each set of variables is an important determinant of financial constraints. Moreover, all underlying model assumptions are met and coefficient results remain stable using hold-out samples. Although not all the tested hypotheses could be confirmed, the overall empirical approach is generally supported. Capital endowment is a significant and relevant factor of impact on financial constraints. As such, it indirectly links financial constraints to firm development. The third hypothesis is clearly supported by the data set. Moreover, the analysis supports the financial economics perspective and reveals that firms' innovativeness tends to exacerbate financial constraints while firms' growth tends to alleviate them. While hypotheses 4a and b are supported by the data set, hypotheses 5 and 6 are partially rejected. First, policy support seems to effectively reduce constraints only if it is implemented via equity-based instruments. Subsidized loans, other debt-based measures and demand-pull systems seem to have no easing effect on financial constraints. Indeed, it appears that firms exposed to pull programs experience greater financial constraints than their non-exposed competitors. With respect to firm-based activities, it appears that the best thing a firm can do is to cultivate its relationships with banks. To put that another way, good banking relationships significantly reduce the likelihood of a firm experiencing financial constraints. Collateral and even retained earnings seem to have less of an effect. On the whole, the results of the third model support the proposed approach to integrate the findings of literature on innovation, growth and financial economics. This inclusive approach is supported in particular by the first, second and third hypotheses. The fourth hypothesis further substantiates the financial economics perspective. Innovation activity and growth at a certain firm determine that firm's propensity to access certain forms of capital and, hence, the likelihood that it will experience financial constraints.

6 FINAL DISCUSSION AND IMPLICATIONS

6.1 Overall summary of results

The theoretical framework proposed the greentech financial innovation growth cycle as a way to integrate three distinct bodies of literature, in addition to political considerations. The framework was shown to clearly present the collected data. However, comparative analysis reveals a number of lessons to be learned. First, capital provided by friends and family seems to play a more important role than was originally anticipated. Second, inside finance remains the most important financing means throughout all phases, not only in the early days when a firm is becoming established. Third, the distinctions drawn between types of innovation did not yield sufficient insights. It does, however, appear that overall (incremental and non-incremental) innovation output increases up to the expansion state and diminishes thereafter. Fourth, firms in the early growth state and start-up firms experience the strongest firm growth. And lastly, government funds not only support early-state firms, but are also extensively used by expanding and even mature firms. Integrating these findings yields a modified financial innovation and growth cycle for the greentech industry that is summarized in Figure 6-1:

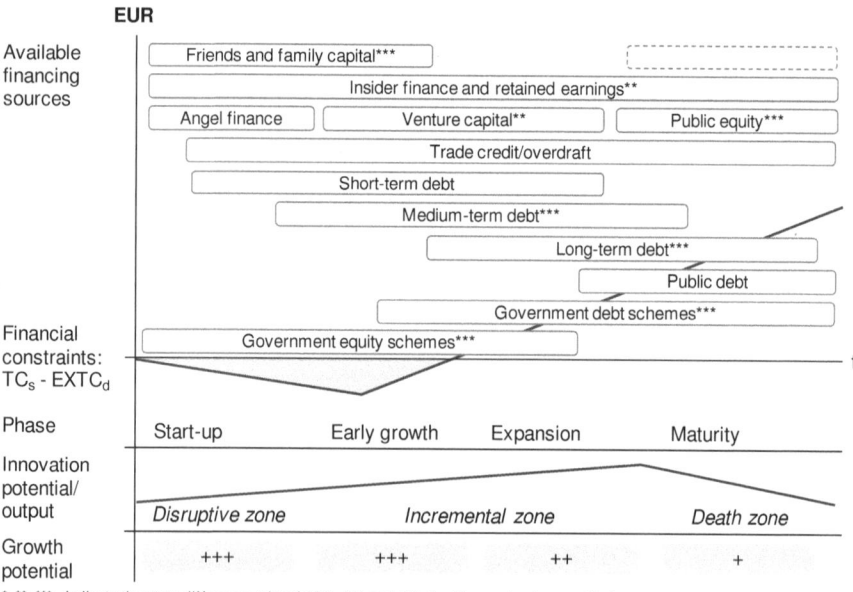

Figure 6-1: The greentech financial innovation and growth cycle (modified)

The figure shows several additional modifications to the original framework presented in chapter 4 and includes the four main lessons pointed out above. Above all, the financial instruments are denoted differently and are adapted to the German greentech market and, hence, to empirical analysis. Inside finance, for example, is enriched by retained earnings; debt is broken down into short-, medium- and long-term debt; and mezzanine capital is eliminated, as it was not used by any of the firms investigated. Additionally, the confidence levels of the comparative analysis are indicated for each capital instrument. In essence, it appears that the use of debt increases over time and reaches its peak – representing 20% of the financial basket – when a firm is in the expansion state. Government capital peaks in the same state and accounts for 12%. Venture capital and business angel capital is underrepresented. At 3%, it is highest for start-up firms. Internal funds provided by shareholders, friends and family and retained earnings constitute the most important financing source for greentech firms throughout all states of development. The only exception is that, during the expansion state, firms predominantly rely on external financing sources. In line with the framework formulated at the outset, financial constraints appear to be harshest for early growth firms and start-up firms, as demonstrated by the shaded area.

The dissertation further tested the theoretical framework by deriving testable hypotheses and using three regression models that focus on the three constructs of interest: innovation, growth and financial constraints. It thereby developed a unique line of argumentation. First, relevant literature reveals that the relationship between the three constructs is at best unclear. For example, while literature on innovation finds that financial constraints have a negative impact on innovation, financial economics literature suggests that innovation has a positive impact on financial constraints. One major problem in interdisciplinary research efforts is therefore the question of causality. Second, existing studies analyzing the impact of financial constraints on innovation or growth suffer from the fact that most compare the level of constraints and performance across different firms. The conclusion thus depends on the chosen sample. For instance, the conclusion that financial constraints reduce firms' growth based on the finding that constrained firms grow more slowly may be false, because constrained firms could have structurally lower performance due to a lower performance potential than their non-constrained peers. A more appropriate research design would have to compare a firm's potential performance to its actual performance subject to due consideration for its level of financial constraints. This, however, is simply not possible given that the performance potential cannot be observed and measured. Third, to resolve these problems, the model proposed herein introduces another construct – capital endowment – to the analysis. On the one hand, capital endowment is a proxy for the relative capital supply to a certain firm which, according to literature, enables innovation and growth. On the other hand, it is also a proxy for capital demand which, in theory, can lead to certain financial constraints when it is subtracted from the relative capital supply. Capital endowment is used instead of relative supply and demand because neither is observable, whereas capital endowment is the observable realization of either supply or demand for a certain firm. In other words, capital endowment equals the minimum of relative capital supply or demand. Essentially, capital endowment is not exposed to the problems discussed above. Potential simultaneity issues can be put aside as capital endowment – unlike financial constraints – is actively chosen by a firm and leads to a corresponding performance outcome. Moreover, analysis of the relationship between capital endowment and financial constraints reveals a structural pattern that has been ignored by current research. This pattern shows that firms with greater capital usage and intensity experience more constraints than those that use less capital. This finding reveals a relationship that has been overlooked in studies to date: the fact that capital-intensive firms are more likely to face financial constraints. At the same time, it indicates that financial constraints indeed prevent the most innovative and fastest-growing firms from exploiting their full performance potential, because capital endowment correlates positively to innovation and growth. Finally, the model adopts a financial economics perspective, assuming that a firm is innovative and growing before it faces financial constraints. Introduction of the capital endowment variable thus helps the analysis to avoid circular relationships.

The overall model is presented in Figure 6-2. The shaded area stands for the capital endowment variable, which represents the minimum of capital demand and supply. Controlling for other factors, capital endowment is regressed toward innovation (model 1) and growth (model 2), as implied by literature on both of these subjects. Capital endowment is also regressed, controlling for relevant factors, toward financial constraints (model 3), indirectly linking them to firms' performance (models 1 and 2) and directly integrating innovation and growth as firm characteristics. The third model thus reflects a view that has been common in literature on financial economics.

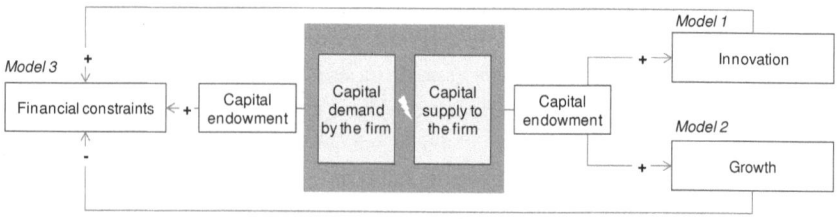

Figure 6-2: Empirical model (confirmed)

All three models yielded the expected relationships and gave substantial support to the first, second, third and fourth research hypotheses. Capital endowment was shown to correlate positively to innovation and growth, representing capital supply in particular. It was also shown to correlate positively to financial constraints, representing capital demand in particular. Since innovative firms seem to have greater informational opacity, innovation increases the likelihood that they will experience financial constraints. Firms' growth, on the other hand, improves their endowment with internal funds and appears to be appreciated by investors. Firms' growth was therefore shown to correlate negatively to financial constraints. Hypotheses 5 and 6 targeted policy-related and firm-based measures that reduce financial constraints. The empirical model did not deliver full support for these hypotheses, however. First, government support correlates negatively to financial constraints only when it is implemented via equity-based funds. Second, only good banking relationships, but not collateral or retained earnings, show a significant decreasing effect on the level of financial constraints faced by firms. The results of this dissertation add to several bodies of literature in various ways. The contributions are discussed in the following section.

6.2 Theoretical contributions

The contributions to literature on innovation, growth and finance were discussed in detail when the findings were presented, as these contributions are directly comparable to existing literature. Accordingly, the major contributions are summarized here only on a high level. With respect to greentech literature, the contributions must be evaluated differently because contemporary research has, up to now, barely even touched on a firm-level perspective. These contributions are therefore discussed in more detail, taking into account three out of the five identified research perspectives in greentech business research (see section 2.2).

6.2.1 Contributions to general business research

The most important contribution made by the present work to general business research is that it proposes a way to integrate various findings from three distinct bodies of literature: (1) innovation, (2) firms' growth and development, and (3) financial economics. It thereby summarizes the status quo in each field (see chapter 3), shows similarities and, most importantly, identifies causal contradictions (see section 4.1). The dissertation contributes to theoretical as well as empirical research. First, it creates a novel theoretical framework based on indicative and graphical analysis. The objective of the framework is to advance the understanding of how innovation, growth and financing are related and why potential financial constraints may occur. It thus combines the three bodies of literature by using the evolutionary segmentation of firms that led to the formulation of the innovation-finance-growth cycle (see section 4.4). The theoretical framework advances the seminal work by Berger and Udell (1998) by integrating state-of-the-art knowledge from literature on innovation and growth. It also pushes general business research in a broader, more interdisciplinary direction. Second, it suggests an empirical model (see section 5.3.1) with which to validate theoretical propositions and put the main findings into practice. It thus finds a way to circumvent causal contradictions, thereby avoiding associated simultaneity issues such as those experienced by Savignac (2008). This is done by introducing another variable: capital endowment. The overall findings support financial economics research, postulating that innovation and growth are determinants of financial constraints (Carreira and Silva (2010)) on the one hand. The findings also support innovation and growth research, however, arguing that financial resources drive firms' development on the other hand (Harhoff (1998)). Innovation, growth and financial constraints are interrelated in both causal directions; and the approach developed in this dissertation produces results that can coexist without logical flaws. It can thus be used in other contexts and may serve as a basis for future interdisciplinary research.

6.2.2   Contributions to greentech research

The dissertation contributes to greentech business research in two main ways. First, it conducts a broad review of literature on green technology, identifying five prevailing perspectives. Second, it advances both theoretical and empirical research by introducing a financial economics and firm-level perspective to the overall discussion. The theoretical and empirical results primarily contribute to three of the five identified perspectives in greentech business research: (1) innovation and diffusion, (2) policy and regulation and (3) finance.

6.2.2.1   Contributions to the innovation and diffusion perspective

With respect to innovation and diffusion, it was shown that the evolutionary approach adopted by various scholars (e.g., Balachandra, Kristle Nathan, and Reddy (2010); Chertow (2000); Grubb (2004); Neuhoff (2005); Wüstenhagen, Markard, and Truffer (2003)) also proves helpful at the level of the firm. The dissertation advances these studies by showing that greentech firms too pass through different states of development during which their characteristics, innovation and growth patterns and financing strategies differ significantly. The many insights provided in this study also point to the need to shift from a technology focus such as the one adopted by the cited authors to a firm-level focus. Essentially, the dissertation provides evidence that innovation in particular is delivered by firms, and that firms are the main vehicle of greentech market development. The identified existence of financial constraints also lends further credence to the claim that greentech market mechanisms are still weak (Grubb (2004); Rennings (2000); Chertow (2000)), creating barriers to innovation and the diffusion of sustainable technologies. The dissertation adds to these studies by demonstrating that weak market mechanisms lead in particular to a lack of financing. This lack is found to be most pronounced for early-state firms that, at least theoretically, have the potential to change existing technology trajectories. The existence of financial constraints benefits larger players, which in turn strengthens the trajectories of existing technologies and thereby locks out new and potentially superior ones. Firm-level evidence thus supports institutional considerations with regard to technology lock-outs (Dismukes, Miller, and Bers (2009)). It further demonstrates that technology lock-out is not necessarily induced by demand markets, as claimed by Unruh (2000) and Jacobsson and Johnson (2000), but that it may arise from firms' inability to commercialize new technologies due to financial constraints.

6.2.2.2 Contributions to the policy and regulation perspective

With respect to government policy and regulation, it seems that policy action eases financial constraints only to a limited extent. However, let us start from the beginning. Some literature argues that spill-over effects from R&D tend to divert profits away from innovators, and that greentech-oriented policy should support firms' R&D activity. (e.g., Nemet (2009); Rausser and Papineau (2010)) The dissertation supports the first argument, as R&D is found to correlate positively to financial constraints. However, push programs were found to ease these constraints only when they are equity-related. Debt-related programs, such as subsidized loans, are not found to have a significant effect. Fortunately, it seems that government's push instruments do not substitute for private investments that would otherwise have been made, as pointed out by Dinica (2006) and Kasemir, Toth, and Masing (2000). A potential crowding-out effect would thus have caused push programs to have a positive effect on financial constraints. Other literature argues that demand pull activities are most effective with respect to greentech market development. (e.g., Couture and Gagnon (2010); Lipp (2007); Loiter and Norberg-Bohm (1999)) Although existing empirical evidence on installed capacity is undeniable, the present study finds pull programs to significantly reinforce financial constraints. In line with the notion that pull activities promote diffusion but not necessarily at the lowest cost (Lund (2007)), the present study finds that pull programs push away private investment. A perceived dependence on "unstable" policy initiatives appears to be one of the main drivers. This finding also contradicts the case study results presented by Bürer (2008). The negative impact of pull programs is shown to be most likely to effect younger firms' innovation and growth. Large-scale demand-pull programs therefore imply the necessity to introduce further regulation, and for governments to intervene in order to offset negative externalities at the very beginning of the innovation value chain. This view supports the emerging consensus that the best policy is a mix that focuses on the technology state. (e.g., Foxon et al. (2005); Krozer and Nentjes (2008); Nill and Kemp (2009)) There is, however, one important difference: The dissertation adopts a firm-level perspective and thus argues that push policies in particular must target the firm. For example, programs that target an early-state technology may allocate funds to a large, expanding conglomerate that has a product portfolio approach and generates sufficient internal funds to finance R&D. At the same time, a young and innovative firm may be working on a completely new technology that is under the government's radar. In this case, funds would be allocated to the first firm but not to the second firm, which might thus be forced to exit the market. This assumption is strongly supported by the present study, which shows that push instruments reach neither early-state firms, nor the firms that experience the greatest financial constraints, nor the most innovative firms. There is one additional implication, namely that government support is a digital function. Either it is implemented for the whole industry or it is not implemented at all.

6.2.2.3 Contributions to the finance perspective

The dissertation also makes direct contributions to the financial perspective in greentech business research. In essence, it provides the first empirical evidence regarding financial instruments used by greentech firms as well as an insight into these firms' average financing basket. The discussion initiated by Diefendorf (2000) and still ongoing now (e.g., Kenney (2009); O'Rourke (2009)) about whether green venture capital is the ultimate enabler of greentech development has thus been elevated to a more concrete level. Firm-level evidence shows that VC is indeed underrepresented, never exceeding more than 3% of firms' overall average financing basket. This is in line with previous findings focusing on the investment side and claiming that a mere fraction of total VC investment is channeled into greentech. (e.g., Hamilton (2007); Moore and Wüstenhagen (2004)) The low number of firms that apply for it also seems to indicate that VC criteria do not fully match the requirements of greentech firms. This assertion lends firm-level support to Kenney (2009), who argues that greentech development requires other means of financing than venture capital.

The dissertation results also provide a basis on which to broaden the scope of discussion in literature. It provides insight into how greentech firms are actually financed. First, it shows not only that VC is of generally low importance, but also that internal funds provided by the owners and their social network account for the main source of financing. Second, it shows that policy support programs are also important at firm level and can be evaluated on the basis of firms' characteristics, such as performance and the degree of financial constraints. Third, it shows that many firms cannot find sufficient capital and often lack even basic instruments such as bank debt. Followed the recommendations of Dinica (2006), the present study therefore moves analysis in the direction of a financial economics perspective. Moreover, the dissertation provides empirical evidence of a financing gap in the industry and advances the theoretical work done by Chertow (2000). However, it also demonstrates that the gap exists in corporate finance, not in technology markets. Steps to bridge the financing gap must therefore target not the development states of technologies but the development states of firms. Apparently, the gap is particularly wide for early growth firms that are attempting to commercialize their products and technologies. The empirical evidence therefore supports theoretical propositions made by Burtis (2004), Burtis (2006) and Murphy and Edwards (2003). It also confirms comments made by study participants, one of whom, for example, argued that "*support instruments focus too much on research and development*". Another study participant claimed that "*money is particularly rare when it comes to commercialization*". Besides theoretical contributions, the dissertation is therefore of high practical value as it contains direct implications for greentech managers and policymakers. These implications are discussed in the following section.

6.3   Managerial implications

The findings of the dissertation have several implications for greentech managers. However, most of them depend on the specific context of and problems faced by the firm in question. It is therefore difficult to draw conclusions that are of broad relevance to the overall firm population. Generally speaking, managers have to innovate in order to grow, and they have to grow in order to receive financing. Clearly, this is a difficult endeavor considering that innovation and growth both require sufficient funds, while attaining funds itself presupposes an adequate growth history – which in turn requires a certain level of innovation. Managers who know about these relationships can use them to their advantage and act accordingly. In this context, the following seven general principles can be derived:

1. Invest in innovation and grow with it!
The first implication is based on the finding that innovative firms grow three times faster than non-innovative firms. Innovation is a strong differentiating factor and a competitive threat in the greentech industry. At the same time, it appears that less than a quarter of all firms engage in process optimization and cost reduction, both of which will most probably become more important in the long run.

2. Don't stick to a business model for too long if it is not growing!
The second implication refers to the empirical finding that age clearly has a negative impact on the performance of small firms. If a certain business model is still not developing after several years, managers should switch to other activities.

3. Internationalize!
The third implication aims at increasing international activity and is born of low current exposure. Only 31% of all firms currently engage in export activities, for example. And only 22% have foreign joint ventures, while only 16% manage international affiliates. Moreover, only 28% of all firms plan to ramp up their international presence in future. Becoming more

international opens up several key advantages such as the ability to increase a firm's size, diversify its risk exposure and make it more competitive.

4. If you need capital, apply for it!

The fourth implication points to two things. First, only a small number of firms actually apply for all the various forms of capital. For example, less than 10% of all firms in this study applied for venture capital. Second, rejection rates range from an average of 15% for long-term debt to a devastating 89% for government debt. Actually obtaining capital thus requires several attempts, assuming that every try increases the likelihood that a firm will succeed.

5. If you still can't find capital, apply for medium-term debt!

The fifth implication refers to medium-term debt. Analysis shows that medium-term debt has the lowest rejection rates at reasonable financing cost. The likelihood of receiving it seems fairly high.

6. Improve and actively work on your relationships with banks!

The sixth implication again relates to financing. Banking relationships have been shown to be the most effective way to reduce financial constraints. It thus seems very worthwhile for firms to invest time and effort in cultivating a long-term, trust-based relationship with their house bank. When dealing with banks, however, it is important to sell a firm in the right way.

7. Sell banks your firm's history and tangibility, not your innovation story!

The seventh implication refers to the problem of credit rationing and looks at the business model operated by banks. In essence, banks have limited upside potential (interest rates) and are thus more risk-averse than equity investors. Tangible assets are attractive to them as they can be liquidated, thereby reducing banks' risk exposure. In addition, a solid tradition and historic growth trajectory make banks confident that a firm can pay back its debt in the future. The rejection rate for debt capital decreases in a straight line with a firm's age and growth. Innovation, on the other hand, makes banks suspicious and should not be a dominant part of the proposal.

6.4   Policy implications

"The inventor produces ideas, the entrepreneur "gets things done," which may but need not embody anything that is scientifically new"
Schumpeter (1947)

Schumpeter's statement reflects an important observation that supports the approach adopted in this dissertation. Push policies are often understood as R&D support programs that aim to generate new products and services. However, an invention alone is often not enough to push markets in a more sustainable direction. In this context, the dissertation finds that start-up and early growth firms in particular experience financial constraints. The main barrier thus appears to be commercialization rather than invention. Policymakers should therefore focus on firms and entrepreneurs, and not so much on technologies. Besides these general perspectives on firms, several specific principles for policymakers can also be derived from this dissertation:

1. Decide whether support is really needed. If so, use an integrated policy mix!
The first principle refers to the question of whether government support is really necessary. While the dissertation has no answer to this dilemma, it does reveal the paramount importance of considering that demand-pull initiatives increase the entry barriers to innovation and tend to lock in existing technologies. Although pull instruments lead to the fast and vigorous diffusion of more sustainable technologies, it appears that existing technologies still suffer significant disadvantages. Examples include limited acceptance in society and regional consumption. Moreover, no support is better than wrong support. For example, if support funds are channeled exclusively into selected technologies that are inferior to other ones, there will not be sufficient private funds available to offset the overall effect. Essentially, the dissertation shows that existing pull policies actually lead to heavier financial constraints at the level of the firm.

2. Use long-term instruments firms can rely on!
The second point refers to the finding that government action must be long-term, stable and reliable, as investors appear to get frightened off by forceful government intervention. This issue ultimately begins when firms apply for capital based on their business planning and expectations. The dissertation reveals that many greentech firms do not regard government actions as stable and would therefore have problems integrating potential government support in their own strategic planning process. This also implies that firms will only target sustainability if they perceive it to be a long-term political goal that adds real value.

3. Adapt support mechanisms to the development state of firms!
The third recommendation is consistent with the central theme of this dissertation. We have seen that firms in the start-up, early growth, expansion and maturity phases possess statistically significant differences in characteristics that should be borne in mind when designing support programs and regulatory frameworks. For example, start-up firms may find initial guidance on how to commercialize a product to be most helpful, together with loan guarantees, while early growth firms are most desperate for cash and will basically take any

type of financial instrument. Expanding firms may be best supported by fostering international expansion, while mature firms may be encouraged to innovate by appropriate networks or institutional cooperation. More generally, the dissertation finds that many programs only foster research and development, whereas financial constraints are greatest in relation to commercialization activities.

4. Support capital-intensive firms!

The fourth point refers to findings of the empirical model. Capital intensity is not only a definite driver of innovation and growth, but also of financial constraints. Any effective support program should therefore target capital-intensive firms and should not be put off by their already high capital endowment.

5. Allocate directly and not via house banks!

The fifth argument targets the process of government fund allocation. There are two main issues. First, the present study finds evidence that the allocation of support resources is suboptimal. In particular, subsidized loans are used intensively by expanding firms that could also find capital on private markets. Moreover, it is shown that government debt programs do not alleviate firms' financial constraints, as they are used by firms that do not face such constraints. Direct allocation via government agencies, as opposed to house banks, would make it easier to control allocation, improving the way in which government's goals are implemented at financial and firm level. Second, the dissertation finds anecdotal evidence of a principal-agent problem between the KfW and local house banks. Study participants complain that they cannot access subsidized loans due to conflicts of interest with their house banks.

6. Use professional bridge fund constructs!

The sixth point builds on the previous one and suggests that professional, market-oriented agents such as the German Gründerfonds should be used to manage funds directly and integrate government's support strategies, objectives and mechanisms. Another way to improve the effectiveness and efficiency of support programs could be to step up the engagement of independent financial intermediaries.

7. Do not underestimate the innovative potential of small and young firms!

The seventh principle refers to the under-researched area of innovation types and how they relate to firms' size. Although the dissertation confirms what has been a dominant theme in literature, namely that innovation output increases with firms' size, it also shows that age correlates negatively with innovation. Small firms that are also young may thus be a creative source of new ideas and innovations with potentially groundbreaking effects.

8. Do not provide financial support for innovation and growth at small and old firms!
The eighth principle builds on the previous one by pointing out that not all small firms are similar. Small and old firms in particular have by far the lowest potential to innovate and grow. Support programs that target innovation and corporate development should therefore avoid this cohort. More generally, policymakers should assess the innovation and growth potential of a given firm before actually providing support funds. This point indirectly refers to previous principles too, as it requires direct contact between government agents and potential recipients of support. The dissertation provides three equations that may be used to assess innovation and growth potential as well as the level of financial constraints with respect to main firm demographics and the corresponding environmental setting.

9. Consider the big greentech picture, not just renewable energy!
Finally, the dissertation shows that greentech is more than just renewable energy. The challenge of sustainable development will not be met solely by solving the problem of future energy supply, but dealing with the wider issue of general resource supplies. Sustainability happens when resource consumption matches resource regeneration. Although this may not fully be achieved by eco-efficiency and the implementation of greentech, the broader industry scope proposed herein captures a larger chunk of the fundamental "problem". Greentech should also be considered as a whole, as it is impossible to set long-term priorities that choose one technology over the other. From a government perspective, for example, reducing emissions seems just as important as water conservation. Establishing a homogenous greentech industry provides fertile soil in which to grow a better understanding of industry mechanisms and potential contributions to sustainability.

6.5   Limitations and outlook

The dissertation uses a mixed method design and aims to provide an insight into the innovation activity, development and financial situation of greentech firms. Several problems had to be resolved during the course of analysis. The greatest obstacle was to resolve the issue of causal circularity in light of the interdisciplinary approach. The dissertation had to develop its own specific theoretical framework, because existing frameworks did not seem to sufficiently account for this need. The dissertation thus contributes to theory building and proposes an integrated framework that has also been tested empirically. However, the approach has five main limitations that could be resolved in future work. First, the theoretical framework is based on several assumptions (e.g. non-consecutive states of development and non-predictive development paths) to account for criticisms leveled at existing lifecycle models. Future works may add further detail and formalize the approach, thereby reducing the number of assumptions. Second, the capital endowment construct introduced herein represents a proxy for two distinct but non-observable variables, i.e. capital demand and capital supply. Although it connects innovation, growth and financial constraints rather well, it remains unclear whether capital supply or demand is a relevant driver. Future research may thus find better ways to actually measure capital demand and capital supply for a certain firm with specific characteristics. Moreover, future research may split capital endowment into sub-types such as equity endowment and debt endowment for the purposes of further rigorous empirical testing. Third, a rather common problem of most empirical studies is the survivorship bias that was not accounted for in the given analysis. Hadjimanolis (2003), for example, points out that studies on barriers to innovation often include a bias as they do not include firms that failed and disappeared from the market. Fourth, there is a general lack of empirical research on innovation types and their determinants. The gap is not fully addressed by this study as the distinctions drawn between different types of innovation were shown to be insufficient. The main problem is to distinguish between incremental and disruptive innovations. Future studies may track certain innovations over time and analyze their impact on the market space for the purposes of classification. Fifth, and as has been pointed out by previous empirical research into firms' growth, most surveys suffer from the fact that measures of growth only compare two points in time or are calculated as averages (Weinzimmer, Nystrom, and Freeman (1998); Delmar, Davidsson, and Gartner (2003)). The same problem occurs in this dissertation, as growth was measured as an average over a three-year period. Structural irregularities could therefore not be considered. Overall, it appears that interdisciplinary research combining several bodies of literature, industries and methodologies is still in its infancy. It is very likely that interdisciplinary attempts in other fields will also yield contradictory results and reveal unclear causal relationships. Hence this study is not only a study of greentech, innovation, growth and finance, but also a plea for increasing interdisciplinary efforts in business science.

7   EXECUTIVE SUMMARY

1. The dissertation was guided by four main research questions that were successfully answered during the course of analysis. The dissertation used a mixed-method design aiming at theory building and empirical validation. This summary reflects the overall structure of the dissertation by introducing the general topic, discussing the context of analysis, reviewing and integrating relevant literature in a theoretical framework, conducting a comparative analysis across different states in firms' development, testing theoretical assertions via statistical modeling and deriving managerial implications.

2. The world is going green. Sustainability has become a major strategic turning point for many firms. The greentech industry is a cornerstone of this development, providing products that potentially reduce environmental degradation without harming economic wealth and development.

3. Existing research has adopted five key perspectives to improve our understanding of the greentech industry: a sustainability perspective, an entrepreneurial perspective, an innovation and diffusion perspective, a policy and regulatory perspective, and a finance perspective. It has thus shown that, while greentech cannot ultimately achieve sustainability, it can make an essential contribution. Entrepreneurial activity, innovation and the diffusion of green technologies are the means to pursue this path. However, this largely depends on the availability of sufficient financial resources, which are particularly scarce in the greentech industry due to persisting market imperfections. Policymakers are therefore trying to offset these imperfections by running various support programs whose effectiveness and efficiency remains the subject of lively discussion.

4. However, existing research is scant when it comes to evidence at firm level. The dissertation fills this gap by integrating findings from general business research in the areas of innovation, growth and financial economics and adapting it to the greentech sector. It thereby shows that, like technologies, firms too can be structured by development states (start-up, early growth, expansion and maturity). Each state possesses distinct characteristics that moderate innovation and growth potential as well as access to certain financing sources. The study thereby contributes to general business research by integrating the three streams of literature in one exhaustive, interdisciplinary framework. By introducing the variable "capital endowment" – a novel step in this context – as a proxy for firms' capital demand and supply, the dissertation overcomes existing antagonisms across the three literature bodies of interest.

5. Comparative analysis revealed that each development state cohort is characterized by distinct firm characteristics:

- *Start-up* firms are smallest but, like early growth firms, have the highest growth potential. Although they are less innovative in terms of overall innovation output, their impact may, at least theoretically, be disruptive and render existing technologies obsolete. Although depending on mostly internal funds, start-up firms are exposed to less financial constraints than early growth firms due to their lower capital requirements. An important external financing source is the German government.

- *Early growth firms* are larger than start-up firms and grow comparatively fast. Their overall innovation output is substantial and comparable to that of expanding firms. In line with theoretical considerations, it appears that early growth firms experience the greatest financial constraints, with a share of 26% of all firms. They therefore remain heavily dependent on internal funds and have limited access to external debt and equity sources. Government programs do not seem to alleviate this situation, which reflects a lack of programs focusing on firms' commercializing activities.

- *Expansion firms* are the largest firms. Generating more than EUR 50 m in average sales, they are significantly older than start-up and early growth firms. They also have the greatest overall innovation output and the lowest propensity to experience financial constraints, as expected from theory. Expanding firms seem to have the best access to external financing sources. This is the case for debt capital in particular but, surprisingly, also for government funds. However, their relative growth potential is lower than that of the first two state cohorts.

- *Maturity firms* are smaller than expanding firms and larger than start-up and early growth firms, but have the lowest innovation output and growth potential. They are more likely than expanding firms but less likely than start-up and early growth firms to experience financial constraints, presumably due to their longer track record. Surprisingly, mature firms make intensive use of government debt funds. It appears that these firms need to foster innovation in order to regain competitiveness.

6. Comparative analysis was amplified by a statistical model that comprises three main parts: innovation, growth and financial constraints. The model was applied to a unique data set of more than 500 German greentech firms. The evidence suggests a good fit with the theoretical framework and the proposed hypotheses. There were seven main findings. First, capital endowment – and hence capital supply – is a major driver of innovation and growth. Second, capital endowment – and hence capital demand – is a powerful determinant of financial constraints, implying that constrained firms cannot

fully exploit their performance potential.[21] Third, innovation activity gives firms greater information opacity. This discourages banks and investors, leading to a higher level of financial constraints. Fourth, past firm growth does the opposite as track record boosts investors' sentiment. Fifth, government support programs ease firms' financial constraints only if they are equity based, whereas debt-based instruments have no significant impact. Sixth, although government pull programs are broadly accepted as strong drivers of technology diffusion, they seem to have a negative impact at the front end of the innovation value chain, i.e. in financial markets. In other words, firms exposed to pull policies are more likely to face financial constraints than their non-exposed counterparts. Lastly, firms themselves have only limited options to reduce financial constraints. The best thing they can do is to engage in and constantly improve their relationships to banks.

7. The elaborated results provide useful advice not only to greentech firms, but also to policymakers. Firms should invest in innovation, abandon their business model if it does not lead to growth, become more international, step up their efforts to apply for necessary capital, use medium-term debt if nothing else seems to work, improve their relationships with banks and "sell" their track record and tangibility when applying for debt. Policymakers should first decide whether support is really needed, only use long-term and reliable instruments, adapt support programs to the development states of firms rather than focusing on technologies, allocate funds directly, use professional but independent intermediaries and financial constructs, not underestimate the innovation potential of small firms and consider greentech as a whole, not just clean energy.

---

[21] Analysis showed that a direct link between financial constraints and innovation or growth is misleading, because firms subject to financial constraints, while not exploiting their full innovation and growth potential, may still be more innovative and grow faster than their less ambitious but also less constrained competitors.

APPENDIX

A. Government support programs (database)

Nr.	Name	Regional scope	Policy type	Instrument type	Impact on greentech	Life cycle	Issuing institute	Policy goals	Details	Amount (max)	Years (max)
1	Agrar- und Ernährungswirtschaft - Umwelt- und Verbraucherschutz	National	Push	Subsidized Loan	Indirect pull	Expansion	Landwirtschaftliche Rentenbank (LR)	Improve efficiency in the food industry	Subsidized loan for technology investment	EUR 10m	10
2	BMU-Umweltinnovationsprogramm	National	Push	Subsidized Loan	Indirect pull	Early Growth	KfW Bankengruppe	Fostering of demonstration projects that would not realize without government support	Subsidized loan; First five years are not amortized	70% of investment	30
3	ERP-Umwelt- und Energieeffizienzprogramm	National	Push	Hybrid	push; indirect pull	Seed, Start-up; Early Growth; Expansion	KfW Bankengruppe	Reduction of energy consumption; increase energy efficiency	Subsidized loan; First three years are not amortized; investment must fulfill efficiency targets; Loans for risk capital companies and funds	EUR 2-10m	20
4	Förderung der Sicherheit und der Umwelt in Unternehmen des Güterkraftverkehrs mit schweren Nutzfahrzeugen	National	Push	Grant, allowance	Indirect pull	Expansion	Bundesamt für Güterverkehr (BAG)	Increase of energy efficiency in the mobility and transport sector	Grant per car	EUR 33,000 per firm	-
5	Förderung des nachträglichen Einbaus von Partikelminderungssystemen bei Personenkraftwagen mit Selbstzündungsmotor (Diesel)	National	Push	Grant, allowance	Indirect pull	Expansion	Bundesamt für Wirtschaft und Ausfuhrkontrolle (BAFA)	Decrease emission from Diesel combustion in cars	Grant per car	EUR 330 per car	-
6	Forschung für nachhaltige Entwicklungen (Fachprogramm)	National	Push	Grant, allowance	Push	Seed, Start-up	Bundesministerium für Bildung und Forschung (BMBF), Deutsches Zentrum für Luft- und Raumfahrt e.V. (DLR)	Sustainable business; R&D	Solution to a sustainability problem; EUR 2bn until 2015; Applicant must bring in at least 50% equity	-	-
7	Modell- und Demonstrationsvorhaben im Bereich der Erhaltung und innovativen nachhaltigen Nutzung der biologischen Vielfalt	National	Push	Grant, allowance	Push	Seed, Start-up	Bundesanstalt für Landwirtschaft und Ernährung (BLE)	Biodiversity; R&D	Demonstration projects	-	-
8	Umweltschutzförderung der Deutschen Bundesstiftung Umwelt	National	Push	Grant, allowance	push	Seed, Start-up; Early Growth	Umweltschutzförderung der Deutschen Bundesstiftung Umwelt	Greentech; R&D; Innovation; Reduction of externalities	Innovation and development of existing knowledge; Foundation capital of more than EUR 1bn from the sale of state owned Salzgitter AG		-
9	Förderprogramm "Nachwachsende Rohstoffe"	National	Push	Grant, allowance	Push	Seed, Start-up	Bundesministerium für Ernährung, Landwirtschaft und Verbraucherschutz (BMELV), Fachagentur Nachwachsende Rohstoffe e.V. (FNR)	Innovation in renewable energies	Grants for R&D; Grants to diminish operating disadvantage towards other technologies	100% of investment	-
10	Innovative Konversionsprozesse für Biogas aus landwirtschaftlichen Substraten	National	Push	Grant, allowance	Push	Seed, Start-up	Bundesministerium für Ernährung, Landwirtschaft und Verbraucherschutz (BMELV), Fachagentur Nachwachsende Rohstoffe e.V. (FNR)	Innovation in renewable energies	Grants for R&D; Grants to diminish operating disadvantage towards other technologies; Profitability should not rise above 6%	100% of investment	-
11	Energieeffiziente IKT für Mittelstand, Verwaltung und Wohnen (IT2Green)	National	Push	Grant, allowance	Push	Seed, Start-up	Bundesministerium für Wirtschaft und Technologie (BMWi); Deutsches Zentrum für Luft- und Raumfahrt e.V.	Energy efficient IT	Grants for R&D in energy efficient IT innovation	50% of investment	-
12	Forschung und Entwicklung im Bereich erneuerbare Energien	National	Push	Grant, allowance	Push	Seed, Start-up	Bundesministerium für Umwelt, Naturschutz und Reaktorsicherheit (BMU); Projektträger Jülich (PTJ)	Renewable energy innovation	Grants for R&D in renewable energies; overall volume of EUR 20m	50% of investment	3
13	Innovation und neue Energietechnologien (5. Energieforschungsprogramm)	National	Push	Grant, allowance	Push	Seed, Start-up	Bundesministerium für Wirtschaft und Technologie; various national public institutions	Renewable energy innovation; Power plant technology innovation	Grants for R&D in renewable energies; high risk projects and governmental interest required	50% of investment	-

Nr. Name	Regional scope	Policy type	Instrument type	Impact on greentech	Life cycle*	Issuing institute	Policy goals	Details	Amount (max)	Years (max)
14 Mobilität und Verkehrstechnologien (3. Verkehrsforschungsprogramm)	National	Push	Grant, allowance	Push	Seed, Start-up	Bundesministerium für Verkehr, Bau und Stadtentwicklung (BMVBS); TÜV Rheinland Consulting GmbH	Sustainable mobility technologies	Grants for R&D with focus of developing state-of-the-art, focus on e-mobility and technologies that are not marketed yet; 50% equity required	50% of investment	-
15 Nationales Innovationsprogramm Wasserstoff- und Brennstoffzellentechnologie (NIP)	National	Push	Grant, allowance	Push	Seed, Start-up	Bundesministerium für Verkehr, Bau und Stadtentwicklung (BMVBS); Nationale Organisation Wasserstoff- und Brennstoffzellentechnologie NOW GmbH	Improve marketability of fuel cell technologies	Grants for R&D and commercialization projects	50% of investment	-
16 Technologien für Nachhaltigkeit und Klimaschutz - Chemische Prozesse und stoffliche Nutzung von CO2	National	Push	Grant, allowance	Push	Seed, Start-up	Bundesministerium für Bildung und Forschung (BMBF); Deutsches Zentrum für Luft- und Raumfahrt e.V. (DLR)	Reduction and usage of CO_2-Emmissions	Grants for R&D	50% of investment	-
17 Erneuerbare-Energien-Gesetz (EEG)	National	Pull	Grant, allowance	Pull	Expansion	Bundesministerium für Umwelt, Naturschutz und Reaktorsicherheit (BMU); Bundesamt für Wirtschaft und Ausfuhrkontrolle (BAFA)	Renewable energy commercialization	Grants to diminish renewable energy cost disadvantages toward conventional power technologies	-	-
18 Exportinitiative Energieeffizienz	National	Push	Grant, allowance	Push	Expansion	Bundesministerium für Wirtschaft und Technologie; Germany Trade and Invest	Internationalization of German efficiency technologies	Grants for international expansion, meetings, light-house projects etc.	-	-
19 Exportinitiative Erneuerbare Energien	National	Push	Grant, allowance	Push	Expansion	Bundesministerium für Wirtschaft und Technologie; Germany Trade and Invest	Internationalization of German renewable energy technologies	Grants for international expansion, meetings, light-house projects etc.	-	-
20 Förderprogramm für emissionsärmere Dieselmotoren von Binnenschiffen	National	Push	Grant, allowance	Indirect pull	Expansion	Bundesministerium für Verkehr, Bau und Stadtentwicklung (BMVBS); Wasser- und Schifffahrtsdirektion West	Modernization of the inland water transportation	Grants for modernization of engines	40% of investment	-
21 Grundlagenforschung Energie 2020+	National	Push	Grant, allowance	Push	Expansion	Bundesministerium für Bildung und Forschung; Projektträger Jülich (PTJ)	Renewable and nuclear energy innovation	Grants for R&D, pre-market development projects	50% of investment	-
22 KfW-Programm Erneuerbare Energien	National	Push	Subsidized Loan	Indirect pull	Expansion	KfW Bankengruppe	Development of renewable energy technologies	Subsidized loan for renewable energy technology investment	EUR 10m	-
23 Klimaschutzinitiative - Maßnahmen an gewerblichen Kälteanlagen	National	Push	Grant, allowance	Indirect pull	Expansion	Bundesministerium für Umwelt, Naturschutz und Reaktorsicherheit (BMU); Bundesamt für Wirtschaft und Ausfuhrkontrolle (BAFA)	Improvement of energy efficiency	Modernization of industrial cooling systems; implementation of innovative technologies	75% of investment	-
24 Klimaschutzinitiative – Mini-KWK-Anlagen	National	Push	Grant, allowance	Indirect pull	Expansion	Bundesministerium für Umwelt, Naturschutz und Reaktorsicherheit (BMU); Bundesamt für Wirtschaft und Ausfuhrkontrolle (BAFA)	Development of small-cogeneration technologies	Development of small-congeneration markets (<50kW)	-	-
25 KMU-innovativ: Ressourcen- und Energieeffizienz	National	Push	Subsidized Loan	Indirect pull	Expansion	Bundesministerium für Bildung und Forschung (BMBF); Projektträger Jülich (PTJ)	Reduction of energy consumption; increase energy efficiency	Environmental improvement of production processes	50% of investment	-
26 Kraft-Wärme-Kopplungsgesetz (KWK)	National	Pull	Grant, allowance	Pull	Expansion	Bundesministerium für Umwelt, Naturschutz und Reaktorsicherheit (BMU); Bundesamt für Wirtschaft und Ausfuhrkontrolle (BAFA)	Cogeneration technology commercialization	Grants to diminish renewable energy cost disadvantages toward conventional power technologies	-	-
27 Maßnahmen zur Nutzung erneuerbarer Energien im Wärmemarkt (Marktanreizprogramm)	National	Push	Grant, allowance	Indirect pull	Expansion	KfW Bankengruppe	Development of renewable energy technologies	Subsidized loans to develop the construction of renewable energy production facilities	EUR 10m	-
28 High Tech Gründerfonds	National	Push	Equity share	Push	Early Growth	Bundesministerium für Wirtschaft und Technologie	R&D, innovation and demonstration projects; market development	Venture capital fund; 20% equity required; 500,000 during first round	EUR 1m	7

Nr.	Name	Regional scope	Policy type	Instrument type	Impact on greentech	Life cycle*	Issuing institute	Policy goals	Details	Amount (max)	Years (max)
29	Agrar- und Ernährungswirtschaft – Umwelt- und Verbraucherschutz	Baden-Württemberg	Push	Subsidized Loan	Indirect pull	Expansion	Staatsbank für Baden-Württemberg	Improvement efficiency in the food industry	Subsidized loan for technology investment	EUR 10m	10
30	Umweltschutz- und Energiesparförderprogramm	Baden-Württemberg	Push	Hybrid	Indirect pull	Early Growth; Expansion	KfW Bankengruppe	Fostering of demonstration projects that would not realize without government support	Subsidized loan; Bail/Absolute gurantee	70% of investment	
31	Umwelttechnik (RWB-EFRE)	Baden-Württemberg	Push	Grant, allowance	Push	Seed, Start-up; Early Growth	Ministerium für Umwelt, Naturschutz und Verkehr Baden-Württemberg; Karlsruher Institut für Technologie; Staatsbank für Baden-Württemberg	Securing regional employment through Greentech; R&D; Innovation	Project related grants, allowances	EUR 500,000	1.5
32	Bioenergiewettbewerb	Baden-Württemberg	Push	Grant, allowance	Push	Seed, Start-up; Early Growth	Wirtschaftsministerium Baden-Württemberg	R&D; Improvement of emmissions, effectiveness of combustion and cost base	Grants to enhance existing bioenergy technologies and increase competitiveness	EUR 250,000	
33	Demonstrationsvorhaben der rationellen Energieverwendung und der Nutzung erneuerer Energieträger	Baden-Württemberg	Push	Grant, allowance	Push	Seed, Start-up; Early Growth	Wirtschaftsministerium Baden-Württemberg	R&D; Improvement of emmissions, effectiveness of combustion and cost base	Grants to enhance existing renewable energy technologies and increase competitiveness (min. investment EUR 40,000)	40% of investment	
34	Förderung von Bioenergiedörfern	Baden-Württemberg	Push	Grant, allowance	Indirect pull	Expansion	Wirtschaftsministerium Baden-Württemberg	Market development	Grants for the rural implementation of cogeneration and regenerative heating technologies	EUR 100,000	
35	Heizen und Wärmenetze mit regenerativen Energien (EFRE)	Baden-Württemberg	Push	Grant, allowance	Indirect pull	Expansion	Ministerium für Umwelt, Naturschutz und Verkehr Baden-Württemberg; KEA Klimaschutz- und Energieagentur Baden-Württemberg GmbH	Market development	Grants for the rural implementation of geothermal and regenerative heating technologies	EUR 200,000	
36	Neue Energien - Energie vom Land	Baden-Württemberg	Push	Subsidized Loan	Indirect pull	Expansion	Staatsbank für Baden-Württemberg	Develop renewable energy technologies in the agricultural- and food industry	Subsidized loan for technology investment	EUR 10m	
37	Umweltschutz- und Energiesparförderprogramm	Baden-Württemberg	Push	Subsidized Loan	Indirect pull	Expansion	Staatsbank für Baden-Württemberg	Increase of energy efficiency	Subsidized loan for energy efficiency and production process improvement	75% of investment	
38	Bayerisches Umweltberatungs- und Auditprogramm	Bavaria	Push	Grant, allowance	Indirect pull	Expansion	Bayerisches Staatsministerium für Umwelt und Gesundheit; Landesgewerbeanstalt Bayern (LGA)	Increase of energy efficiency	Grants for consulting fees towards environmental improvement	EUR 2,700	
39	Bayerisches Umweltkreditprogramm / Ökokredit	Bavaria	Push	Subsidized Loan	Indirect pull	Expansion	LfA Förderbank Bayern	Environmental protection	Loans for investing into environmental protection; 20% equity required	EUR 12.5m	
40	Förderprogramm "Elektromobilität"	Bavaria	Push	Grant, allowance	Push	Seed, Start-up	Bayerisches Staatsministerium für Wirtschaft, Infrastruktur, Verkehr und Technologie	Innovation in electric mobility technologies	Grants for high-risk innovation projects with focus on industrial R&D	50% of investment	
41	Förderung der CO2-Vermeidung durch Biomasseheizanlagen (BioKlima)	Bavaria	Push	Grant, allowance	Indirect pull; Push	Seed, Start-up; Early Growth	Bayerisches Staatsministerium für Ernährung, Landwirtschaft und Forsten; Technologie- und Förderzentrum (TFZ)	Reduction of CO2-Emmissions	Market development for Biogas technologies	EUR 200,000	8
42	Förderung von Tiefengeothermie-Wärmenetzen	Bavaria	Push	Hybrid	Indirect pull; Push	Expansion	Bayerisches Staatsministerium für Wirtschaft, Infrastruktur, Verkehr und Technologie; LfA Förderbank Bayern	Development of the geothermal infrastructure	Grants or subsidized loans for set-up, construction and initiation	EUR 1.5m	
43	Rationelle Energiegewinnung und -verwendung (BayREV)	Bavaria	Push	Grant, allowance	Indirect pull; Push	Expansion	Bayerisches Staatsministeriums für Wirtschaft, Infrastruktur, Verkehr und Technologie	Reduction of CO2-Emmissions	Grants for projects that either increase energy efficiency or renewable energy usage	50% of investment	
44	Umweltentlastungsprogramm (UEP II)	Bavaria	Push	Grant, allowance	Indirect pull; Push	Seed, Start-up; Early Growth; Expansion	Senatsverwaltung für Gesundheit, Umwelt und Verbraucherschutz; B.&S.U. Beratungs- und Service-Gesellschaft Umwelt mbH	Reduction of CO2-Emmissions	Grants for projects that reduce CO2-emmissions	80% of investment	

Nr.	Name	Regional scope	Policy type	Instrument type	Impact on greentech	Life cycle*	Issuing institute	Policy goals	Details	Amount (max)	Years (max)
45	Umweltentlastungsprogramm (UEP II)	Berlin	Push	Grant, allowance	Indirect pull; Push	Expansion	Staatsverwaltung für Gesundheit, Umwelt und Verbraucherschutz; B. & S. U.	Reduction of externalities in industrial context	Investment must fulfill efficiency targets	35% of investment	-
46	Brandenburg-Kredit für den Ländlichen Raum	Brandenburg	Push	Subsidized Loan	Indirect pull	Expansion	InvestitionsBank des Landes Brandenburg (ILB)	Reduction of externalities in rural areas	Subsidized loan	EUR 10m	20
47	Förderrichtlinie Umweltschutz	Brandenburg	Push	Grant, allowance	Indirect pull	Expansion	InvestitionsBank des Landes Brandenburg (ILB)	CO2 reduction, waste managment, noise reduction	Investment must fulfill the interest of government and has a societal benefit	50% of investment	-
48	Brandenburg Capital	Brandenburg	Push	Equity share	Push	Seed, Start-up; Early Growth	Brandenburg Capital	Innovation and economic growth	Venture capital fond; EUR 150m under management	-	-
49	Energieeffizienz und Nutzung erneuerbarer Energien (REN-Programm)	Brandenburg	Push	Grant, allowance	Push; Indirect pull	Seed, Start-up; Early Growth; Expansion	Ministerium für Wirtschaft; InvestitionsBank des Landes Brandenburg (ILB)	Reduction of CO2-Emissions	Grants for projects that reduce CO2-emmisions or introduce market innovations	EUR 500,000	-
50	Förderung der betrieblichen ökologischen Effizienz und des verantwortlichen Wirtschaftens	Bremen	Push	Grant, allowance	Indirect pull	Expansion	Senator für Umwelt, Bau, Verkehr und Europa	Improvement of efficiency through consulting and certification	Grant for up to 60% of consulting fees	EUR 12,000	-
51	Förderung von Investitionen für den Aufbau einer Kreislaufwirtschafts-Infrastruktur	Bremen	Push	Grant, allowance	Indirect pull	Expansion	Senator für Umwelt, Bau, Verkehr und Europa	Recycling infrastructure	Grant with respect to environmental target	-	-
52	Programm zur Förderung anwendungsnaher Umwelttechniken (PFAU): Markterschließungen	Bremen	Push	Grant, allowance	Push	Early Growth	WFB Wirtschaftsförderung Bremen GmbH	Market development	Grant; 50% equity required	EUR 50,000	-
53	Programm zur Förderung anwendungsnaher Umwelttechniken (PFAU): Pilotprojekte	Bremen	Push	Grant, allowance	Push	Seed, Start-up	WFB Wirtschaftsförderung Bremen GmbH	R&D; Demonstration projects	Grant; 50% equity required	50% of investment	-
54	Sparsame und rationelle Energienutzung und -umwandlung in Industrie und Gewerbe (REN-Richtlinie)	Bremen	Push	Grant, allowance	Indirect pull	Expansion	Senator für Umwelt, Bau, Verkehr und Europa	Reduction of CO2-Emmisions	Grants for projects that either increase energy efficiency or renewable energy usage	EUR 15,000	-
55	Windkraftnutzung im Land Bremen nach §9 BremEG	Bremen	Pull	Grant, allowance	Pull	Expansion	Senator für Umwelt, Bau, Verkehr und Europa	Development of wind energy technologies	Grants to diminish operating disadvantage towards other technologies	-	12
56	Förderrichtlinie Erneuerbare Energien	Hamburg	Push	Hybrid	Indirect pull; Push	Early Growth; Expansion	Behörde für Stadtentwicklung und Umwelt	Reduction of externalities by introducing green technologies	Grants or subsidized loans for set-up, construction and initiation	-	-
57	Klimaschutzkredite	Hamburg	Push	Subsidized Loan	Indirect pull	Expansion	Hamburgische Wohnungsbaukreditanstalt	Reduction of externalities in all sectors	Investment must reduce resurce usage and environmental impacts	EUR 100,000	-
58	Unternehmen für Ressourcenschutz	Hamburg	Push	Grant, allowance	Indirect pull	Expansion	Hamburgische Wohnungsbaukreditanstalt	Reduction of externalities in all sectors	Investment must reduce resurce usage and environmental impacts; 70% equity required	EUR 100,000	-
59	Klimaschutzprogramm Bioenergie	Hamburg	Push	Grant, allowance	Indirect pull; Push	Early Growth	Behörde für Stadtentwicklung und Umwelt	Reduction of emmissions; Development of Biogas technologies	Grants for development of bioenergy facilities	EUR 75 per kW	-
60	Klimaschutzprogramm Photovoltaik	Hamburg	Push	Grant, allowance	Indirect pull; Push	Early Growth	Behörde für Stadtentwicklung und Umwelt	Reduction of emmissions; Development of Photovoltaik technologies	Grants for development of photovoltaik facilities	EUR 250 per kWp	-
61	Klimaschutzprogramm Solarthermie und Heizung	Hamburg	Push	Grant, allowance	Indirect pull; Push	Early Growth	Behörde für Stadtentwicklung und Umwelt	Reduction of emmissions; Development of solar thermal technologies	Grants for development of solar thermal facilities	EUR 60 per sqm	-

Nr.	Name	Regional scope	Policy type	Instrument type	Impact on greentech	Life cycle*	Issuing institute	Policy goals	Details	Amount (max)	Years (max)
62	Unternehmen für Ressourcenschutz	Hamburg	Push	Grant, allowance	Indirect pull	Expansion	Senator für Umwelt, Bau, Verkehr und Europa	Reduction of CO2-Emmissions	Grants for projects that either increase energy efficiency or renewable energy usage	EUR 100,000	-
63	Technologieprogramm Klimaschutz und Energieeffizienz	Hesse	Push	Grant, allowance	Push	Seed, Start-up	Hessisches Ministerium für Wirtschaft, Verkehr und Landesentwicklung (HMWVL); Wirtschafts- und Infrastrukturbank Hessen (WIBank)	Renewable energy innovation	Grants for R&D; pre-market development projects	60% of investment	-
64	Niedersächsisches Innovationsförderprogramm	Lower saxony	Push	Grant, allowance	Push	Seed, Start-up	Ministerium für Wirtschaft, Arbeit und Verkehr; Ministerium für Umwelt und Klimaschutz; Investitions- und Förderbank Niedersachsen (NBank)	Greentech; R&D; Innovation	Grants to enhance existing technologies and increase competitiveness	45% of investment	-
65	Klimaschutz-Förderrichtlinie	Mecklenburg-Western Pomerania	Push	Grant, allowance	Indirect pull	Expansion	Landesförderinstitut Mecklenburg-Vorpommern (LFI)	Reduction of CO2-Emmissions	Grants for projects that either increase energy efficiency or renewable energy usage	40% of investment	-
66	Investitionsprogramm Abwasser NRW	North Rhine-Westphalia	Push	Hybrid	Indirect pull	Early Growth	NRW.BANK	Reduction of externalities in water management	Subsidized loan; Grants, allowances; Funding must be secured also without governmental support	-	-
67	Maßnahmen des Wasserbaus einschließlich Talsperren	North Rhine-Westphalia	Push	Grant, allowance	Indirect pull	Expansion	Ministerium für Umwelt und Naturschutz, Landwirtschaft und Verbraucherschutz	Improvement of water infrastructure	Project must be relevant and environmentally oriented	-	-
68	Ressourceneffizienz-Programm des Landes Nordrhein-Westfalen	North Rhine-Westphalia	Push	Grant, allowance	Push; indirect pull	Seed, Start-up; Early Growth	Landesamt für Natur, Umwelt und Verbraucherschutz Nordrhein-Westfalen (LANUV)	Improvement of resource efficiency; R&D; Innovation	R&D receives highest grants; Other areas include process improvement and consultancy services	EUR 7.5m	-
69	progres.nrw - Programm für Rationelle Energieverwendung, Regenerative Energien und Energiesparen - Programmbereiche Innovation; Markteinführung	North Rhine-Westphalia	Push	Grant, allowance	Push	Seed, Start-up	Ministerium für Wirtschaft, Mittelstand und Energie; Projektträger Energie, Technologie, Nachhaltigkeit (PT ETN)	Energy innovation and commercialization of innovative technologies	Grants for R&D (-85%); Commercialization projects (-35%); Energy concepts and consulting	85% of investment	-
70	NRW.BANK.Seed.Fonds.Initiative	North Rhine-Westphalia	Push	Equity share	Push	Seed, Start-up	NRW.BANK	R&D; Innovation and commercialization of greentech products	Focused funds that invest exclusively into greentech	EUR 500,000	-
71	Zinszuschüsse für Investitionen im Bereich der Energieeffizienz und der Energieversorgung	Rhineland-Palatinate	Push	Subsidized Loan	Indirect pull	Expansion	Ministerium für Umwelt, Forsten und Verbraucherschutz; EffizienzOffensive Energie Rheinland-Pfalz (EOR) e. V.	Improve efficiency in the industry; Devolopment of renewable energy technologies	Subsidized loan for technology investment	EUR 5m	-
72	Saarländisches Umweltmanagement-Förderprogramm	Saarland	Push	Grant, allowance	Indirect pull	Expansion	Ministerium für Umwelt, Energie und Verkehr	Development of environmental management	Grants for implementing environmental certification	45.5% of investment	-
73	Förderung von Maßnahmen im Zukunftsenergieprogramm (ZEP-Tech)	Saarland	Push	Grant, allowance	Push; Indirect pull	Seed, Start-up; Early Growth	Ministerium für Umwelt, Energie und Verkehr	R&D; Improvement of emmissions, Increase decentralized energy technologies	Grants to enhance existing renewable energy technologies and increase competitiveness; R&D	EUR 100,000	-
74	Energieeffizienz und Klimaschutz	Saxony	Push	Hybrid	Push; Indirect pull	Seed, Start-up; Early Growth	Staatsministerium für Umwelt und Landwirtschaft; Staatsministerium für Wirtschaft, Arbeit und Verkehr; Sächsische Aufbaubank – Förderbank – (SAB)	Reduction of CO2-Emmissions	Grants to develope existing technologies and improve energy efficiency; R&D and innovation (first 40,000 as grant)	EUR 40,000	-

Nr.	Name	Regional scope	Policy type	Instrument type	Impact on greentech	Life cycle*	Issuing institute*	Policy goals	Details	Amount (max)	Years (max)
75	Mittelstandsförderung – Umweltmanagement	Saxony	Push	Grant, allowance	Indirect pull	Expansion	Sächsischen Staatsministeriums für Wirtschaft, Arbeit und Verkehr	Development of environmental management	Grants for implementing environmental certification	50% of investment	-
76	Förderung abfallwirtschaftlicher Maßnahmen	Saxony-Anhalt	Push	Hybrid	Push; indirect pull	Seed, Start-up; Early Growth; Expansion	Investitionsbank Sachsen-Anhalt	Development and innovation in the waste management sector	Subsidized loan; Grants, allowances; Funding must be secured also without governmental support; Track record of supported company	40% of investment	-
77	Zukunftsfonds Entsorgungswirtschaft Sachsen-Anhalt	Saxony-Anhalt	Push	Hybrid	Push	Seed, Start-up; Early Growth	Investitionsbank Sachsen-Anhalt	Innovation and innovative projects in the waste management sector	Subsidized loan; Grants, allowances; Funding must be secured also without governmental support	EUR 1.5m	-
78	Förderung von Umweltinnovationen (UI-Richtlinie)	Schleswig-Holstein	Push	Grant, allowance	Push; indirect pull	Seed, Start-up; Early Growth	Wirtschaftsförderung und Technologietransfer Schleswig-Holstein GmbH (WTSH)	Improvement of energy efficiency, material efficiency and waste management	Grants for technology supplier and companies that implement new processes or projects	70% of investment	-
79	Energetische Nutzung von Biomasse im ländlichen Raum	Schleswig-Holstein	Push	Grant, allowance	Push; indirect pull	Seed, Start-up; Early Growth	Landesamt für Landwirtschaft, Umwelt und ländliche Räume	Development of bioenergy technologies	Grants for set-up, construction and development of biofuel combustion technologies	40% of investment	2
80	Förderung von Kleinkläranlagen im Freistaat Thüringen	Thuringia	Push	Grant, allowance	Indirect pull	Expansion	Thüringer Aufbaubank (TAB)	Improvement of water infrastructure	Grants for modernization of existing sewage water systems	EUR 750 per "Einwohner wert"	-
81	Nachhaltige Entwicklung	Thuringia	Push	Grant, allowance	Indirect pull	Expansion	Thüringer Ministerium für Landwirtschaft, Forsten, Naturschutz und Umwelt	Sustainable business	Grants for consulting fees towards environmental improvement	70% of investment	-
82	Marco Polo II – Verbesserung der Umweltfreundlichkeit des Güterverkehrssystems	Europe	Push	Grant, allowance	Indirect pull	Early Growth	Europäische Kommission Generaldirektion Energie und Verkehr	Sustainable development in the logistics and transport industry	Grants for projects that reduce structural barriers and improve current transport technologies	35% of investment	-
83	Programm zur Konjunkturbelebung durch eine finanzielle Unterstützung der Gemeinschaft zugunsten von Vorhaben im Energiebereich	Europe	Push	Grant, allowance	Indirect pull	Early Growth; Expansion	Europäische Kommission Generaldirektion Energie und Verkehr	Modernization of existing energy facilities; Development of offshore-wind parks	Grants for construction and operation of offshore-wind parks, CCS and gas/power networks	80% of investment	-
84	Rahmenprogramm für Wettbewerbsfähigkeit und Innovation (CIP) (2007-2013)	Europe	Push	Hybrid	push	Seed, Start-up; Early Growth	Europäische Kommission Generaldirektion Energie und Verkehr	R&D; Innovation and demonstration projects; market development	Grants and absolute guarantees for innovative development and commercialization projects with focus on greentech	60% of investment	-
85	Gemeinsame Technologieinitiative für Brennstoffzellen und Wasserstoff - Gemeinsames Unternehmen FCH	Europe	Push	Grant, allowance	Push	Seed, Start-up	Europäische Kommission Generaldirektion Forschung	Improve marketability of fuel cell technologies	Grants for R&D and commercialization projects	50% of investment	-
86	Gemeinsame Technologieinitiative "Clean Sky" - Gemeinsames Unternehmen "Clean Sky"	Europe	Push	Grant, allowance	Push	Seed, Start-up	Europäische Kommission Generaldirektion Forschung	R&D; Innovation and demonstration projects; market development	Grants for innovative development and commercialization projects with focus on sustainable aviation	-	-

B. Questionnaire

Vielen Dank, dass Sie sich zur Teilnahme an der Studie **"Greentech: Innovation, Wachstum und Finanzierung"** entschieden haben. Die Studie wird im Rahmen der **Doktorarbeit von Philipp Hoff**, innerhalb eines **Forschungsprojektes der Universität St. Gallen** unter der Leitung von **Professor Dr. Thomas Berndt** und mit Unterstützung der Unternehmensberatung **Roland Berger** durchgeführt. Das Erheben der Daten geschieht anonym und sie werden aggregiert ausgewertet. Der erhobene Datensatz wird nicht an Dritte weitergegeben.
Der folgende Fragebogen umfasst knapp **30 Ankreuzfragen und dauert etwa 20 Minuten**. Am Ende des Fragebogens haben Sie die Möglichkeit eine **Greentech-Förderprogrammübersicht und die Studienergebnisse anzufordern**, sowie an der **Verlosung eines Apple iPad** teilzunehmen.
Sie können jederzeit die Befragung unterbrechen und zu einem späteren Zeitpunkt fortfahren - klicken Sie dazu einfach auf den Link aus der E-mail.
Sämtliche Fragen beziehen sich auf Ihr gesamtes Unternehmen. Sollte Ihr Unternehmen eine diversifizierte Struktur mit Nicht-Greentech-Aktivitäten aufweisen, bitten wir Sie (sofern möglich) die Fragen auf die relevante Geschäftseinheit zu beziehen. Der Fragebogen gliedert sich in fünf Teile: (1) Strategie, (2) Innovation, (3) Finanzierung, (4) Förderung und (5) Unternehmenscharakteristika. Pflichtfelder sind mit einem * gekennzeichnet
Die Umsetzung des Projekts wird von Herrn Philipp Hoff verantwortet, der Ihnen gerne jederzeit für Fragen, Anmerkungen und Diskussionen unter philipp.hoff@student.unisg.ch oder +49(160) 744 6193 zur Verfügung steht.
Weitere Informationen erhalten Sie hier .

Strategie und Schwerpunkte

1a. Welche **Rechtsform** besitzt Ihre Firma:
(Bitte auswählen, eine Nennung)

○ Personengesellschaft (z.B. OHG, GbR, KG)
○ Kapitalgesellschaft (z.B. GmbH, GmbH & Co. KG, AG)

1b. Ist Ihre Firma **Teil einer Unternehmensgruppe oder eines Konzerns**?*

○ Ja
○ Nein

2. Was ist der wesentliche **Geschäftszweck** Ihres Unternehmens?*
(Bitte Schwerpunkt(e) auswählen)

☐ Entwicklung, Herstellung und Vermarktung von **Technologien und physischen Produkten**
☐ Erbringung von **Dienstleistungen**
☐ **Handel** mit Technologien und Produkten

3. In welchen **Leitmärkten/Technologiefeldern** ist Ihr Unternehmen schwerpunktmäßig aktiv?*
(Bitte Schwerpunkt(e) auswählen)

☐ Erneuerbare Energietechnologien und -dienstleistungen
☐ Energieeffizienztechnologien und -dienstleistungen
☐ Nachhaltige Wasserwirtschaft
☐ Nachhaltige Mobilitätstechnologien
☐ Nachhaltige Materialien und Materialeffizienz
☐ Abfallwirtschaft, -aufbereitung und -recycling
☐ Sonstige *(Bitte benennen)*:

4. Welche der folgenden **Strategien** trifft auf Ihr Unternehmen am besten zu?*
(Bitte auswählen, eine Nennung)

○ Innovativste Technologien und Dienstleistungen am Markt
○ Kostenführerschaft und Profitabilität durch Effizienz
○ Konstantes Einkommen, Stabilität und Substanzerhaltung

5. Welchen **Stellenwert** haben Nachhaltigkeitsaspekte (sozial und ökologisch) in Ihrem Unternehmen?*
(Bitte auswählen, eine Nennung)

○ Nachhaltigkeitsziele sind **wichtiger als** Gewinn- und Umsatzziele
○ Nachhaltigkeitsziele sind **genauso wichtig** wie Gewinn- und Umsatzziele
○ Nachhaltigkeitsziele sind **wichtig, aber** Gewinn- und Umsatzzielen **untergeordnet**
○ Nachhaltigkeitsziele werden **nicht aktiv** formuliert, aber wenn möglich umgesetzt
○ Nachhaltigkeitsziele spielen **keine Rolle**

6.   Ist Ihr Unternehmen **international aktiv**?*

(Bitte auswählen)

	Keine Aktivität 1	2	3	4	5	Hohe Aktivität 6
Import/Export	O	O	O	O	O	O
Internationale(s) Mutterunternehmen/Tochterunternehmen	O	O	O	O	O	O
Internationale Kooperationen (Joint Ventures)	O	O	O	O	O	O
Internationale Kapitalbeschaffung	O	O	O	O	O	O
Internationale Aktivitäten in Vorbereitung	O	O	O	O	O	O

7a. In welchem Jahr wurde Ihr **Unternehmen gegründet?***
(Bitte Jahr angeben, z.B. 2002)

7b. In welcher **Phase befindet sich Ihr Unternehmen?***
(Bitte auswählen, eine Nennung)

○ **Gründung:** z.B. Beginn der Geschäftsaktivitäten; erste Anbahnung von Kundenkontakten; Ausstattung der Räumlichkeiten; Aufbau von Produktionsanlagen, Rechnern etc.; Investorensuche; Technologieentwicklung

○ **Start-up:** z.B. Aufbau eines Kernkundensegments; flexible Prozesse; Informelle Organisationsstruktur bei hohem Zentralisierungsgrad; generalistische Ausrichtung der Mitarbeiter; lokaler und regionaler Fokus der Verkaufsaktivitäten

○ **Frühes Wachstum:** z.B. Stärkere Formalisierung der Prozesse, schnelles Wachstum der Kundenbasis; schnelles Wachstum der Organisation; Ausweitung der Aktivitäten auf neue Regionen, Produkte und Kundensegmente; Einrichtung von spezialisierten Funktionen im Unternehmen

○ **Expansion:** z.B. Starke Marktposition; moderates bis schnelles Wachstum; hohe Durchdringung der potenziellen Kundenbasis; verstärkte Aktivitäten zur Einsparung von Kosten; Optimierung der Prozesse; hoher Grad der Spezialisierung der Unternehmensfunktionen

○ **Erhaltung:** z.B. Erwirtschaftung von konstanten Erträgen; hoher Anteil Stammkunden; keine Ausweitung der Kundenbasis oder Produktsegmente; kein oder sehr wenig Wachstum; Unternehmen kann sowohl klein als auch groß sein

○ **Keine** der genannten Phasen, sondern:

*Auf die mit * markierten Fragen folgen Pflichteingabefelder.*

Innovation und Investition

8a. Hat Ihr Unternehmen in den letzten drei Jahren (2008-2010)...*
(Bitte auswählen)

	Ja	Nein
...ein innovatives und neues Produkt/Dienstleistung auf den Markt gebracht?	O	O
...ein Produkt/Dienstleistung wesentlich weiterentwickelt?	O	O
...einen innovativen und völlig neuen Prozess eingeführt?	O	O
...einen Prozess wesentlich weiterentwickelt?	O	O

8b. Hat Ihr Unternehmen **derzeit** ein oder mehrere laufende Projekte zur...*
(Bitte auswählen)

	Ja	Nein
...Produkt-/ Dienstleistungsentwicklung oder -verbesserung?	O	O
...Prozessentwicklung oder Verbesserung?	O	O

9. Wie viel **Prozent des Umsatzes** hat Ihr Unternehmen im **letzten Jahr (2009) in Forschung und Entwicklung** von neuen Produkten und Dienstleistungen investiert?*
(Bitte Angabe in % vom Umsatz 2009 z.B. 5)

10. In welche der unten genannten Bereiche wird Ihr Unternehmen voraussichtlich in den **nächsten drei Jahren investieren** (2011-2013)?*
(Bitte auswählen)

	Keine Investition 1	2	3	4	5	Hohe Investition 6
Forschung und Entwicklung von Produkten und Dienstleistungen	O	O	O	O	O	O
Vertriebskanäle und Marketing	O	O	O	O	O	O

Internationalisierung	○	○	○	○	○	○
Verbesserung von Prozessen	○	○	○	○	○	○
Anlagen, Maschinen und Gebäude	○	○	○	○	○	○
Andere:	○	○	○	○	○	○

11. Welche der folgenden Aussagen trifft bezüglich **der letzten drei Jahre (2008-2010)** auf Ihr Unternehmen zu?*
    (Bitte auswählen)

(Investitions-)Projekte wurden **langsamer, später oder gar nicht umgesetzt,**...	Trifft nicht zu 1	2	3	4	5	Trifft voll zu 6
...da das **ökonomische Risiko zu hoch** war...	○	○	○	○	○	○
...da auch zu höheren Zinssätzen **nicht ausreichend finanzielle Mittel** zur Verfügung standen...	○	○	○	○	○	○
...da die **notwendige Finanzierung zu teuer** gewesen wäre...	○	○	○	○	○	○
...da der **Finanzierungsprozess zu viel Zeit** in Anspruch nimmt...	○	○	○	○	○	○
...da es nur **unzureichende oder unpassende Förderprogramme** gibt...	○	○	○	○	○	○
Andere:	○	○	○	○	○	○

*Auf die mit * markierten Fragen folgen Pflichteingabefelder.*

Finanzierung

12. Welche **Finanzierungsinstrumente** nutzt Ihr Unternehmen?*
    (Bitte auswählen)

Eigenkapital	Keine Nutzung 1	2	3	4	5	Häufige Nutzung 6
Kapital von Freunden/Familie	○	○	○	○	○	○
Eigenes, firmeninternes Eigenkapital (Gesellschafter, Gründer)	○	○	○	○	○	○
Externes Eigenkapital (Investoren, Aktien)	○	○	○	○	○	○
Thesaurierte/einbehaltene Gewinne	○	○	○	○	○	○
Business Angel Kapital	○	○	○	○	○	○
Venture Capital	○	○	○	○	○	○
Förderzuschüsse (z.B. KfW, Bund, Land)	○	○	○	○	○	○
Fremdkapital						
Überziehungskredite bei Banken	○	○	○	○	○	○
Kreditkarten	○	○	○	○	○	○
Gesellschafterkredite	○	○	○	○	○	○
Kredite von Freunden/Bekannten/Familie	○	○	○	○	○	○
Kurzfristige Bank- und Lieferantenkredite	○	○	○	○	○	○
Mittelfristige Bankkredite	○	○	○	○	○	○
Langfristige Bankkredite	○	○	○	○	○	○
Förderdarlehen (z.B. KfW, NRW Bank)	○	○	○	○	○	○
Externe Darlehen (Bonds, Schuldscheine etc.)	○	○	○	○	○	○

13. Hat Ihr Unternehmen für die Aufnahme der Bankkredite **Sicherheiten** erbracht, die über das Haftungskapital der Firma hinausgingen?
    (Bitte auswählen, Mehrfachnennung möglich)

☐ Ja, persönliches Vermögen der Firmeneigner
☐ Ja, Haftungserklärung durch Dritte
☐ Ja, andere:
☐ Nein

14. Nutzen Sie den **Verkauf von Forderungen (Factoring)** als Finanzierungsinstrument?*
    (Bitte auswählen)

○ Ja
○ Nein

15a. Hat sich Ihr Unternehmen in der Vergangenheit für eine der folgenden **Finanzierungsformen beworben und wurde abgelehnt?***
(Bitte auswählen)

	Ja	Nein
Eigenkapital		
Externes Eigenkapital (Investoren, Aktien)	○	○
Business Angel Kapital	○	○
Venture Capital	○	○
Förderzuschüsse (z.B. KfW, Bund, Land)	○	○
Fremdkapital		
Kurzfristige Bank- und Lieferantenkredite	○	○
Mittelfristige Bankkredite	○	○
Langfristige Bankkredite	○	○
Förderdarlehen (z.B. KfW, NRW Bank)	○	○
Externe Darlehen (Sonstige Kapitalgeber, Schuldscheine)	○	○

15b. Falls eine Bewerbung abgelehnt wurde, **warum** wurde Ihr Unternehmen abgelehnt?
(Bitte auswählen, Mehrfachnennung möglich)

☐ Unzureichendes Marktverständnis der Geldgeber
☐ Zu hohes Risiko aus Sicht der Geldgeber
☐ Unzulänglicher Business Plan aus Sicht der Geldgeber
☐ Nicht ausreichend Sicherheiten vorhanden
☐ Andere Gründe _____

16a. Werden Ihrem Unternehmen bei der Bezahlung von Lieferungen und Leistungen regelmäßig **Skontorabatte** gewährt?***
(Bitte auswählen)

○ Ja
○ Nein

16b. Wenn ja, nimmt Ihre Firma die Skonto-Angebote in Anspruch?
(Bitte auswählen)

	Nein, nie 1	2	3	4	5	Ja, immer 6
	○	○	○	○	○	○

17a. Wie schätzen Sie das **Verhältnis zu Ihre(n) Hausbank(en)** ein?***
(Bitte auswählen)

	Sehr schlecht 1	2	3	4	5	Sehr gut 6
	○	○	○	○	○	○

17b. Mit wie vielen Banken unterhalten Sie **Geschäftsbeziehungen?**
(Bitte auswählen)

○ 1-2
○ 3-4
○ mehr als 4

18. Wie hoch war Ihr durchschnittlicher **Zinssatz für Fremdkapital im Jahr 2009?**
(Bitte Angabe in % p.a., z.B. 5)

[%]

Langfristige Kredite ☐
Kurz- und mittelfristige Kredite ☐

19. Wie schätzen Sie die **künftige Finanzierungslage** ein?***
(Bitte auswählen)

	Viel schwerer 1	2	3	4	5	Viel leichter 6
Finanzierung wird...						

...für alle Unternehmen	O	O	O	O	O	O
...für Greentech-Unternehmen	O	O	O	O	O	O
...für Ihr Unternehmen	O	O	O	O	O	O

Auf die mit * markierten Fragen folgen Pflichteingabefelder.

Förderung

20a. Hat Ihr Unternehmen in den letzten drei Jahren (2008-2010) **staatliche Förderprogramme in Anspruch genommen?***
(Bitte auswählen)

Greentech- und Nachhaltigkeitsförderung (spezifische Programme)	Keine Nutzung 1	2	3	4	5	Häufige Nutzung 6
Zuschüsse (z.B. KWK-Zuschuss)	O	O	O	O	O	O
Subventionierte Kredite (z.B. KFW erneuerbare Energien, LfA Ökokredit)	O	O	O	O	O	O
Staatliches Eigenkapital/Venture Capital (z.B. High-tech Gründerfonds)	O	O	O	O	O	O
Mischformen (z.B. Kredite mit Zuschüssen)	O	O	O	O	O	O
Sonstige Förderprogramme ohne Greentech-Fokus (z.B. KMU-Förderkredit, Gründerzuschuss)	O	O	O	O	O	O

20b. Falls Ihr Unternehmen **Förderung** in Anspruch genommen hat, wären Ihre **Investitionen ohne Förderung niedriger** ausgefallen?
(Bitte auswählen)

	Nein, keinesfalls 1	2	3	4	5	Ja, sehr 6
	O	O	O	O	O	O

20c. Falls Ihr Unternehmen **keine Förderung** in Anspruch genommen hat, **was waren die Gründe dafür?**
(Bitte auswählen)

	Trifft nicht zu 1	2	3	4	5	Trifft voll zu 6
Unzureichende Kenntnis hinsichtlich der Förderprogramme	O	O	O	O	O	O
Keine Erfüllung der Förderkriterien	O	O	O	O	O	O
Zu hoher Aufwand/Komplexität der Antragstellung/Bewerbung	O	O	O	O	O	O
Andere:	O	O	O	O	O	O

21. Ist Ihre Geschäftstätigkeit auf **staatliche Anreizprogramme** angewiesen, die auf die **Erhöhung der Nachfrage** Ihrer Produkte abzielen?
(Bitte auswählen)

	Nein, keinesfalls 1	2	3	4	5	Ja, sehr 6
Erneuerbare-Energien-Gesetz (EEG)	O	O	O	O	O	O
Kraft-Wärme-Kopplungs-Gesetz (KWK)	O	O	O	O	O	O
Andere:	O	O	O	O	O	O

Auf die mit * markierten Fragen folgen Pflichteingabefelder.

Unternehmenscharakteristika

22a.Wie viele **Mitarbeiter** hat Ihr Unternehmen **Ende 2009** beschäftigt?*
(Bitte Anzahl Mitarbeiter angeben, z.B. 25)

22b.Wie viel **Prozent Ihrer Mitarbeiter** besitzen schätzungsweise einen **(Fach-)Hochschulabschluss?**
(Bitte in Prozent der Gesamtbelegschaft schätzen, z.B. 10)

23.Wie hoch war **der Jahresumsatz** Ihres Unternehmens im Jahr 2009?*
(Bitte auswählen)

○ Bis 250,000 Euro
○ Mehr als 250,000 Euro bis 1 Mio. Euro
○ Mehr als 1 Mio. Euro bis 2 Mio. Euro
○ Mehr als 2 Mio. Euro bis 10 Mio. Euro
○ Mehr als 10 Mio. Euro bis 25 Mio. Euro
○ Mehr als 25 Mio. Euro bis 50 Mio. Euro
○ Mehr als 50 Mio. Euro bis 250 Mio. Euro
○ Mehr als 250 Mio. Euro

24a.Wie hat sich der **Umsatz** Ihres Unternehmens im **Durchschnitt in den letzten drei Jahren entwickelt?***
(Bitte in % pro Jahr angeben, z.B. -/+5% p.a.)

24b.Welches Umsatzwachstum erwarten Sie für die **nächsten drei Jahre** (Durchschnitt)?*
(Bitte in % pro Jahr schätzen, z.B. -/+5% p.a.)

25a.Wie hoch war die **Umsatzrendite vor Steuern** (Betriebsergebnis als % vom Umsatz) Ihres Unternehmens **im Jahr 2009?**
(Bitte auswählen)

○ Kleiner 0%
○ 0 bis 3%
○ Mehr als 3% bis 5%
○ Mehr als 5% bis 10%
○ Mehr als 10% bis 15%
○ Mehr als 15% bis 20%
○ Mehr als 20%

25b.Wie schätzen Sie die **Entwicklung Ihrer Umsatzrendite** für die nächsten drei Jahre (2010-2012) ein?*
(Bitte auswählen, eine Nennung)

○ Stark steigend
○ Steigend
○ Gleich bleibend
○ Leicht fallend
○ Fallend

26a.Wie hoch war **Ende 2009** die **Bilanzsumme** Ihres Unternehmens?
(Bitte in Euro angeben)

26b. Wie war **Ende 2009** das **Vermögen** Ihres Unternehmens strukturiert (Bilanzposten)?*
(Bitte auswählen)

Anteil...an der Bilanz	Sehr niedriger Anteil 1	2	3	4	5	Sehr hoher Anteil 6
...Materielles Anlagevermögen (z.B. Maschinen, Gebäude)	○	○	○	○	○	○
...Finanzielles Anlagevermögen (z.B. Aktien, Bundesanleihen)	○	○	○	○	○	○
...Immaterielles Anlagevermögen (z.B. Software, Patente)	○	○	○	○	○	○
...Umlaufvermögen (z.B. Vorräte, Forderungen)	○	○	○	○	○	○

27.Sollten aus Ihrer Sicht die politischen Rahmenbedingungen für Greentech-Unternehmen in Deutschland verändert werden? Wenn, ja welche Änderungen würden Sie sich wünschen?
(Bitte formulieren Sie Stichpunkte, hier können Sie auch weitere Anmerkungen und Themen platzieren)

*Auf die mit * markierten Fragen folgen Pflichteingabefelder.*

Haben Sie vielen Dank für Ihre Teilnahme!

Wie bereits erwähnt werden Ihre Angaben **streng vertraulich behandelt** und nur in **aggregierter Form ausgewertet**. Selbstverständlich werden die gesammelten Daten auch nicht an Dritte weitergegeben. Ihr Firmenname wird weder in den Auswertungen noch im Zusammenhang mit den von Ihnen angegebenen Daten erwähnt. Eine Ausnahme bildet das Unternehmensverzeichnis in der Dissertation, sofern Sie mit der Erwähnung im Folgenden einverstanden sind (ausschließlich Name, keine Daten).

A. Bitte geben Sie an, ob Sie im **Unternehmensverzeichnis der Dissertation** namentlich genannt werden möchten.*

    ◯ Ja, mit Firmennamen
    ◯ Ja, mit Firmennamen und persönlichem
       Namen
    ◯ Nein, ich möchte anonym bleiben

B.   Als Dank für Ihre Teilnahme möchten wir Ihnen die **Ergebnisse der Studie** zukommen lassen. Darüber hinaus stellen wir Ihnen bei Interesse eine übersichtliche **Darstellung der knapp 100 deutschen Förderprogramme** mit Greentech-Fokus zur Verfügung. Außerdem verlosen wir unter allen Studienteilnehmern ein **Apple iPad.***

(Bitte geben Sie an, ob Sie hieran Interesse haben und mit der notwendigen Speicherung Ihrer persönlichen Daten einverstanden sind)

Ich bin mit der Speicherung/Nutzung meiner persönlichen Daten einverstanden,...

	Ja	Nein
...um die Ergebnisse und exklusiven Auswertungen der Studie zugesendet zu bekommen	◯	◯
...um an der Verlosung des Apple iPad teilzunehmen	◯	◯

Kontaktdaten

(E-Mail Angabe notwendig, wenn Sie Ergebnisse der Studie wünschen oder an der iPad-Verlosung teilnehmen möchten)

Firmenname:

[   ]

Personalien des Ausfüllers:

Name:

[   ]

Vorname:

[   ]

E-Mail (*) :

[   ]

Telefon:

[   ]

*Auf die mit * markierten Fragen folgen Pflichteingabefelder.*

Vielen Dank für Ihre Teilnahme!

Sie können dieses Fenster nun schließen .

REFERENCES

Achleitner, Ann-Kristin, and Margarita Tchouvakhina, 2006, Der deutsche Beteiligungs-markt. Entwicklung des Anbieterverhaltens, *Finanz Betrieb Aufsätze*, 538.

Acs, Zoltan J., and David B. Audretsch, 1988, Innovation and firm size in manufacturing, *Technovation* 7, 197–210.

Acs, Zoltan J., and David J. Storey, 2004, Introduction: Entrepreneurship and Economic Development, *Regional Studies* 38, 871–877.

Adizes, Ichak, 1988, *Corporate lifecycles. How and why corporations grow and die and what to do about it* (Prentice-Hall, Englewood Cliffs NJ).

Agarwal, Rajshree, and Michael Gort, 2002, Firm and Product Life Cycles and Firm Survival, *American Economic Review* 92, 184–190.

Agarwal, Rajshree, M. B. Sarkar, and Raj Echambadi, 2002, The Conditioning Effect of Time on Firm Survival: An Industry Life Cycle Approach, *The Academy of Management Journal* 45, 971–994.

Akerlof, George A., 1970, The Market for "Lemons": Quality Uncertainty and the Market Mechanism, *Quarterly Journal of Economics* 84, 488–500.

Anderson, Terry L., and Donald Leal, 2001, *Free market environmentalism* (St. Martin's Press, New York, NY).

Armstrong, J. S., and Terry S. Overton, 1977, Estimating Nonresponse Bias in Mail Surveys, *Journal of Marketing Research* 14, 396–402.

Arrow, Kenneth J., and Rand Corporation., 1959, *Economic welfare and the allocation of resources for invention* (Rand Corp., Santa Monica, CA).

Audretsch, David B., and Zoltan J. Acs, 1991, Innovation and Size at the Firm Level, *Southern Economic Journal* 57, 739–744.

Audretsch, David B., and Julie A. Elston, 2002, Does firm size matter? Evidence on the impact of liquidity constraints on firm investment behavior in Germany, *International Journal of Industrial Organization* 20, 1–17.

Awerbuch, Shimon, 2000, Investing in photovoltaics: risk, accounting and the value of new technology, *Energy Policy* 28, 1023–1035.

Ayres, Robert, 1991, Evolutionary economics and environmental imperatives, *Structural Change and Economic Dynamics* 2, 255–273.

Backhaus, Klaus, 2008, *Multivariate Analysemethoden. Eine anwendungsorientierte Einführung* (Springer, Berlin).

Bah, Rahim, and Pascal Dumontier, 2001, R&D Intensity and Corporate Financial Policy: Some International Evidence, *Journal of Business Finance & Accounting* 28.

Bahadir, S. C., Sundar Bharadwaj, and Michael Parzen, 2009, A meta-analysis of the determinants of organic sales growth, *International Journal of Research in Marketing* 26, 263–275.

Balachandra, P., Hippu S. Kristle Nathan, and B. S. Reddy, 2010, Commercialization of sustainable energy technologies, *Renewable Energy* 35, 1842–1851.

Barron, David N., Elizabeth West, and Michael T. Hannan, 1994, A Time to Grow and a Time to Die: Growth and Mortality of Credit Unions in New York City, 1914-1990, *The American Journal of Sociology* 100, 381–421.

Barrow, Colin, 2006, *Financial management for the small business* (Kogan Page, London).

Bass, Frank M., 1969, A New Product Growth for Model Consumer Durables, *Management Science* 15, 215–227.

Basu, Anuradha, Asbjorn Osland, and Michael Solt, 2009, *A New Course on Sustainability Entrepreneurship*, Working Paper (accessed February 5, 2010).

Becchetti, Leonardo, and Giovanni Trovato, 2002, The Determinants of Growth for Small and Medium Sized Firms. The Role of the Availability of External Finance, *Small Business Economics* 19, 291–306.

Beck, Thorsten, and Asli Demirguc-Kunt, 2006, Small and medium-size enterprises: Access to finance as a growth constraint, *Journal of Banking & Finance* 30, 2931–2943.

Berger, Allen N., and Gregory F. Udell, 1992, Some evidence on the empirical significance of credit rationing, *Journal of Political Economy* 100, 1047.

Berger, Allen N., and Gregory F. Udell, 1998, The economics of small business finance: The roles of private equity and debt markets in the financial growth cycle, *Journal of Banking & Finance*, 613–673.

Bhaird, Ciarán M., 2010, *Resourcing Small and Medium Sized Enterprises. A Financial Growth Life Cycle Approach* (Springer-Verlag Berlin Heidelberg, Heidelberg).

BMU, 2009, *Copenhagen Accord 2009. Decision -/CP.15 (Advance unedited version)* (Copenhagen, Denmark).

Bollingtoft, Anne, John P. Ulhoi, Henning Madsen, and Helle Neergaard, 2003, The Effect of Financial Factors on the Performance of New Venture Companies in High Tech and Knowledge-Intensive Industries: An Empirical Study in Denmark, *International Journal of Management* 20, 535–547.

Börner, Christoph J., Grichnik, Dietmar, eds., 2005, *Entrepreneurial Finance. Kompendium der Gründungs- und Wachstumsfinanzierung* (Physica-Verlag Heidelberg, Heidelberg).

Börner, Christoph J., and Dietmar Grichnik, 2010, Finanzierungsentscheidungen mittelständischer Unternehmer der Einflussfaktoren der Fremdfinanzierung deutscher KMU, *Zeitschrift für betriebswirtschaftliche Forschung*, 227–250.

Bortz, Jürgen, Gustav A. Lienert, and Klaus Boehnke, 2008, *Verteilungsfreie Methoden in der Biostatistik. Mit 247 Tabellen* (Springer Medizin Verlag Heidelberg, Berlin, Heidelberg).

Bougheas, Spiros, Holger Görg, and Eric Strobl, 2003, Is R&D Financially Constrained? Theory and Evidence from Irish Manufacturing, *Review of Industrial Organization* 22, 159–174.

Bozkaya, Ant, and Bruno P. de van Potterie, 2008, Who Funds Technology-Based Small Firms? Evidence from Belgium, *Econ. Innov. New Techn.* 17, 97–122.

Braun, Christoph-Friedrich von, 1997, *The innovation war* (Prentice Hall, Upper Saddle River NJ).

Brechin, and Steven R., 2003, Comparative public opinion and knowledge on global climatic change and the Kyoto protocol: The U.S. versus the world?, *International Journal of Sociology and Social Policy* 23, 106–132.

Brundtland, Gro H., 1991, *Our common future*, World Commission on Environment and Development (Univ. Press, Oxford).

Bruno, Albert V., and Tyzoon T. Tyebjee, 1985, The entrepreneur's search for capital, *Journal of Business Venturing* 1, 61–74.

Brush, Candida G., Dennis J. Ceru, and Robert Blackburn, 2009, Pathways to entrepreneurial growth: The influence of management, marketing and money, *Business Horizonz*, 481–491.

Büchele, Ralph, Henzelmann, Torsten, Emrich, Arnim, Zelt, Thilo, eds., 2007, *GreenTech made in Germany. Umwelttechnologie-Atlas für Deutschland* (Vahlen, München).

Büchele, Ralph, Henzelmann, Torsten, Hoff, Philipp, Engel, Michael, eds., 2009, *GreenTech made in Germany 2.0. Umwelttechnik-Atlas für Deutschland* (Vahlen, München).

Bürer, Mary J., 2008, *Public Policy and Clean Energy Private Equity Investment*, PhD Dissertation, University of St. Gallen (HSG) (Zürich).

Bürer, Mary J., and Rolf Wüstenhagen, 2009, Which renewable energy policy is a venture capitalist's best friend? Empirical evidence from a survey of international cleantech investors, *Energy Policy* 37, 4997–5006.

Burtis, Patrick R., 2004, *Creating the California cleantech cluster. How innovation and investment can promote job growth and a healthy environment.* http://72.32.110.154/air/energy/cleantech/cleantech.pdf (accessed February 17, 2010).

Burtis, Patrick R., 2006, *Creating the California cleantech cluster*, 2004 Update. http://www.globalurban.org/Creating_Cleantech_Clusters.pdf (accessed February 17, 2010).

Buttner, E. H., and Benson Rosen, 1992, Rejection in the loan application process: Male and female entrepreneurs' and subsequent intentions, *Journal of Small Business Management* 30, 58–65.

2007, *BVK Statistik. Das Jahr in Zahlen 2006*, Bundesverband Deutscher Kapitalbeteiligungsgesellschaften (Berlin).

2008, *BVK Statistik. Das Jahr in Zahlen 2007*, Bundesverband Deutscher Kapitalbeteiligungsgesellschaften (Berlin).

2009, *BVK Statistik. Das Jahr in Zahlen 2008*, Bundesverband Deutscher Kapitalbeteiligungsgesellschaften (Berlin).

2010, *BVK Statistik. Das Jahr in Zahlen 2009*, Bundesverband Deutscher Kapitalbeteiligungsgesellschaften (Berlin).

Canepa, Alessandra, and Paul Stoneman, 2008, Financial constraints to innovation in the UK: evidence from CIS2 and CIS3, *Oxford Economic Papers* 60, 711–730.

Carayannis, Elias G., Edgar Gonzalez, and John Wetter, 2003, The Nature and Dynamics of Discontinuous and Disruptive Innovations from a Learning and Knowledge Management Perspective, in Larisa V. Shavinina, ed.: *The International Handbook on Innovation* (Pergamon, Oxford).

Carpenter, Robert E., and Bruce C. Petersen, 2002, Is the growth of small firms constrained by internal finance?, *Review of Economics & Statistics* 84, 298–309.

Carreira, Carlos, and Filipe Silva, 2010, No deep pockets: Some stylized empirical results on firm's financial constraints, *Journal of Economic Surveys* 24, 731–753.

Cassar, Gavin, 2004, The financing of business start-ups, *Journal of Business Venturing*, 261–283.

Cassar, Gavin, and Scott Holmes, 2003, Capital structure and financing of SMEs: Australian evidence, *Accounting & Finance* 43, 123–147.

Casson, Peter D., Roderick Martin, and Tahir M. Nisar, 2008, The financing decisions of innovative firms, *Research in International Business and Finance* 22, 208–221.

Chakraborty, Atreya, and Charles X. Hu, 2006, Lending relationships in line-of-credit and nonline-of-credit loans: Evidence from collateral use in small business, *Journal of Financial Intermediation* 15, 86–107.

Chan, Yolande E., Niraj Bhargava, and Christopher T. Street, 2006, Having Arrived: The Homogeneity of High-Growth Small Firms, *Journal of Small Business Management* 44, 426–440.

Chandler, Alfred D., 2001, *Strategy and structure. Chapters in the history of the industrial enterprise* (MIT Press, Cambridge, Mass.).

Chertow, Marian R., 2000, *Accelerating commercialization of environmental technology in the U.S.: Theory and case studies*, PhD Dissertation, Graduate School of Yale University (Yale University, New Haven, CT).

Chittenden, Francis, Graham Hall, and Patrick Hutchinson, 1996, Small firm growth, access to capital markets and financial structure: Review of issues and an empirical investigation, *Small Business Economics* 8, 59–67.

Choi, David Y., and Edmund R. Gray, 2008, The venture development processes of "sustainable" entrepreneurs, *Management Research News* 31, 558–569.

Christensen, Clayton M., 2008, *The innovator's dilemma. The revolutionary book that will change the way you do business* (Collins Business Essentials, New York, NY).

Christensen, Jesper L., ed., 2009, *Greens rush in?: CleanTech venture capital investments - prospects or hype? DRUID summer conference: Innovation, Strategy and Knowledge*, Conference report (Copenhagen, Denmark).

Christmann, Petra, 2000, Effects of best practices of environmental management on cost advantage: the role of complementary assets, *Academy of Management Journal* 43, 663–680.

Churchill, Neil C., and Virginia L. Lewis, 1983, The five stages of small business growth, *Harvard Business Review* 61, 30.

Clulow, Robert, 1999, Financing for wind energy. Renewable Energy Energy Efficiency, Policy and the Environment, *Renewable Energy* 16, 858–862.

Coco, Giuseppe, 2000, On the Use of Collateral, *Journal of Economic Surveys* 14, 191.

Cohen, Boyd, and Monika I. Winn, 2007, Market imperfections, opportunity and sustainable entrepreneurship, *Journal of Business Venturing* 22, 29–49.

Colombo, Massimo, and Luca Grilli, 2007, Funding Gaps? Access To Bank Loans By High-Tech Start-Ups, *Small Business Economics* 29, 25–46.

Conover, W. J., and Ronald L. Iman, 1981, Rank Transformations as a Bridge Between Parametric and Nonparametric Statistics, *The American Statistician* 35, 124–129.

Correa, Paulo G., Ana M. Fernandes, and Chris J. Uregian, 2010, Technology Adoption and the Investment Climate: Firm-Level Evidence for Eastern Europe and Central Asia, *World Bank Econ Rev* 24, 121–147.

Couture, Toby, and Yves Gagnon, 2010, An analysis of feed-in tariff remuneration models: Implications for renewable energy investment, *Energy Policy* 38, 955–965.

Cowan, Kelly R., and Turgrul Daim, 2009, Comparative technological road-mapping for renewable energy, *Technology in Society* 31, 333–341.

Cowan, Robin, and Staffan Hultén, 1996, Escaping lock-in: The case of the electric vehicle. Technology and the Environment, *Technological Forecasting and Social Change* 53, 61–79.

Cowling, Marc, and Peter Mitchell, 2003, Is the Small Firms Loan Guarantee Scheme Hazardous for Banks or Helpful to Small Business?, *Small Business Economics* 21, 63–71.

Cronan, Timothy P., Donald R. Epley, and Larry G. Perry, 1986, The use of rank transformation and multiple regression analysis in estimating residential property values with a small sample, *Journal of Real Estate Research* 1, 19.

Dahlqvist, Jonas, Per Davidsson, and Johan Wiklund, 2000, Initial Conditions as Predictors of New Venture Performance: A Replication and Extension of the Cooper et al.study, *Enterprise & Innovation Management Studies* 1.

Damanpour, Fariborz, 1992, Organizational Size and Innovation, *Organization Studies* 13, 375–402.

Dasgupta, Parta, 1986, The Theory of Technological Competition, in Joseph E. Stiglitz, ed.: *New developments in the analysis of market structure. Proceedings of a conference held by the International Economic Association in Ottawa, Canada* (MIT Press, Cambridge, Mass.).

Davila, Antonio, George Foster, and Mahendra Gupta, 2003, Venture capital financing and the growth of startup firms, *Journal of Business Venturing* 18, 689.

Dean, Thomas J., and Jeffery S. McMullen, 2007, Toward a theory of sustainable entrepreneurship: Reducing environmental degradation through entrepreneurial action, *Journal of Business Venturing* 22, 50–76.

Del Monte, Alfredo, and Erasmo Papagni, 2003, R&D and the growth of firms: empirical analysis of a panel of Italian firms, *Research Policy* 32, 1003–1014.

Delmar, Frédéric, Per Davidsson, and William B. Gartner, 2003, Arriving at the high-growth firm, *Journal of Business Venturing* 18, 189–216.

Destatis, 2009, *Umwelt. Umsatz mit Waren, Bau- und Dienstleistungen für den Umweltschutz* (Wiesbaden).

Diefendorf, Sarah, 2000, Venture capital & the environmental industry, *Corporate Environmental Strategy* 7, 388–399.

Diekmann, Andreas, 2010, *Empirische Sozialforschung. Grundlagen, Methoden, Anwendungen* (Rowohlt-Taschenbuch-Verl., Reinbek bei Hamburg).

Dillman, Don A., 2000, *Mail and internet surveys. The tailored design method* (Wiley, New York, NY).

Dinica, Valentina, 2006, Support systems for the diffusion of renewable energy technologies: an investor perspective, *Energy Policy* 34, 461–480.

Dismukes, John P., Lawrence K. Miller, and John A. Bers, 2009, The industrial life cycle of wind energy electrical power generation: ARI methodology modeling of life cycle dynamics. Knowledge Driven Planning Tools for Emerging and Converging Technologies, *Technological Forecasting and Social Change* 76, 178–191.

Dodge, H. R., and John E. Robbins, 1992, An empirical investigation of the organizational life cycle model for small business development and survival, *Journal of Small Business Management* 30, 27–37.

Durbin, J., and G. S. Watson, 1951, Testing for serial correlation in least squares regression. II, *Biometrika* 38, 159–178.

Egeln, Jurgen, and Georg Licht, 1997, Firm foundations and the role of financial constraints, *Small Business Economics* 9, 137.

Eggers, John H., and Kim T. Leahy, 1994, Stages of small business growth revisted: Insights into growth path and needed leadership/management skills in low and high growth companies, *Frontiers of Entrepreneurship Research, Babson College Edition*, 131–144.

Elkington, John, 1998, Partnerships from Cannibals with Forks: The Triple Bottom Line of 21st-Century Business, *Environmental Quality Management* 8, 37–51.

Elsayed, Khaled, and David Paton, 2009, The impact of financial performance on environmental policy: does firm life cycle matter?, *Business Strategy and the Environment* 18, 397–413.

Elyasiani, Elyas, and Lawrence G. Goldberg, 2004, Relationship lending: a survey of the literature. Research Perspectives Special Issue, *Journal of Economics and Business* 56, 315–330.

Enzensberger, N., W. Fichtner, and O. Rentz, 2003, Financing renewable energy projects via closed-end funds--a German case study, *Renewable Energy* 28, 2023–2036.

Enzensberger, N., M. Wietschel, and O. Rentz, 2002, Policy instruments fostering wind energy projects -- a multi-perspective evaluation approach, *Energy Policy* 30, 793–801.

Evans, David S., 1987, Tests of Alternative Theories of Firm Growth, *The Journal of Political Economy* 95, 657–674.

Fahrmeir, Ludwig, Thomas Kneib, and Stefan Lang, 2009, *Regression. Modelle, Methoden und Anwendungen* (Springer, Berlin).

Fama, Eugene F., and Kenneth R. French, 2002, Testing Trade-Off and Pecking Order Predictions about Dividends and Debt, *The Review of Financial Studies* 15, 1–33.

Fazzari, Steven M., R. G. Hubbard, and Bruce C. Petersen, 1988, Financing Constraints and Corporate Investment, *Brookings Papers on Economic Activity*, 141–206.

Fazzari, Steven M., R. G. Hubbard, and Bruce C. Petersen, 2000, Investment-cash flow sensitivities are useful: A comment on Kaplan and Zingales, *Quarterly Journal of Economics* 115, 695–705.

Federal Environment Agency, 2009, *Data on the Environment. Edition 2009* (Berlin).

Fluck, Z., D. Holtz-Eakin, and H. S. Rosen, 1998, *Where does the money come from? The Financing of Small Entrepreneurial Enterprises* (New York University, Leornard N. Stern School Finance Department).

Fourt, Louis A., and Joseph W. Woodlock, 1960, Early Prediction of Market Success for New Grocery Products, *The Journal of Marketing* 25, 31–38.

Foxon, T. J., R. Gross, A. Chase, J. Howes, A. Arnall, and D. Anderson, 2005, UK innovation systems for new and renewable energy technologies: drivers, barriers and systems failures, *Energy Policy* 33, 2123–2137.

Freel, Mark S., 1999, The financing of small firm product innovation within the UK, *Technovation* 19, 707–719.

Freeman, Christopher, and Luc Soete, 2004, *The economics of industrial innovation* (Routledge, London).

Galbraith, John K., 1997, *American capitalism. The concept of countervailing power* (Transaction Publ., New Brunswick, NJ).

Galende, Jesús, and Juan M. de La Fuente, 2003, Internal factors determining a firm's innovative behaviour, *Research Policy* 32, 715–736.

Garnsey, Elizabeth, Erik Stam, and Paul Heffernan, 2006, New Firm Growth: Exploring Processes and Paths, *Industry & Innovation* 13, 1–20.

Garud, Raghu, and Peter Karnøe, 2003, Bricolage versus breakthrough: distributed and embedded agency in technology entrepreneurship. Special Issue on Technology Entrepreneurship and Contact Information for corresponding authors, *Research Policy* 32, 277–300.

Ge, Wenxia, and G. Whitmore, 2010, Binary response and logistic regression in recent accounting research publications: a methodological note 34, 81–93.

Geroski, P. A., 2000, Models of technology diffusion, *Research Policy* 29, 603–625.

Geroski, Paul, 2003, *The Evolution of New Markets* (Oxford University Press, Oxford).

Gibrat, R., 1931, *Les Inegalite´s Economiques* (Librairie du Recueil Sirey).

Gil, Pedro M., 2010, Stylised facts and other empirical evidence on firm dynamics, business cycle and growth, *Research in Economics* 64, 73–80.

Gilbert, Brett A., Patricia P. McDougall, and David B. Audretsch, 2006, New Venture Growth: A Review and Extension, *Journal of Management* 32, 926–950.

Giudici, Giancarlo, and Stefano Paleari, 2000, The Provision of Finance to Innovation: A Survey Conducted among Italian Technology-based Small Firms, *Small Business Economics* 14, 37–53.

Goldenberg, Jacob, Barak Libai, and Eitan Muller, 2002, Riding the Saddle: How Cross-Market Communications Can Create a Major Slump in Sales, *The Journal of Marketing* 66, 1–16.

Golder, Peter N., and Gerard J. Tellis, 1997, Will It Ever Fly? Modeling the Takeoff of Really New Consumer Durables, *Marketing Science* 16, 256–270.

Golder, Peter N., and Gerard J. Tellis, 2004, Growing, Growing, Gone: Cascades, Diffusion, and Turning Points in the Product Life Cycle, *Marketing Science* 23, 207–218.

Gompers, Paul A., 1996, Grandstanding in the venture capital industry, *Journal of Financial Economics* 42, 133–156.

Gompers, Paul A., and Joshua Lerner, 2001, *The money of invention. How venture capital creates new wealth* (Harvard Business School Press, Boston, Mass.).

González, Xulia, and Consuelo Pazó, 2008, Do public subsidies stimulate private R&D spending?, *Research Policy* 37, 371–389.

Gottschalk, Sandra, Helmut Fryges, Georg Metzger, Diana Heger, and Georg Licht, 2007, *Start-ups zwischen Forschung und Finanzierung: Hightech-Gründungen in Deutschland*, ZEW (Mannheim).

Grabowski, Henry G., and Dennis C. Mueller, 1975, Life-cycle Effects on Corporate Returns on Retentions, *Review of Economics & Statistics* 57, 400.

Graves, S. B., and N. S. Langowitz, 1993, Innovative productivity and returns to scale in the pharmaceutical industry, *Strat. Mgmt. J.* 14, 593–605.

Greenland, Sander, James J. Schlesselmann, and Michael H. Criqui, 1986, The fallacy of employing standardized regression coefficients and correlations as measure of effect, *American Journal of Epidemiology* 123, 203–208.

Gregory, Brian T., Matthew W. Rutherford, Sharon Oswald, and Lorraine Gardiner, 2005, An Empirical Investigation of the Growth Cycle Theory of Small Firm Financing, *Journal of Small Business Management* 43, 382–392.

Greiner, Larry E., 1998, Evolution and revolution as organiations grow, *Harvard Business Review* 76, 55–68.

Griliches, Zvi, 1960, Hybrid Corn and the Economics of Innovation, *Science* 132, 275–280.

Grubb, Michael, 2004, Technology Innovation and Climate Change Policy: An Overview of Issues and Options, *Keio Journal of Economics* 41, 103–132.

Grübler, Arnulf, 1996, Time for a Change: On the Patterns of Diffusion of Innovation, *Daedalus* 125, 19–42.

Guiso, Luigi, 1998, High-tech firms and credit rationing, *Journal of Economic Behavior & Organization* 35, 39.

Haan, Leo de, and Jeroen Hinloopen, 2003, Preference hierarchies for internal finance, bank loans, bond, and share issues: evidence for Dutch firms, *Journal of Empirical Finance* 10, 661–681.

Hadjimanolis, Athanasios, 2003, The Barriers Approach to Innovation, in Larisa V. Shavinina, ed.: *The International Handbook on Innovation* (Pergamon, Oxford).

Hair, Joseph F., 2006, *Multivariate data analysis* (Pearson/Prentice Hall, Upper Saddle River, NJ).

Hall, Bronwyn H., 2002, The Financing of Research and Development, *Oxford Review Economic Policy*, 35–51.

Hall, Bronwyn H., 2007, Innovation and Diffusion, in Jan Fagerberg, David C. Mowery, and Richard R. Nelson, eds.: *The Oxford handbook of innovation* (Oxford Univ. Press, Oxford).

Hall, Graham C., Patrick J. Hutchinson, and Nicos Michaelas, 2004, Determinants of the Capital Structures of European SMEs, *Journal of Business Finance & Accounting* 31, 711–728.

Hall, Jeremy, and Harrie Vredenburg, 2003, The Challenges of Innovating for Sustainable Development, *MIT Sloan Management Review* 45, 61–68.

Hall, Jeremy K., Gregory A. Daneke, and Michael J. Lenox, 2010, Sustainable development and entrepreneurship: Past contributions and future directions, *Journal of Business Venturing* In Press, Corrected Proof.

Hamilton, Barton H., and Jackson A. Nickerson, 2003, Correcting for Endogeneity in Strategic Management Research, *Strategic Organization* 1, 51–78.

Hamilton, Ian, 2007, *Why the climate change debate has not created more cleantech funds in Sweden* (Umea).

Hamilton, Kirsty, 2009, *Unlocking Finance for Clean Energy: The Need for 'Investment Grade' Policy* .

Hanks, Steven H., Collin J. Watson, Erik Jansen, and Gaylen N. Chandler, 1993, Tightening the Life-Cycle Construct: A Taxonomic Study of Growth Stage Configurations in High-Technology Organizations, *Entrepreneurship: Theory & Practice* 18, 5–30.

Hansen, John A., 1992, Innovation, firm size, and firm age, *Small Business Economics* 4, 37–44.

Harhoff, Dietmar, 1998, Are There Financing Constraints for R&D and Investment in German Manufacturing Firms?, *Annales d'Économie et de Statistique*, 421–456.

Harris, Robert S., and Susan Chaplinsky, 2006, *Capital Structure Theory: A Current Perspective* .

Hart, Stuart L., 1995, natural resource-based view of the firm, *Academy of Management Review* 20, 986–1014.

Hart, Stuart L., and Mark B. Milstein, 1999, Global Sustainability and the Creative Destruction of Industries, *Sloan Management Review* 41, 23–33.

Hauser, John, Gerard J. Tellis, and Abbie Griffin, 2006, Research on Innovation: A Review and Agenda for Marketing Science, *Marketing Science* 25, 687–717.

Headd, Brian, and Bruce Kirchhoff, 2009, The Growth, Decline and Survival of Small Businesses: An Exploratory Study of Life Cycles, *Journal of Small Business Management* 47, 531–550.

Hellström, Thomas, 2007, Dimensions of environmentally sustainable innovation: the structure of eco-innovation concepts, *Sustainable Development* 15, 148–159.

Heyman, Dries, Marc Deloof, and Hubert Ooghe, 2008, The Financial Structure of Private Held Belgian Firms, *Small Business Economics* 30, 301–313.

Himmelberg, Charles P., and Bruce C. Petersen, 1994, R&D and internal finance: A panel study of small firms in high-tech industries, *The review of Economics and Statistics* 76, 38–51.

Hite, Julie M., and William S. Hesterly, 2001, The Evolution of Firm Networks: From Emergence to Early Growth of the Firm, *Strategic Management Journal* 22, 275–286.

Hockerts, Kai, and Rolf Wüstenhagen, 2009, Greening Goliaths versus emerging Davids — Theorizing about the role of incumbents and new entrants in sustainable entrepreneurship, *Journal of Business Venturing*.

Hoetker, Glenn, 2007, The use of logit and probit models in strategic management research: Critical issues, *Strat. Mgmt. J.* 28, 331 343.

Hoffman, Andrew J., and Marc J. Ventresca, 1999, The Institutional Framing of Policy Debates, *American Behavioral Scientist* 42, 1368.

Hogan, Teresa, and Elaine Hutson, 2005, Capital structure in new technology-based firms: Evidence from the Irish software sector. Special Issue, *Global Finance Journal* 15, 369–387.

Honjo, Yuji, and Nobuyuki Harada, 2006, SME Policy, Financial Structure and Firm Growth: Evidence From Japan, *Small Business Economics* 27, 289–300.

Horbach, Jens, 2008a, Determinants of environmental innovation--New evidence from German panel data sources, *Research Policy* 37, 163–173.

Horbach, Jens, 2008b, *The impact of Innovation Activities on Employment in the Environmental Sector. Empirical Results for Germany at the Firm Level* .

Horrell, James F., and V. P. Lessig, 1975, A note on a multivariate generalization of the Kruskal-Wallis test, *Decision Sciences* 6, 135–141.

Hosmer, David W., and Stanley Lemeshow, 1989, *Applied logistic regression* (Wiley, New York, NY).

Hoy, Frank, 2006, The Complicating Factor of Life Cycles in Corporate Venturing, *Entrepreneurship: Theory & Practice* 30, 831–836.

Huber, Joseph, 1986, *Die verlorene Unschuld der Ökologie. Neue Technologien und superindustrielle Entwicklung* (Fischer, Frankfurt a.M.).

Huber, Joseph, 2004, *New technologies and environmental innovation* (Elgar, Cheltenham).

Huergo, Elena, and Jordi Jaumandreu, 2004, How Does Probability of Innovation Change with Firm Age?, *Small Business Economics* 22, 193–207.

Hyytinen, Ari, and Mika Pajarinen, 2008, Opacity of young businesses: Evidence from rating disagreements, *Journal of Banking & Finance* 32, 1234–1241.

Hyytinen, Ari, and Otto Toivanen, 2005, Do financial constraints hold back innovation and growth?: Evidence on the role of public policy, *Research Policy* 34, 1385–1403.

IHK, 2009, *UMFIS database*. http://www.umfis.de/ (accessed November 15, 2009).

Iman, Ronald L., and W. J. Conover, 1979, The Use of the Rank Transform in Regression, *Technometrics* 21, 499.

Islam, Towhidul, and Nigel Meade, 1997, The diffusion of successive generations of a technology: A more general model, *Technological Forecasting and Social Change* 56, 49–60.

Islam, Towhidul, and Nigel Meade, 2006, Modelling and forecasting the diffusion of innovation - A 25-year review. Twenty five years of forecasting, *International Journal of Forecasting* 22, 519–545.

Jacobsson, Staffan, and Anna Johnson, 2000, The diffusion of renewable energy technology: an analytical framework and key issues for research, *Energy Policy* 28, 625–640.

Jacobsson, Staffan, and Volkmar Lauber, 2006, The politics and policy of energy system transformation: explaining the German diffusion of renewable energy technology, *Energy Policy* 34, 256–276.

Jensen, Michael C., and William H. Meckling, 1976, Theory of the firm: Managerial behavior, agency costs and ownership structure, *Journal of Financial Economics* 3, 305–360.

Jovanovic, Boyan, 1982a, Selection and Evolution of Industry, *Econometrica*, 649–670.

Jovanovic, Boyan, 1982b, Selection and the Evolution of Industry, *Econometrica* 50, 649–670.

Jovanovic, Boyan, and Saul Lach, 1989, Entry, Exit, and Diffusion with Learning by Doing, *The American Economic Review* 79, 690–699.

Kammlott, C., and Dirk Schiereck, 2009, Finanzierungsengpässe im Bereich der erneuerbaren Energien, *Immobilien und Finanzierung - Der langfristige Kredit*, 260–262.

Kang, Jae, Almas Heshmati, and Gyoung-Gyu Choi, 2008, Effect of credit guarantee policy on survival and performance of SMEs in Republic of Korea, *Small Business Economics* 31, 443-443.

Kaplan, Steven N., and Luigi Zingales, 1997, Do investment-cash flow sensitivities provide useful measures of financing constraints?, *Quarterly Journal of Economics* 112, 169–215.

Kaplan, Steven N., and Luigi Zingales, 2000, Investment-Cash Flow Sensitivities are not Valid Measures of Financing Constraints*, *Quarterly Journal of Economics* 115, 707–712.

Kasemir, Bernd, Ferenc Toth, and Vanessa Masing, 2000, Climate Policy, Venture Capital and European Integration, *Journal of Common Market Studies* 38, 891.

Kazanjian, Robert K., and Robert Drazin, 1989, An empirical test of a stage of growth progression model, *Management Science* 35, 1489–1503.

Kazanjian, Robert K., and Robert Drazin, 1990, A stage-contingent model of design and growth for technology based new ventures, *Journal of Business Venturing* 5, 137–150.

Kemp, René, and Luc Soete, 1992, The greening of technological progress. An evolutionary perspective, *Futures* 24, 437–457.

Kenney, Martin, 2009, *Venture Capital Investment in the Greentech Industries: A Provocative Essay*, Department of Human and Community Development; University of California, Davis .

Khan, Arshad M., and V. Manopichetwattana, 1989, Innovative and Noninnovative Small Firms: Types and Characteristics, *Management Science* 35, 597–606.

King, Andrew, and Michael Lenox, 2002, Exploring the Locus of Profitable Pollution Reduction, *Management Science* 48, 289–299.

Klassen, Kenneth J., Jeffrey A. Pittman, and Margaret P. Reed, 2004, A Cross-national Comparison of R&D Expenditure Decisions: Tax Incentives and Financial Constraints, *Contemporary Accounting Research* 21, 639–680.

Klassen, Robert D., and Curtis P. McLaughlin, 1996, The impact of environmental management on firm performance, *Management Science* 42, 1199–1214.

Kleinknecht, Alfred H., and Pierre A. Mohnen, 2002, *Innovation and firm performance. Econometric explorations of survey data* (Palgrave, Basingstoke, Hampshire).

Klepper, Steven, 1996, Entry, Exit, Growth, and Innovation over the Product Life Cycle, *The American Economic Review* 86, 562–583.

Kline, Steven J., and Nathan Rosenberg, 1986, An Overview of Innovation, in Ralph Landau, ed.: *The positive sum strategy. Harnessing technology for economic growth* (Nat. Acad. Pr., Washington, DC).

Koberg, Christine S., Nikolaus Uhlenbruck, and Yolanda Sarason, 1996, Facilitators of organizational innovation: The role of life-cycle stage, *Journal of Business Venturing* 11, 133–149.

Kohn, Karsten, 2009, Marktversagen und Gründungshemmnisse - Was können wir aus der empirischen Literatur lernen?, *Finanz Betrieb Aufsätze*, 678.

Krozer, Yoram, and Andries Nentjes, 2008, Environmental policy and innovations, *Business Strategy and the Environment* 17, 219–229.

Kruskal, William H., and W. A. Wallis, 1952, Use of Ranks in One-Criterion Variance Analysis, *Journal of the American Statistical Association* 47, 583–621.

Kuckertz, Andreas, and Marcus Wagner, 2010, The influence of sustainability orientation on entrepreneurial intentions -- Investigating the role of business experience, *Journal of Business Venturing* In Press, Corrected Proof.

La Rocca, Maurizio, and Tiziana La Rocca, 2007, *Capital structure and corporate strategy: An overview*, University of Calabria .

La Rocca, Maurizio, Tiziana La Rocca, and Alfio Cariola, 2009, Capital Structure Decisions During a Firm's Life Cycle, *Small Business Economics*.

Lamont, Owen, Christopher Polk, and Jesús Saá-Requejo, 2001, Financial Constraints and Stock Returns, *The Review of Financial Studies* 14, 529–554.

Lee, Chang-Yang, 2010, A theory of firm growth: Learning capability, knowledge threshold, and patterns of growth, *Research Policy* 39, 278–289.

Lee, Choonwoo, Kyungmook Lee, and Johannes M. Pennings, 2001, Internal Capabilities, External Networks, and Performance: A Study on Technology-Based Ventures, *Strategic Management Journal* 22, 615.

Leland, Hayne E., and David H. Pyle, 1977, Informational asymmetries, financial structure and financial intermediation, *Journal of Finance* 32, 371–387.

Lester, Donald L., John A. Parnell, William ". Crandall, and Michael L. Menefee, 2008, Organizational life cycle and performance among SMEs, *International Journal of Commerce & Management* 18, 313–330.

Levie, Jonathan, and Benyamin B. Lichtenstein, 2008, *From "Stages" of Business Growth to a Dynamic States Model of Entrepreneurial Growth and Change*, Hunter center for entrepreneurship (Boston, Mass.).

Levitt, Theodore, 1965, Exploit the Product Life Cycle, *Harvard Business Review* 43, 81–94.

Lewis, Joanna I., and Ryan H. Wiser, 2007, Fostering a renewable energy technology industry: An international comparison of wind industry policy support mechanisms, *Energy Policy* 35, 1844–1857.

Lichtenstein, Benyamin B., Jonathan Levie, and Michael Hay, 2007, *Stage Theory Is Dead! Long Live the New Stages Theory of Organizational Change*, UMASS College of Management (Boston, Mass.).

Lipp, Judith, 2007, Lessons for effective renewable electricity policy from Denmark, Germany and the United Kingdom, *Energy Policy* 35, 5481–5495.

Loiter, Jeffrey M., and Vicki Norberg-Bohm, 1999, Technology policy and renewable energy: public roles in the development of new energy technologies, *Energy Policy* 27, 85–97.

López-Gracia, José, and Francisco Sogorb-Mira, 2008, Testing trade-off and pecking order theories financing SMEs, *Small Business Economics* 31, 117–136.

Love, James H., and Brian Ashcroft, 1999, Market versus Corporate Structure in Plant-Level Innovation Performance, *Small Business Economics* 13, 97–109.

Lund, P. D., 2007, Effectiveness of policy measures in transforming the energy system, *Energy Policy* 35, 627–639.

Macpherson, Allan, and Robin Holt, 2007, Knowledge, learning and small firm growth: A systematic review of the evidence, *Research Policy* 36, 172–192.

Magri, Silvia, 2007, *The financing of small innovative firms: The Italian case*, Banca d'Italia .

Mahajan, Vijay, and Eitan Muller, 1996, Timing, diffusion, and substitution of successive generations of technological innovations: The IBM mainframe case, *Technological Forecasting and Social Change* 51, 109–132.

Mahajan, Vijay, and Eitan Muller, 1998, When Is It Worthwhile Targeting the Majority Instead of the Innovators in a New Product Launch?, *Journal of Marketing Research* 35, 488–495.

Mahajan, Vijay, Eitan Muller, and Rajendra K. Srivastava, 1990, Determination of Adopter Categories by Using Innovation Diffusion Models, *Journal of Marketing Research* 27, 37–50.

Mahajan, Vijay, and Robert A. Peterson, 1996, *Models for innovation diffusion* (Sage, Newbury Park, Calif.).

Mansfield, Edwin, 1961, Technical Change and the Rate of Imitation, *Econometrica* 29, 741–766.

Mansfield, Edwin, 1968, *Industrial research and technological innovation. An econometric analysis*, Cowles Foundation for Research in Economics (Norton & Co., New York, NY).

Marinova, Dora, and John Phillimore, 2003, Models of Innovation, in Larisa V. Shavinina, ed.: *The International Handbook on Innovation* (Pergamon, Oxford).

Matell, Michael S., and Jacob Jacoby, 1972, Is there an opitmal number of alternatives for likert-scale items?, *Journal of Applied Psychology* 56, 506–509.

McKelvie, Alexander, and Johan Wiklund, 2010, Advancing Firm Growth Research: A Focus on Growth Mode Instead of Growth Rate, *Entrepreneurship: Theory & Practice* 34, 261–288.

Meek, William R., Desirée F. Pacheco, and Jeffrey G. York, 2010, The impact of social norms on entrepreneurial action: Evidence from the environmental entrepreneurship context, *Journal of Business Venturing* In Press, Corrected Proof.

Menanteau, Philippe, Dominique Finon, and Marie-Laure Lamy, 2003, Prices versus quantities: choosing policies for promoting the development of renewable energy, *Energy Policy* 31, 799–812.

Metzger, Georg, Diana Heger, Daniel Höwer, and Georg Licht, 2010, *High-Tech-Gründungen in Deutschland. Zum Mythos des jungen High-Tech-Gründers*. ftp://ftp.zew.de/pub/zew-docs/gutachten/hightechgruendungen10.pdf (accessed February 17, 2010).

Miller, Danny, and Peter H. Friesen, 1984, A Longitudinal Study of the Corporate Life Cycle, *Management Science* 30, 1161–1183.

Mitchell, C., D. Bauknecht, and P. M. Connor, 2006, Effectiveness through risk reduction: a comparison of the renewable obligation in England and Wales and the feed-in system in Germany. Renewable Energy Policies in the European Union, *Energy Policy* 34, 297–305.

Modigliani, Franco, and Merton H. Miller, 1958, The Cost of Capital, Corporation Finance and the Theory of Investment, *The American Economic Review* 48, 261–297.

Modigliani, Franco, and Merton H. Miller, 1963, Corporate Income Taxes and the Cost of Capital: A Correction, *The American Economic Review* 53, 433–443.

Moore, Bill, and Rolf Wüstenhagen, 2004, Innovative and Sustainable Energy Technologies: The Role of Venture Capital, *Business Strategy and the Environment*, 235–245.

Moore, Geoffrey A., 2006, *Crossing the chasm. Marketing and selling disruptive products to mainstream customers* (Collins Business Essentials, New York, NY).

Müller, Elisabeth, and Volker Zimmermann, 2009, The importance of equity finance for R&D activity, *Small Business Economics* 33, 303–318.

Müller-Christ, Georg, 2001, *Nachhaltiges Ressourcenmanagement. Eine wirtschaftsökologische Fundierung*, Univ., Habil.-Schr.--Bayreuth, 2001. (Metropolis-Verl., Marburg).

Murphy, L. M., and P. L. Edwards, 2003, *Bridging the Valley of Death: Transitioning from Public to Private Sector Financing*, National Renewable Energy Laboratory (NREL).

Musso, Patrick, and Stefano Schiavo, 2008, The impact of financial constraints on firm survival and growth, *Journal of Evolutionary Economics* 18, 135–149.

Myers, Stewart C., 1977, Determinants of corporate borrowing, *Journal of Financial Economics* 5, 147–175.

Myers, Stewart C., 1984, The Capital Structure Puzzle, *Journal of Finance* 39, 575–592.

Nelson, Richard R., 1959, The Simple Economics of Basic Scientific Research, *The Journal of Political Economy* 67, 297–306.

Nelson, Richard R., Alexander Peterhansl, and Bhaven Sampat, 2004, Why and how innovations get adopted: a tale of four models, *Ind Corp Change* 13, 679–699.

Nemet, Gregory F., 2009, Demand-pull, technology-push, and government-led incentives for non-incremental technical change, *Research Policy* 38, 700–709.

Neuhoff, Karsten, 2005, Large-scale deployment of renewables for electricity generation, *Oxford Review of Economic Policy* 21, 88–110.

Nill, Jan, and René Kemp, 2009, Evolutionary approaches for sustainable innovation policies: From niche to paradigm? Special Issue: Emerging Challenges for Science, Technology and Innovation Policy Research: A Reflexive Overview, *Research Policy* 38, 668–680.

Nitzsch, Rüdiger von, Christian Rouette, and Olaf Stotz, 2005, Kapitalstrukturentscheidungen junger Unternehmen, in Christoph J. Börner, and Dietmar Grichnik, eds.: *Entrepreneurial Finance. Kompendium der Gründungs- und Wachstumsfinanzierung* (Physica-Verlag Heidelberg, Heidelberg).

Norton, John A., and Frank M. Bass, 1987, A Diffusion Theory Model of Adoption and Substitution for Successive Generations of High-Technology Products, *Management Science* 33, 1069–1086.

O' Sullivan, Mary, 2007, Finance and Innovation, in Jan Fagerberg, David C. Mowery, and Richard R. Nelson, eds.: *The Oxford handbook of innovation* (Oxford Univ. Press, Oxford).

OECD, 2005, *The Measurement of Scientific and Technological Activities. Proposed Guidelines for Collecting and Interpreting Technological Innovation Data*, Oslo Manual .

Oltmanns, Torsten, ed., 2011, *Green growth, green profit. How green transformation boosts business* (Palgrave Macmillan, Basingstoke).

O'Rourke, Anastasia R., 2009, *The Emergence of Cleantech*, PhD Dissertation, Granduate School of Yale University (New Haven, CT).

Pasanen, Mika, 2007, SME growth strategies: organic or non-organic, *Journal of Enterprising Culture* 15, 317–338.

Peneder, Michael, 2008, The problem of private under-investment in innovation: A policy mind map, *Technovation* 28, 518–530.

Peng, Chao-Ying J., Kuk L. Lee, and Gary M. Ingersoll, 2002, An Introduction to Logistic Regression Analysis and Reporting, *The Journal of Educational Research* 96, 3–14.

Penrose, Edith T., 1997, *The theory of the growth of the firm* (Oxford Univ. Press, Oxford).

Peres, Renana, Eitan Muller, and Vijay Mahajan, 2010, Innovation diffusion and new product growth models: A critical review and research directions, *International Journal of Research in Marketing* 27, 91–106.

Phelps, Robert, Richard Adams, and John Bessant, 2007, Life cycles of growing organizations: A review with implications for knowledge and learning, *International Journal of Management Reviews* 9, 1–30.

Piga, Claudio A., and Gianfranco Atzeni, 2007, R&D investment, credit rationing and sample selection, *Bulletin of Economic Research* 59, 149–178.

Pissarides, Francesca, 1999, Is lack of funds the main obstacle to growth? ebrd's experience with small- and medium-sized businesses in central and eastern europe, *Journal of Business Venturing* 14, 519–539.

Ploetscher, Claudia, and Horst Rottmann, 2002, Investment Behavior and Financing Constraints in German Manufacturing and Construction Firms: A Bivariate Ordered Probit Estimation, *Ifo-Studien : Zeitschrift für Empirische Wirtschaftsforschung* 48, 383–400.

Pohl, Elke, 2010, Aktion "Klima und Finanzen", *Versicherungswirtschaft* 65, 369.

Porter, Michael E., and Claas van der Linde, 1995, Green and Competitive: Ending the Stalemate, *Harvard Business Review* 73, 120–134.

Poutziouris, Panikkos, 2003, The strategic orientation of owner-managers of small ventures: Evidence from the UK small business economy, *International Journal of Entrepreneurial Behaviour & Research* 9, 185–214.

Randjelovic, Jelena, Anastasia R. O'Rourke, and Renato J. Orsato, 2003, The emergence of "green" venture capital, *Business Strategy and the Environment* 12, 240–253.

Randolph, Justus J., 2009, A Guide to Writing the Dissertation Literature Review, *Practical Assessment, Research and Evaluation* 14.

Rausser, Gordon C., and Maya Papineau, 2010, *Managing R&D Risk in Renewable Energy*, Department of Agricultural and Resource Economics, UCB, UC Berkeley .

Rave, Klaus, 1998, Finance and banking for wind energy. Renewable Energy Energy Efficiency, Policy and the Environment, *Renewable Energy* 16, 855–857.

Rennings, Klaus, 2000, Redefining innovation -- eco-innovation research and the contribution from ecological economics, *Ecological Economics* 32, 319–332.

Rivera, Jorge, 2001, Does it pay to be green in the developing world? Participation in a Costa Rican voluntary environmental program and its impact on hotels' competitive advantage, *Academy of Management Proceedings & Membership Directory*, C1-C6.

Rogers, Everett M., 2003, *Diffusion of innovations* (Free Press, New York, NY).

Romano, Claudio A., George A. Tanewski, and Kosmas X. Smyrnios, 2001, Capital structure decision making: A model for family business, *Journal of Business Venturing* 16, 285–310.

Ross, Stephen A., 1977, The Determination of Financial Structure: The Incentive-Signalling Approach, *The Bell Journal of Economics* 8, 23–40.

Rostow, W. W., 1959, The Stages of Economic Growth, *The Economic History Review* 12, 1–16.

Russo, Michael V., 2003, The emergence of sustainable industries: building on natural capital, *Strategic Management Journal* 24, 317–331.

Rutherford, Matthew W., Paul F. Buller, and Patrick R. McMullen, 2003, Human Resource Management Problems over the Life Cycle of Small to Medium-Sized Firms, *Human Resource Management* 42, 321–335.

Savignac, Frédérique, 2008, Impact of financial constraints on innovation: What can be learned from a direct measure?, *Economics of Innovation & New Technology* 17, 553–569.

Scellato, Giuseppe, 2007, Patents, firm size and financial constraints: an empirical analysis for a panel of Italian manufacturing firms, *Cambridge Journal of Economics* 31, 55–76.

Schendera, Christian F. G., 2008, *Regressionsanalyse mit SPSS* (Oldenbourg, München).

Scherer, F. M., 1965, Firm Size, Market Structure, Opportunity, and the Output of Patented Inventions, *The American Economic Review* 55, 1097–1125.

Schulte, Reinhard, 2006, Finanzierungsstrategien kleiner und mittlerer Unternehmen, in Albert Martin, ed.: *Managementstrategien von kleinen und mittleren Unternehmen. Stand der theoretischen und empirischen Forschung* (Hampp, München).

Schumpeter, Joseph A., 1943, *Capitalism, socialism and democracy* (Harperperennial Modern Thought, New York N.Y.).

Schumpeter, Joseph A., 1947, The Creative Response in Economic History, *The Journal of Economic History* 7, 149–159.

Schumpeter, Joseph A., 2007, *The theory of economic development. An inquiry into profits, capital, credit, interest, and the business cycle* (Transaction Publ., New Brunswick, NJ).

Sharma, Sanjay, and Harrie Vredenburg, 1998, Proactive corporate environmental strategy and the development of competitively valuable organizational capabilities, *Strategic Management Journal* 19, 729.

Shavinina, Larisa V., and Kavita L. Seeratan, 2003, On the Nature of Individual Innovation, in Larisa V. Shavinina, ed.: *The International Handbook on Innovation* (Pergamon, Oxford).

Shepherd, Dean, and Johan Wiklund, 2009, Are We Comparing Apples With Apples or Apples With Oranges? Appropriateness of Knowledge Accumulation Across Growth Studies, *Entrepreneurship Theory and Practice* 33, 105–123.

Shrivastava, Paul, 1995, Environmental Technologies and Competitive Advantage, *Strategic Management Journal* 16, 183–200.

Siegel, Sidney, and N. J. Castellan, 2003, *Nonparametric statistics for the behavioral sciences* (McGraw-Hill, Boston, Mass.).

Smith, Keith, 2007, Measuring Innovation, in Jan Fagerberg, David C. Mowery, and Richard R. Nelson, eds.: *The Oxford handbook of innovation* (Oxford Univ. Press, Oxford).

Soerensen, Jesper B., and Toby E. Stuart, 2000, Aging, Obsolescence, and Organizational Innovation, *Administrative Science Quarterly* 45, 81–112.

Sogorb-Mira, Francisco, 2005, How SME Uniqueness Affects Capital Structure: Evidence From A 1994–1998 Spanish Data Panel, *Small Business Economics* 25, 447–457.

Souitaris, Vangelis, 2003, Determinants of Technological Innovation: Current Research Trends and Future Prospects, in Larisa V. Shavinina, ed.: *The International Handbook on Innovation* (Pergamon, Oxford).

Sovacool, Benjamin K., 2009, The cultural barriers to renewable energy and energy efficiency in the United States, *Technology in Society* 31, 365–373.

Spence, Michael, 1984, Cost Reduction, Competition, and Industry Performance, *Econometrica* 52, 101–121.

Spicer, John, 2005, *Making sense of multivariate data analysis* (Sage, Thousand Oaks, Calif.).

Srinivasan, Sunderasan, 2007, The Indian solar photovoltaic industry: a life cycle analysis, *Renewable and Sustainable Energy Reviews* 11, 133–147.

Steinmetz, Lawrence L., 1969, Critical stages of small business growth, *Business Horizons* 12, 29.

Stern, Nicholas, 2008, *The economics of climate change. The Stern review* (Cambridge Univ. Press, Cambridge).

Sternberg, Robert J., Jean E. Pretz, and James C. Kaufman, 2003, Types of Innovations, in Larisa V. Shavinina, ed.: *The International Handbook on Innovation* (Pergamon, Oxford).

Stiglitz, Joseph E., and Andrew Weiss, 1981, Credit Rationing in Markets with Imperfect Information, *American Economic Review* 71, 393.

Storey, David J., 1996, *Understanding the small business sector* (Routledge, London).

Storey, John, 2000, The Management of Innovation Problem, *International Journal of Innovation Management* 4, 347.

Sutton, John, 1997, Gibrat's Legacy, *Journal of Economic Literature* 35, 40–59.

Svensson, Roger, 2007, Commercialization of patents and external financing during the R&D phase, *Research Policy* 36, 1052–1069.

Taylor, Margaret, 2008, Beyond technology-push and demand-pull: Lessons from California's solar policy. Technological Change and the Environment, *Energy Economics* 30, 2829–2854.

Tensie Steijvers, and Wim Voordeckers, 2009, Collateral and Credit Rationing: Review of Recent Empirical Studies as a Guide for Future Research, *Journal of Economic Surveys* 23, 924–946.

Teppo, Tarja, 2006, *Financing Clean Energy Market Creation. Clean Energy Ventures, Venture Capitalists and other Investors*, PhD Dissertation, Helsinki University of Technology Development and Management in Industry (Helsinki University of Technology, Espoo, Finnland).

Thornhill, Stewart, Guy Gellatly, and Allan Riding, 2004, Growth history, knowledge intensity and capital structure in small firms, *Venture Capital* 6, 73–89.

Tsoutsos, Theocharis D., and Yeoryios A. Stamboulis, 2005, The sustainable diffusion of renewable energy technologies as an example of an innovation-focused policy, *Technovation* 25, 753–761.

Unruh, Gregory C., 2000, Understanding carbon lock-in, *Energy Policy* 28, 817–830.

Unruh, Gregory C., 2002, Escaping carbon lock-in, *Energy Policy* 30, 317–325.

Utterback, James M., 1971, The Process of Technological Innovation Within the Firm, *Academy of Management Journal* 14, 75–88.

Utterback, James M., and William J. Abernathy, 1975, A dynamic model of process and product innovation, *Omega* 3, 639–656.

Utterback, James M., and Fernando F. Suárez, 1993, Innovation, competition, and industry structure, *Research Policy* 22, 1–21.

Vaona, Andrea, and Mario Pianta, 2008, Firm Size and Innovation in European Manufacturing, *Small Business Economics* 30, 283–299.

Vincent, Leslie H., Sundar G. Bharadwaj, and Goutam N. Challagalla, 2004, *Does innovation mediate firm performance? A meta-analysis of determinants and consequences of Organizational innovation*, Working paper, Georgia Institute of Technology (Atlanta, GA).

Vos, Ed, Andy J.-Y. Yeh, Sara Carter, and Stephen Tagg, 2007, The happy story of small business financing, *Journal of Banking & Finance* 31, 2648–2672.

Voulgaris, F., D. Asteriou, and G. Agiomirgianakis, 2004, Size and Determinants of Capital Structure in the Greek Manufacturing Sector, *International Review of Applied Economics* 18, 247–262.

Watson, Robert, and Nick Wilson, 2002, Small and Medium Size Enterprise Financing: A Note on Some of the Empirical Implications of a Pecking Order, *Journal of Business Finance & Accounting* 29, 557.

Weber, Christiana, and Ariane B. Antal, 2001, The Role of Time in Organizational Learning, *Handbook of Organizational Learning & Knowledge*, 351–368.

Weinzimmer, Laurence G., Paul C. Nystrom, and Sarah J. Freeman, 1998, Measuring Organizational Growth: Issues, Consequences and Guidelines, *Journal of Management* 24, 235–262.

Westhead, Paul, and David J. Storey, 1997, Financial constraints on the growth of high technology small firms in the United Kingdom, *Applied Financial Economics* 7, 197–201.

Wijewardena, Hema, and Shiran Cooray, 1995, Determinants of Growth in Small Japanese Manufacturing Firms: Survey Evidence from Kobe, *Journal of Small Business Management* 33, 87–92.

Witt, Peter, and Andreas Hack, 2008, Staatliche Gründungsfinanzierung: Stand der Forschung und offene Fragen, *Journal für Betriebswirtschaft* 58, 55–79.

Wolf, Björn, 2006, *Empirische Untersuchung zu den Einflussfaktoren der Finanzierungsprobleme junger Unternehmen in Deutschland und deren Auswirkungen auf die Wirtschaftspolitik* .

Wolff, Philipp, ed., 2009, *Deutsches CleanTech Jahrbuch 2009/ 2010: Beiträge aus Wirtschaft, Wissenschaft und Praxis. Eine Bestandsaufnahme (Gebundene Ausgabe)* (DCTI).

Wu, Yonghong, D. Popp, and S. Bretschneider, 2007, The Effects of Innovation Policies on Business R&D: A Cross-National Empirical Study, *Economics of Innovation & New Technology* 16, 237–253.

Wüstenhagen, Rolf, 2008, *Sustainable innovation and entrepreneurship* (Elgar, Cheltenham).

Wüstenhagen, Rolf, and Michael Bilharz, 2006, Green energy market development in Germany: effective public policy and emerging customer demand, *Energy Policy* 34, 1681–1696.

Wüstenhagen, Rolf, Jochen Markard, and Bernhard Truffer, 2003, Diffusion of green power products in Switzerland, *Energy Policy* 31, 621.

Wüstenhagen, Rolf, and Tarja Teppo, 2006, Do venture capitalists really invest in good industries? Risk-return perceptions and path dependence in the emerging European energy VC market, *Int. J. Technology Management* Vol. 34, 63–87.

Wüstenhagen, Rolf, Maarten Wolsink, and Mary J. Bürer, 2007, Social acceptance of renewable energy innovation: An introduction to the concept, *Energy Policy* 35, 2683–2691.

WWF, 2009, *Low Carbon Jobs for Europe. Current Opportunities and Future Prospects* (World Wide Fund for Nature, London).

Yarow, Jay, 2009, *Cleantech Industry Collapses In 2009*. http://www.businessinsider.com/cleantech-thus-far-2009-7 (accessed February 16, 2010).

York, Jeffrey G., and S. Venkataraman, 2010, The entrepreneur-environment nexus: Uncertainty, innovation, and allocation, *Journal of Business Venturing* In Press, Corrected Proof.

Zecchini, Salvatore, and Marco Ventura, 2009, The impact of public guarantees on credit to SMEs, *Small Business Economics* 32, 191–206.

GABLER RESEARCH

„Schriften zum europäischen Management"
Herausgeber: Roland Berger Strategy Consultants – Academic Network
zuletzt erschienen:

Rainer Bizenberger
Informal Private Debt

Simon Plankenhorn
Innovation Offshoring

Richard Federowski
Unternehmensroutinen im Turnaroundmanagement

Christian Neuner
**Kofiguration internationaler Produktionsnetzwerke
unter Berücksichtigung von Unsicherheit**

Christoph M. Auerbach
Fusionen deutscher Kreditinstitute

Fabian Sommerrock
Erfolgreiche Post-Merger-Integration bei öffentlichen Institutionen

Julia Däcke
Nutzung virtueller Welten zur Kundenintegration in die Neuproduktentwicklung

Christina Welsch
**Organisationale Trägheit und ihre Wirkung auf die strategische Früherkennung
von Unternehmenskrisen**

Bärbel Fleischer
Einsatz von Erfolgshonoraren in der Unternehmensberatung

Florian Geiger
Mergers & Acquisitions in the Machinery Industry

Sebastian Durst
Strategische Lieferantenentwicklung

Sandra Strohbücker
**Bepreisen von Preis- und Mengenrisiken der Strombeschaffung
unter Berücksichtigung von Portfolioaspekten
bei Großkunden im Strommarkt**

Adele J. Huber
Effective Strategy Implementation

Philipp H. Hoff
Greentech Innovation and Diffusion

Änderungen vorbehalten. Stand: November 2011.
Erhältlich im Buchhandel oder beim Verlag.
Gabler Verlag . Abraham-Lincoln-Str. 46 . 65189 Wiesbaden . www.gabler.de

GABLER